Leviathan and the Air-Pump

STEVEN SHAPIN & SIMON SCHAFFER

LEVIATHAN AND THE AIR-PUMP

HOBBES, BOYLE, AND THE EXPERIMENTAL LIFE

INCLUDING A TRANSLATION OF THOMAS HOBBES,

DIALOGUS PHYSICUS DE NATURA AERIS,

BY SIMON SCHAFFER

Princeton University Press

1985

PUBLISHED BY PRINCETON UNIVERSITY PRESS
41 WILLIAM STREET, PRINCETON, NEW JERSEY 08540
IN THE UNITED KINGDOM: PRINCETON UNIVERSITY PRESS
OXFORD

LIBRARY OF CONGRESS CATALOGING IN PUBLICATION DATA
WILL BE FOUND ON THE LAST PRINTED PAGE OF THIS BOOK
ISBN 0-691-08393-2
ISBN 0-691-02432-4, PBK.

FIRST PRINCETON PAPERBACK PRINTING,
WITH CORRECTIONS, 1989

THIS BOOK HAS BEEN COMPOSED IN LINOTRON BASKERVILLE
CLOTHBOUND EDITIONS OF PRINCETON UNIVERSITY PRESS BOOKS ARE
PRINTED ON ACID-FREE PAPER, AND BINDING MATERIALS ARE
CHOSEN FOR STRENGTH AND DURABILITY. PAPERBACKS,
ALTHOUGH SATISFACTORY FOR PERSONAL COLLECTIONS,
ARE NOT USUALLY SUITABLE FOR LIBRARY REBINDING

PRINTED IN THE UNITED STATES OF AMERICA
BY PRINCETON UNIVERSITY PRESS
PRINCETON, NEW JERSEY

For our Parents

Every afternoon Father Nicanor would sit by the chestnut tree preaching in Latin, but José Arcadio Buendía insisted on rejecting rhetorical tricks and the transmutation of chocolate, and he demanded the daguerreotype of God as the only proof. Father Nicanor then brought him medals and pictures and even a reproduction of the Veronica, but José Arcadio Buendía rejected them as artistic objects without any scientific basis. He was so stubborn that Father Nicanor gave up his attempts at evangelization and continued visiting him out of humanitarian feelings. But then it was José Arcadio Buendía who took the lead and tried to break down the priest's faith with rationalist tricks. On a certain occasion when Father Nicanor brought a checker set to the chestnut tree and invited him to a game, José Arcadio Buendía would not accept, because according to him he could never understand the sense of a contest in which the two adversaries have agreed upon the rules.

GABRIEL GARCÍA MARQUEZ, *One Hundred Years of Solitude*

What a blessing to mankind, in himself and in his writings, was the ingenious, humble, and pious Mr. Boyle; what a common pest to society was the fallacious, proud, and impious Hobbes! Accordingly we find the former bad adieu to this world with the utmost serenity, honour, and hope; while the other went out of it in the dark, with an odium on his name, as well as with terrible apprehensions of an unknown future.

W. DODD, *The Beauties of History; or, Pictures of Virtue and Vice Drawn from Examples of Men, Eminent for Their Virtues or Infamous for Their Vices* (1796)

· *CONTENTS* ·

· *ILLUSTRATIONS* ·

· NOTES ON SOURCES AND CONVENTIONS ·

For citations of sources in footnotes we have adopted an economical convention similar to that employed in Elizabeth Eisenstein's *The Printing Press as an Agent of Change.* Bibliographic information is kept to a minimum in the notes, apart from the occasional addition of date of publication where that information is not given in the text and is germane. Full titles and publication details are provided in the Bibliography. Complete details of unpublished manuscript sources, seventeenth-century periodical articles, and items in state and parliamentary papers are, however, given in the notes and not repeated in the Bibliography.

We have made liberal use of correspondence and other material not published in the seventeenth century. Our major concerns have been with knowledge that was public or designed to be so, and this has affected the extent of our use of such sources. Where we are interested in material that was incompletely public or, possibly, intended to be restricted (as in chapter 6), our use of manuscript material is correspondingly greater.

During the period with which this book is concerned, the British Isles employed a calendar different from that used in most Continental countries, especially Catholic ones. The former used the Julian (old style) calendar, which was ten days behind the Gregorian (new style) calendar employed on the Continent. In addition, the British new year was reckoned to begin on 25 March. Because we deal in some detail with exchanges between England and Continental countries, we give all dates in both old and new style form, but we adjust years to correspond with a new year commencing 1 January. Thus, the English 6 March 1661 is given as 6/16 March 1662; the Dutch (who used the Gregorian calendar even though Protestant) 24 July 1664 is given as 14/24 July 1664; and so forth.

We have endeavoured, within reason, to preserve seventeenth-century orthography, punctuation, and emphases, and have dispensed with *sic* indications, save where absolutely necessary.

In our usage, "Hobbesian" refers to the beliefs and practices of Hobbes as an individual; "Hobbist" to the beliefs and practices of his real or alleged followers. We distinguish between religious Dissent (upper case) and intellectual and political dissent (lower case).

· *ACKNOWLEDGMENTS* ·

Material from this book was presented to seminars at the Science Studies Centre, Bath University; the Department of History and Philosophy of Science, Cambridge University; the Institute for Historical Research, University College, London; Groupe Pandore, Paris; the Department of History and Sociology of Science, University of Pennsylvania; the Program in History of Science, Princeton University; the Institute for the History and Philosophy of Science and Ideas, Tel Aviv University. Talks based on the book were also presented to a joint meeting of the British Society for the History of Science and British Society for the Philosophy of Science at Leicester and to a joint course in the history of design at the Victoria & Albert Museum, London. We are grateful to members of those audiences for much constructive criticism. Portions of the manuscript were read by David Bloor, Harry Collins, Peter Dear, Nicholas Fisher, Jan Golinski, John Henry, Bruno Latour, and Andrew Pickering. We thank them all for their comments. We also wish to acknowledge the careful and sympathetic reports of the readers for Princeton University Press. Our other debts, diffuse and specific, are too numerous to list, but we must mention the encouragement, hospitality, and warm friendship of Yehuda Elkana and the generous bibliographic assistance of Jeffrey Sturchio at the E. F. Smith History of Chemistry Collection at the University of Pennsylvania. We also thank David Edge (for general support), Michael Aaron Dennis (for badges), Moyra Forrest (for proofreading the manuscript), Alice Calaprice (for wise editorial advice), and Dorinda Outram (for telling us not to).

During 1979-1980 Shapin was the recipient of a research fellowship from the John Simon Guggenheim Memorial Foundation. This book partly originated in work done at that time. Shapin would like to express his gratitude for that support and for the hospitality extended during the year by the students and staff of the Department of History and Sociology of Science, University of Pennsylvania. Research for chapter 6 was supported by a grant from the Royal Society of London, and we gratefully acknowledge that assistance.

A version of part of chapter 2 was published as "Pump and Circumstance: Robert Boyle's Literary Technology," in *Social Studies of Science* 14 (1984), 481-520. We thank Sage Publications Ltd. for permission to use this material. For permission to quote from manuscripts in their care we should like to thank the Syndics of Cambridge University Library and the Trustees of the British Library. For permission to reproduce pictorial material in their keeping, we thank the National Portrait Gallery, London (figure 5); the Sutherland Collection of the Ashmolean Museum, Oxford (figure 16); Cambridge University Library (figures 17, 20, 21, and 22); the British Library (figures 2 and 4); and Edinburgh University Library (figures 1, 3, 6, 7, 8, 9, 11, 12, 13, 14, 15, 18, 19, and the diagram in the translation of the *Dialogus physicus*). For permission to use the epigraph to chapter 1, we thank the holders of the original copyright to Umberto Eco's *The Name of the Rose*: Gruppo Editoriale Fabbri, Bompiani, Sonzogno, Etas S.p.A., Milan (American edition published by Harcourt Brace Jovanovich).

January 1985
Aulthucknall, Derbyshire

Leviathan and the Air-Pump

· I ·

Understanding Experiment

Adso:	*"But how does it happen," I said with admiration, "that you were able to solve the mystery of the library looking at it from the outside, and you were unable to solve it when you were inside?"*
William of Baskerville:	*"Thus God knows the world, because He conceived it in His mind, as if from the outside, before it was created, and we do not know its rule, because we live inside it, having found it already made."*

UMBERTO ECO, The Name of the Rose

OUR subject is experiment. We want to understand the nature and status of experimental practices and their intellectual products. These are the questions to which we seek answers: What is an experiment? How is an experiment performed? What are the means by which experiments can be said to produce matters of fact, and what is the relationship between experimental facts and explanatory constructs? How is a successful experiment identified, and how is success distinguished from experimental failure? Behind this series of particular questions lie more general ones: *Why* does one do experiments in order to arrive at scientific truth? Is experiment a privileged means of arriving at consensually agreed knowledge of nature, or are other means possible? What recommends the experimental way in science over alternatives to it?

We want our answers to be historical in character. To that end, we will deal with the historical circumstances in which experiment as a systematic means of generating natural knowledge arose, in which experimental practices became institutionalized, and in which experimentally produced matters of fact were made into the foundations of what counted as proper scientific knowledge. We start, therefore, with that great paradigm of experimental procedure: Robert Boyle's researches in pneumatics and his employment of the air-pump in that enterprise.

Boyle's air-pump experiments have a canonical character in science texts, in science pedagogy, and in the academic discipline of the history of science. Of all subjects in the history of science it might be thought that this would be the one about which least new could be said. It is an oft-told tale and, in the main, a well-told tale. Indeed, there are many aspects of Boyle's experimental work and the setting in which it occurred that have been sufficiently documented and about which we shall have little novel to say: our debt to previous historical writing is too extensive to acknowledge adequately. It is entirely appropriate that an excellent account of Boyle's pneumatic experiments of the 1660s constitutes the first of the celebrated series of *Harvard Case Histories in Experimental Science*.[1] This thirty-five-year-old study admirably establishes our point of departure: it shows that Boyle's air-pump experiments were designed to provide (and have since provided) a heuristic model of how authentic scientific knowledge should be secured.

Interestingly, the Harvard history has itself acquired a canonical status: through its justified place in the teaching of history of science it has provided a concrete exemplar of how to do research in the discipline, what sorts of historical questions are pertinent to ask, what kinds of historical materials are relevant to the inquiry, what sorts are not germane, and what the general form of historical narrative and explanation ought to be. Yet it is now time to move on from the methods, assumptions, and the historical programme embedded in the Harvard case history and other studies like it. We want to look again at the air-pump experiments, to put additional questions to these materials and to rephrase traditional questions. We did not initiate our project with a view to criticizing existing accounts of Boyle's experimental work. In fact, at the outset we were doubtful that we could add much to the work of distinguished Boyle scholars of the past. Yet, as our analysis proceeded, we became increasingly convinced that the questions we wished to have answered had not been systematically posed by previous writers. Why not?

A solution might reside in the distinction between "member's accounts" and "stranger's accounts." Being a member of the culture one seeks to understand has enormous advantages. Indeed, it is difficult to see how one could understand a culture to which one was a complete stranger. Nevertheless, unreflective membership also carries with it serious disadvantages to the search for understanding, and the chief of these might be called "the self-evident

[1] Conant, "Boyle's Experiments in Pneumatics"; idem, *On Understanding Science*, pp. 29-64.

method."[2] One reason why historians have not systematically and searchingly pressed the questions we want to ask about experimental practices is that they have, to a great extent, been producing accounts coloured by the member's self-evident method. In this method the presuppositions of our own culture's routine practices are not regarded as problematic and in need of explanation. Ordinarily, our culture's beliefs and practices are referred to the unambiguous facts of nature or to universal and impersonal criteria of how people just do things (or do them when behaving "rationally"). A lay member of our culture, if asked why he calls an ostrich a bird, will probably tell his inquisitor that ostriches just *are* birds, or he will point to unproblematic criteria of the Linnaean system of classification by which ostriches are so categorized. By contrast, this lay member will think of a range of explanations to bring to bear upon a culture that excludes ostriches from the class of birds.[3] In the case of experimental culture, the self-evident method is particularly noticeable in historians' accounts; and it is easy to see why this should be the case, for historians are in wide agreement in identifying Boyle as a founder of the experimental world in which scientists now live and operate. Thus, historians start with the assumption that they (and modern scientists) share a culture with Robert Boyle, and treat their subject accordingly: the historian and the seventeenth-century experimentalist are both members. The historical career of experimental culture can be enlisted in support of this assumption. Boyle's programme triumphed over alternatives and objections, and in his own country it did so very rapidly, largely aided and abetted by the vigorously partisan publicity of the Royal Society of London. The success of the experimental programme is commonly treated as its own explanation.[4] Even so, the usual way in which the self-evident method presents itself in historical practice is more subtle—not as a set of explicit

[2] See, for example, Douglas, "Self-Evidence."

[3] A classic site for relativist and realist discussions of classification and the natural world is Bulmer, "Why is the Cassowary not a Bird?" Bulmer's account is crucially asymmetrical: only cultures that do not classify the cassowary as a bird arouse his curiosity. For symmetrical treatments of this question, see Bloor, "Durkheim and Mauss Revisited"; idem, *Knowledge and Social Imagery*, chap. 1; Barnes and Bloor, "Relativism, Rationalism and the Sociology of Knowledge," esp. pp. 37-38.

[4] For a powerful nineteenth-century expression of this view, see Herschel, *Preliminary Discourse on the Study of Natural Philosophy*, pp. 115-116. Among many twentieth-century examples, see L. T. More, *Life of Boyle*, p. 239: "[Boyle's] conclusions were universally accepted, disregarding the objections of Linus and Hobbes, and he was immediately proclaimed as the highest authority in science."

claims about the rise, acceptance, and institutionalization of experiment, but as a disposition not to see the point of putting certain questions about the nature of experiment and its status in our overall intellectual map.

The member's account, and its associated self-evident method, have great instinctive appeal; the social forces that protect and sustain them are powerful. The member who poses awkward questions about "what everybody knows" in the shared culture runs a real risk of being dealt with as a troublemaker or an idiot. Indeed, there are few more reliable ways of being expelled from a culture than continuing seriously to query its taken-for-granted intellectual framework.[5] Playing the stranger is therefore a difficult business; yet this is precisely what we need to do with respect to the culture of experiment. We need to *play* the stranger, not to *be* the stranger. A genuine stranger is simply ignorant. We wish to adopt a calculated and an informed suspension of our taken-for-granted perceptions of experimental practice and its products. By playing the stranger we hope to move away from self-evidence. We want to approach "our" culture of experiment as Alfred Schutz suggests a stranger approaches an alien society, "not [as] a shelter but [as] a field of adventure, not a matter of course but a questionable topic of investigation, not an instrument for disentangling problematic situations but a problematic situation itself and one hard to master."[6] If we pretend to be a stranger to experimental culture, we can seek to appropriate one great advantage the stranger has over the member in explaining the beliefs and practices of a specific culture: the stranger is in a position to *know* that there are alternatives to those beliefs and practices.[7] The awareness of alternatives and the pertinence of the explanatory project go together.

Of course, we are not anthropologists but historians. How can the historian play the stranger to experimental culture, a culture we are said to share with a setting in the past and of which one of our subjects is said to be the founder? One means we can use is

[5] See the "experiments" of Harold Garfinkel on questioning taken-for-granted rules of social interaction: *Studies in Ethnomethodology*, esp. chap. 2.

[6] Schutz, *Collected Papers, Vol. II*, p. 104.

[7] The relative advantages of the member's and stranger's perspective have been debated by sociologists undertaking participant observation of modern science. Latour and Woolgar, *Laboratory Life*, chap. 1, are wary of the methodological dangers of identifying with the scientists they study, whereas Collins, "Understanding Science," esp. pp. 373-374, argues that only by becoming a competent member of the community under study can one reliably test one's understanding.

the identification and examination of episodes of *controversy* in the past. Historical instances of controversy over natural phenomena or intellectual practices have two advantages, from our point of view. One is that they often involve disagreements over the reality of entities or propriety of practices whose existence or value are subsequently taken to be unproblematic or settled. In H. M. Collins' metaphor, institutionalized beliefs about the natural world are like the ship in the bottle, whereas instances of scientific controversy offer us the opportunity to see that the ship was once a pile of sticks and string, and that it was once outside the bottle.[8] Another advantage afforded by studying controversy is that historical actors frequently play a role analogous to that of our pretend-stranger: in the course of controversy they attempt to deconstruct the taken-for-granted quality of their antagonists' preferred beliefs and practices, and they do this by trying to display the artifactual and conventional status of those beliefs and practices. Since this is the case, participants in controversy offer the historian resources for playing stranger. It would, of course, be a great mistake for the historian simply to appropriate and validate the analysis of one side to scientific controversy, and this is not what we propose to do. We have found it valuable to note the constructive and deconstructive strategies employed by both sides to the controversy. While we use participants' accounts, we shall not confuse them with our own interpretative work: the historian speaks for himself.

The controversy with which we are concerned took place in England in the 1660s and early 1670s. The protagonists were Robert Boyle (1627-1691) and Thomas Hobbes (1588-1679). Boyle appears as the major practitioner of systematic experimentation and one of the most important propagandists for the value of experimental practices in natural philosophy. Hobbes takes the role of Boyle's most vigorous local opponent, seeking to undermine the particular claims and interpretations produced by Boyle's researches and, crucially, mobilizing powerful arguments why the experimental programme could not produce the sort of knowledge Boyle recommended. There are a number of reasons why the Hobbes-Boyle disputes are particularly intractable ones for the historian to analyze. One reason is the extent to which the figure of Hobbes as a *natural philosopher* has disappeared from the literature. Kargon rightly says that "Hobbes was one of the three most important mechanical philosophers of the mid-seventeenth century,

[8] Collins, "The Seven Sexes"; idem, "Son of Seven Sexes."

along with Descartes and Gassend."[9] There is no lack of evidence of the seriousness with which Hobbes's natural philosophical views were treated in the seventeenth century, especially, but not exclusively, by those who considered them to be seriously flawed. We know that as late as the early eighteenth century Hobbes's natural philosophical tracts formed an important component of the Scottish university curriculum.[10] Yet by the end of the eighteenth century Hobbes had largely been written out of the history of science. The entry on Hobbes in the 1797 third edition of the *Encyclopaedia Britannica* scarcely mentions Hobbes's scientific views and totally ignores the tracts written against Boyle. Much the same is true of the *Encyclopaedia*'s 1842 *Dissertation on the History . . . of Mathematical and Physical Science*: Hobbes is to be remembered as an ethical, political, psychological, and metaphysical philosopher; the unity of those concerns with the philosophy of nature, so insisted upon by Hobbes, has been split up and the science dismissed from consideration. Even Mintz's article on Hobbes in the *Dictionary of Scientific Biography* is biased heavily towards his moral, political, and psychological writings.[11] Fortunately for us, since Brandt's 1928 monograph on Hobbes's mechanical philosophy, this situation has begun to improve. Our indebtedness to recent work on Hobbes's science by scholars such as R. H. Kargon, J.W.N. Watkins, Alan Shapiro, Miriam Reik, and Thomas Spragens will be evident in what follows. Nevertheless, we are still very far from appreciating Hobbes's true place in seventeenth-century natural philosophy, and, if this book stimulates further research, one of its functions will have been fulfilled.

Kargon suggests that one of the reasons for the neglect of Hobbes by historians of science lies in the fact that he disagreed with the hero Boyle and, accordingly, suffered ostracism from the Royal Society of London.[12] There is no doubt that Hobbes's scientific controversies in England, all of which his contemporaries considered he decisively lost, have much to do with his dismissal by historians. Within the tradition of "Whig" history, losing sides have little interest, and in no type of history has this tendency been more

[9] Kargon, *Atomism in England*, p. 54.

[10] Shepherd, "Newtonianism in Scottish Universities," esp. p. 70; idem, *Philosophy and Science in the Scottish Universities*, pp. 8, 116, 153, 167, 215-217.

[11] Anon., "Hobbes"; Mackintosh, "Dissertation Second," pp. 316-323 (on ethical philosophy); Playfair, "Dissertation Third" (on mathematical and physical science, where Hobbes is scarcely mentioned at all); Mintz, "Hobbes."

[12] Kargon, *Atomism in England*, p. 54.

apparent than in classical history of science.[13] This book is concerned with Hobbes's natural philosophical controversies, yet his mathematical disputes with John Wallis and Seth Ward, which we cannot treat in any detail, were lost even more spectacularly and have disappeared from the historical record more thoroughly than the fight with Boyle. In Leslie Stephen's *Dictionary of National Biography* entry, Hobbes's opponents showed his "manifold absurdities"; Croom Robertson's more extended account in the eleventh edition of the *Encyclopaedia Britannica* echoes that judgment; and no historian dissents.[14]

The situation is similar in historians' accounts of Hobbes's controversies with Boyle. There is not very much written about these disputes, and even that little has contained some fundamental errors. For example, one writer has claimed that Hobbes's objections to Boyle's natural philosophy stemmed from Hobbes's belief in the Aristotelian *horror vacui* (which is quite wrong),[15] and another, more sensitive, writer has argued that Hobbes approved of a central role for experimentation in natural philosophy (which we shall be at pains to show to be wrong).[16] It is possible that part of the reason for these errors, and for the general neglect of the Hobbes-Boyle controversies, is documentary. So far as we have been able to determine, only two historians give solid indications that they have opened the crucial text and digested any of its contents: Hobbes's *Dialogus physicus de natura aeris* of 1661.[17] True, Hobbes's *Dialogus*

[13] The Whiggish tendency in the treatment of the disputes between Boyle, Hobbes, and Linus is briefly noted in Brush, *Statistical Physics*, p. 16.

[14] Stephen, "Hobbes," esp. p. 935 (cf. idem, *Hobbes*, pp. 51-54); Robertson, "Hobbes," esp. pp. 549-550 (cf. idem, *Hobbes*, pp. 160-185); A. E. Taylor, *Hobbes*, esp. pp. 18-21, 40-41. See also Scott, "John Wallis," p. 65. For work on Hobbes's geometry and the controversies with the Oxford professors, see Sacksteder, "Hobbes: Geometrical Objects"; idem, "Hobbes: The Art of the Geometricians"; Breidert, "Les mathématiques et la méthode mathématique chez Hobbes"; Scott, *The Mathematical Work of Wallis*, ch. 10.

[15] For the *horror vacui* claim, see Greene, "More and Boyle on the Spirit of Nature," p. 463; for a note pointing out this error, see Applebaum, "Boyle and Hobbes."

[16] Watkins, *Hobbes's System*, p. 70n. This claim is dealt with in detail in chapter 4 below.

[17] The exceptions are Gargani, *Hobbes e la scienza*, pp. 278-285, and Lupoli, "La polemica tra Hobbes e Boyle." Gargani points out that the *Dialogus* "belongs to a fairly advanced stage of Hobbes's philosophical and scientific career." Gargani does not see the *Dialogus* as developing anything original; instead, he reads it as continuous with the plenist physics and the critique of naive experimentalism in earlier writings (notably *De corpore* and the *Short Tract on First Principles*: see pp. 134-138, 271-278). But Gargani *only* cites the two prefatory dedications of Hobbes's *Dialogus*

has never been translated from the Latin original, and this may go some way to explain its neglect. (To remedy this state of affairs, we offer an English translation, by Schaffer, as an appendix to this book.) With these two exceptions, historians have been content to align themselves with the victorious Boyle and his associates, to repeat Boyle's judgment on Hobbes's text, and to keep silent about what Hobbes actually had to say. Even Brandt, who wrote the most detailed study of Hobbes's science, declined engagement with the *Dialogus physicus* and later natural philosophical texts. Brandt, too, accepted Boyle's evaluation of Hobbes's views:

> We will not examine the works subsequent to *De Corpore* [of 1655, six years before the *Dialogus physicus*]. . . . No less than three times during these years Hobbes took up his physics for further elaboration . . . , but it retains exactly the same character as the physics of *De Corpore*. This character becomes especially conspicuous in Hobbes' attack on Boyle's famous "New Experiments touching the Spring of the Aire." Here again Hobbes shows how little he understands the significance of the experiment. In spite of the continual experiments on vacuity,

and pays no attention to the actual text or to the attack on Boyle's air-pump programme. Lupoli gives a full and valuable exposition of Boyle's response to Hobbes in the *Examen*. He places the controversy in the context of the earlier pneumatic trials in Italy and France in the 1640s, notably the Pascal-Noël debate. Lupoli suggests that Hobbes attacked Boyle because of his "disappointment at being excluded from the new scientific association, but above all the disillusion and preoccupation with seeing his foundation of physical science ignored" (p. 324). Lupoli highlights Boyle's prolixity as a response to Hobbes's attack on the "rhetoric of ingenuity," and Boyle's tactic of point-by-point refutation of empirical claims as a means of avoiding a direct confrontation with Hobbes's whole physical programme (p. 329). But Lupoli is much more interested in Boyle's utterances on method and on experimental philosophy, and does not give any detailed account of the sources of Hobbes's own polemic. We are grateful to Agostino Lupoli for a copy of his paper (received after our manuscript was written): it is the only source we have found that cites the *Dialogus* in detail. Other major recent sources for Hobbes's natural philosophy do not treat the controversies with Boyle in any detail, nor do they examine the contents of Hobbes's *Dialogus physicus*; see, for example, Spragens, *The Politics of Motion*, esp. chap. 3; Reik, *The Golden Lands of Hobbes*, chap. 7; Goldsmith, *Hobbes's Science of Politics*, chap. 2, although each of these is valuable in other connections. In addition, there are many allusions to Hobbes's science by mainstream Hobbes scholars. They have tended to mine his philosophy of nature because of the generally high evaluation that historians of ideas have placed upon the significance of Hobbes's political and psychological theories and because of their conviction that there must be an overall pattern in his thought. Historians of science, given their low evaluation of Hobbes's natural philosophy and mathematics, have not tended to search for such a pattern.

> in spite of the invention of the air-pump, Hobbes still adhered to his view of the full world. Hobbes' last years were rather tragic. He did not well understand the great development of English empirical science that took place just at that time. . . . And when the members of the Royal Society adopted the experimental method of research . . . Hobbes could no longer keep abreast of them.[18]

Here we see the germ of a standard historiographic strategy for dealing with the Hobbes-Boyle controversy, and, arguably, for handling rejected knowledge in general. We have a dismissal, the rudiments of a causal explanation of the rejected knowledge (which implicitly acts to justify the dismissal), and an asymmetrical handling of rejected and accepted knowledge. First, it is established that the rejected knowledge is not knowledge at all, but error. This the historian accomplishes by taking the side of accepted knowledge and using the victorious party's causal explanation of their adversaries' position as the historian's own. Since the victors have thus disposed of error, so the historian's dismissal is justified.[19] Thus, L. T. More notes that Hobbes's "sneers" at Boyle were "a farrago of nonsense," and quotes Boyle's decisive riposte without detailing what Hobbes's position was.[20] McKie deals with the disputes simply by saying that "Boyle disposed very competently of Hobbes's arguments and very gracefully of his contentious and splenetic outburst."[21] John Laird concludes that "the essential justice of Boyle's criticisms [of Hobbes] shows . . . that it would be unprofitable to examine much of Hobbian special physics in detail. . . ."[22] Peters claims that Hobbes's criticisms "would have come better from one . . . who had himself done some experiments" (which cannot be the best way of seeking to understand a controversy over the validity and value of experiment),[23] and R. F. Jones concurs.[24] Other his-

[18] Brandt, *Hobbes' Mechanical Conception*, pp. 377-378.

[19] For alternative sociological and historical approaches to rejected knowledge, see the contributions to Wallis, ed., *On the Margins of Science*, and Collins and Pinch, *Frames of Meaning*.

[20] L. T. More, *Life of Boyle*, p. 97. Maddison's more recent *Life of Boyle* (pp. 106-109) has even less to say about the controversy.

[21] McKie, "Introduction," pp. xii*-xiii*.

[22] Laird, *Hobbes*, p. 117.

[23] Peters, *Hobbes*, p. 40.

[24] R. F. Jones, *Ancients and Moderns*, p. 128; de Beer, "Some Letters of Hobbes," p. 197: Hobbes "failed to appreciate . . . the paramount value of experiment in deciding any question of natural philosophy."

torians go further in wiping the historical record clean of significant opposition to the experimental programme: Marie Boas Hall, though without mentioning Hobbes by name, says that "No one but a dedicated Aristotelian" (which Hobbes most certainly was not) "could fail to find Boyle's arguments powerful and convincing,"[25] and Barbara Shapiro, in her admirable account of English empiricism and experimentalism, concludes that "Except for a tiny group of critics who poked fun at the virtuosi" (whose names she does not mention), "there was no serious opposition to the new philosophy."[26]

Pervasively, historians have drawn upon the notion of "misunderstanding" (and the reasons for it) as the basis of their causal accounting and dismissal of Hobbes's position. The *Harvard Case Histories* relate that Hobbes's arguments against Boyle "were based in part on a misunderstanding of Boyle's views."[27] M. A. Stewart refers to Boyle's pneumatics as leading "Hobbes into ill-advised controversy on matters he did not understand."[28] Leslie Stephen and Croom Robertson both attempt to explain Hobbes's misunderstanding by referring to factors that distorted his judgment or made him unfit to appreciate the validity of Boyle's programme: he was ill-qualified in mathematics and physics; he was too old and rigid at the time of his controversies with Boyle; he was temperamentally obstinate and dogmatic; he had ideological axes to grind.[29] (To the best of our knowledge no historian has ever suggested that Boyle may have "misunderstood" Hobbes.)

Since our way of proceeding will dispense with the category of "misunderstanding" and the asymmetries associated with it, some words on method are indicated here. Almost needless to say, our purpose is not evaluative: it is descriptive and explanatory. Nevertheless, questions relating to evaluation do figure centrally in this book, and they do so in several ways. We have said that we shall be setting out by pretending to adopt a "stranger's perspective" with respect to the experimental programme; we shall do this be-

[25] M. B. Hall, "Boyle," p. 379. Her *Boyle and Seventeenth-Century Chemistry* makes no mention of the Boyle-Hobbes disputes; cf. Burtt, *Metaphysical Foundations of Modern Science*, p. 169.

[26] B. Shapiro, *Probability and Certainty*, p. 73; cf. p. 68.

[27] Conant, "Boyle's Experiments in Pneumatics," p. 49.

[28] Stewart, "Introduction," p. xvi. Hobbes's "misunderstanding" of Boyle even creeps into accounts written for young people; see Kuslan and Stone, *Boyle: The Great Experimenter*, p. 26.

[29] Stephen, "Hobbes," p. 937; Robertson, "Hobbes," p. 552.

cause we have set ourselves the historical task of inquiring into *why* experimental practices were accounted proper and *how* such practices were considered to yield reliable knowledge. As part of the same exercise we shall be adopting something close to a "member's account" of Hobbes's anti-experimentalism. That is to say, we want to put ourselves in a position where objections to the experimental programme seem plausible, sensible, and rational. Following Gellner, we shall be offering a "charitable interpretation" of Hobbes's point of view.[30] Our purpose is not to take Hobbes's side, nor even to resuscitate his scientific reputation (though this, in our opinion, has been seriously undervalued). Our goal is to break down the aura of self-evidence surrounding the experimental way of producing knowledge, and "charitable interpretation" of the opposition to experimentalism is a valuable means of accomplishing this. Of course, our ambition is not to rewrite the clear judgment of history: Hobbes's views found little support in the English natural philosophical community. Yet we want to show that there was nothing self-evident or inevitable about the series of historical judgments in that context which yielded a natural philosophical consensus in favour of the experimental programme. Given other circumstances bearing upon that philosophical community, Hobbes's views might well have found a different reception. They were not widely credited or believed—but they were *believable*; they were not counted to be correct—but there was nothing inherent in them that prevented a different evaluation. (True, there were points at which Hobbes's criticisms were less than well-informed, just as there were aspects of Boyle's position that might be regarded as ill-informed and even sloppy. If the historian *wanted* to evaluate the actors by the standards of present-day scientific procedure, he would find both Hobbes and Boyle vulnerable.) On the other hand, our treatment of Boyle's experimentalism will stress the fundamental roles of convention, of practical agreement, and of labour in the creation and positive evaluation of experimental knowledge. We shall try to identify those features of the historical setting that bore upon intellectuals' decisions that these conventions were appropriate, that such agreement was necessary, and that the labour involved in experimental knowledge-production was worthwhile and to be preferred over alternatives.

Far from avoiding questions of "truth," "objectivity," and "proper method," we will be confronting such matters centrally. But we

[30] Gellner, "Concepts and Society"; cf. Collins, "Son of Seven Sexes," pp. 52-54.

shall be treating them in a manner slightly different from that which characterizes some history and much philosophy of science. "Truth," "adequacy," and "objectivity" will be dealt with as accomplishments, as historical products, as actors' judgments and categories. They will be topics for our inquiry, not resources unreflectively to be used in that inquiry. How and why were certain practices and beliefs accounted proper and true? In assessing matters of scientific method we shall be following a similar path. For us, methodology will not be treated solely as a set of formal statements about how to produce knowledge, and not at all as a determinant of intellectual practice. We shall be intermittently concerned with explicit verbal statements about how philosophers should conduct themselves, but such method-statements will invariably be analyzed in relation to the precise setting in which they were produced, in terms of the purposes of those making them, and in reference to the actual nature of contemporary scientific practice.[31] More important to our project is an examination of method understood as real practical activity. For example, we shall devote much attention to such questions as: How is an experimental matter of fact actually produced? What are the practical criteria for judging experimental success or failure? How, and to what extent, are experiments actually replicated, and what is it that enables replication to take place? How is the experimental boundary between fact and theory actually managed? Are there crucial experiments and, if so, on what grounds are they accounted crucial? Further, we shall be endeavouring to broaden our usual appreciations of what scientific method consists of and how method in natural philosophy relates to practical intellectual procedures in other areas of culture and in the wider society. One way we shall try to do this is by situating scientific method, and controversies about it, in a social context.

By adducing "social context" it is routinely understood that one is pointing to the wider society, and, to a very large extent, we shall be concerned to show the connections between the conduct of the natural philosophical community and Restoration society in general. However, we also mean something else when we use the term "social context." We intend to display scientific method as crystallizing forms of social organization and as a means of regulating social interaction within the scientific community. To this end, we

[31] For examples of empirical studies which assess method-statements in these terms, see P. B. Wood, "Methodology and Apologetics"; Miller, "Method and the 'Micropolitics' of Science"; Yeo, "Scientific Method and the Image of Science."

will make liberal, but informal, use of Wittgenstein's notions of a "language-game" and a "form of life." We mean to approach scientific method as integrated into *patterns of activity*. Just as for Wittgenstein "the term 'language-*game*' is meant to bring into prominence the fact that the *speaking* of language is part of an activity or of a form of life," so we shall treat controversies over scientific method as disputes over different patterns of doing things and of organizing men to practical ends.[32] We shall suggest that solutions to the problem of knowledge are embedded within practical solutions to the problem of social order, and that different practical solutions to the problem of social order encapsulate contrasting practical solutions to the problem of knowledge. *That* is what the Hobbes-Boyle controversies were about.

It will not escape our readers' notice that this book is an exercise in the sociology of scientific knowledge. One can either debate the possibility of the sociology of knowledge, or one can get on with the job of doing the thing.[33] We have chosen the latter option. It follows from our decision that we shall be making relatively few references to the theoretical literature in the sociology of knowledge that has been a major and continuing source of inspiration to our project. Nevertheless, we trust that our practical historical procedures will bear sufficient witness to our obligations in that quarter. Our methodological debts also extend in many other directions, and they are too deep and extensive to be adequately acknowledged. Among Hobbes scholars we are especially indebted to J.W.N. Watkins (for his insistence upon the relationships between the natural and civic philosophy), even while we dissent from him on the issue of Hobbes's attitudes to experiment; and to Quentin Skinner (for aspects of his historiography), even while diverging from him over Hobbes's relations with the Royal Society. Among historians of science we have found substantial inspiration in recent studies of the actual nature of experimental practice: we have particularly in mind the work of Robert Frank and John Heilbron. The particular orientation to the understanding of scientific experiment that we have found most valuable derives from the work

[32] Wittgenstein, *Philosophical Investigations*, I, 23; idem, *Blue and Brown Books*, pp. 17, 81; Bloor, *Wittgenstein*, chap. 3. Foucault's "discourse" has a number of interesting similarities with Wittgenstein's "language-game," but we prefer Wittgenstein because of his stress on the primacy of practical *activity*. For Foucauldian usages, see, especially, *The Archaeology of Knowledge*, chaps. 1-2.

[33] The present state of the sociology of scientific knowledge as an empirical practice is examined in Shapin, "History of Science and Its Sociological Reconstructions."

of British and French micro-sociologists of science: H. M. Collins, T. J. Pinch, Bruno Latour, and Andrew Pickering, and from the pioneering Ludwik Fleck.

Since these debts are obvious and evident, it may be of some interest to acknowledge two pieces of empirical history whose connection with our own project may be less readily apparent, but which exemplify similar orientations to those employed here. John Keegan opens his magnificent study of the history of battle with the following confession:

> I have not been in a battle; not near one, nor heard one from afar, nor seen the aftermath. . . . I have read about battles, of course, have talked about battles, have been lectured about battles. . . . But I have never been in a battle. And I grow increasingly convinced that I have very little idea of what a battle can be like.[34]

It is a graceful admission of an ignorance that Keegan recognized in himself as a teacher at Sandhurst and in many military historians. Without this recognition, Keegan would have been unable to write the vivid and moving history that he ultimately produced. As we began the research for this book, we felt ourselves to be in a position similar to Keegan's. We had read much about experiment; we had both even performed a few as students; but we did not feel that we had a satisfactory idea of what an experiment was and how it yielded scientific knowledge. The parallel with Keegan's account of battle extends even farther. Keegan identifies a dominant variety of military history, shaped by Count von Moltke, which he refers to as "General Staff History." In General Staff History, what is of overarching significance is the role of the generals, their strategic planning, their rational decision-making, and their influence on the ultimate course of the battle. What is systematically left out of General Staff History is the contingency and the confusion of actual combat, the role of small groups of soldiers, the relationship between battle on the ground and the planning of the generals. It would not be a flight of fancy to recognize in General Staff History a family resemblance to "rational reconstructionist" tendencies in the history and philosophy of science. The "von Moltkes" of the history of science have shown similar disinclinations to engage with actual scientific practice, preferring idealizations and simplifications

[34] Keegan, *The Face of Battle*, p. 15; see also Keegan's more detailed account of a World War II series of battles, *Six Armies in Normandy*.

to messy contingencies, speech of essences to the identification of conventions, references to unproblematic facts of nature and transcendent criteria of scientific method to the historical work done by real scientific actors.[35] It is too much to think that we have added to the history of experiment a fraction of what Keegan has contributed to military history, but we are happy to be engaged in the same historiographic enterprise.

Our other unexpected model is closer in its empirical focus to our own objects of study: Svetlana Alpers' *The Art of Describing*. Unfortunately for us, Alpers' book was published when our own work was substantially completed, and we have not been able to engage with it as extensively as we would have liked. Nevertheless, the parallels with our project are highly important, and we want briefly to point them out. Alpers is concerned with Dutch descriptive art in the seventeenth century. In particular, she wants to understand the assumptions behind Dutch preferences for descriptive painting and the conventions employed in making such pictures. She writes: "It was a particular assumption of the seventeenth century that finding and making, our discovery of the world and our crafting of it, are presumed to be one."[36] She shows that such assumptions spread across disparate areas of culture: universal language projects, the experimental programme in science, and painting, and that they were particularly pronounced in the Netherlands and in England. Both Dutch descriptive painting and English empiricist science involved a perceptual metaphor for knowledge: "By this I mean a culture that assumes that we know what we know through the mind's mirroring of nature."[37] The basis for certain knowledge was to be nature witnessed. The craft of the painter,

[35] The deep-rooted bias against the study of experimental practice displayed by historians of science has been noted by several writers; see, for example, Eklund, *The Incompleat Chymist*, p. 1. Even philosophers are now beginning to admit the anti-practice and pro-theory prejudices of their discipline; see Hacking, *Representing and Intervening*, chap. 9, esp. pp. 149-150: "History of the natural sciences is now almost always written as a history of theory. Philosophy of science has so much become the philosophy of theory that the very existence of pre-theoretical observations or experiments has been denied."

[36] Alpers, *The Art of Describing*, p. 27. Similar exercises in art history that offer valuable resources to the sociologically inclined historian of science include Baxandall's *Painting and Experience*, his *Limewood Sculptors of Renaissance Germany*, and Edgerton's *The Renaissance Discovery of Linear Perspective*.

[37] Alpers, *The Art of Describing*, pp. 45-46. Alpers alludes to Rorty's important survey of the development of mirror theories of knowledge: *Philosophy and the Mirror of Nature*, esp. chap. 3.

and the art of the experimentalist, was, therefore, to make representations that reliably imitated the act of unmediated seeing.

There are two points in Alpers' account of special interest to us. One is the contrast she draws between Northern (and particularly Dutch) conceptions of the picture and those characteristic of Italian painting. In the latter the painting was conceived primarily as a gloss on a text; in the former the textual meaning of the picture was dispensed with in favour of direct visual apprehension of natural reality. Although the details of the contrast cannot concern us here, Alpers concludes that different theories of picturing expressed different conceptions of knowledge: the text versus the eye. The parallel between the Hobbes-Boyle controversy, and its underlying conflict over theories of knowledge, is far from exact; nevertheless, in the case of conflicts over the propriety of experimental methods we see a quite similar dispute over the reliability of the eye, and of witnessing, as the basis for generating and warranting knowledge. Secondly, Alpers adopts what we have termed a "stranger's perspective" to the nature of realist images. Their "mirroring" of reality is treated as the product of *convention* and of *craft*: "To appear lifelike, a picture has to be carefully made." The craft of realist representation is predicated upon the acceptance of Hooke's conventions for making realist statements in science: the "sincere hand" and the "faithful eye."[38] With the acceptance of this convention for knowledge, and with the execution of the craft of representation, the artful nature of making representations disappears, and they acquire the status of mirrors of reality. Our project, therefore, is the same as Alpers': to display the conventions and the craft.

In the following chapter we examine the form of life that Boyle proposed for experimental philosophy. We identify the technical, literary, and social practices whereby experimental matters of fact were to be generated, validated, and formed into bases for consensus. We pay special attention to the operation of the air-pump and the means by which experiments employing this device could be made to yield what counted as unassailable knowledge. We discuss the social and linguistic practices Boyle recommended to experimentalists; we show how these were important constitutive elements in the making of matters of fact and in protecting such facts from items of knowledge that were thought to generate discord

[38] Alpers, *The Art of Describing*, pp. 72-73 (quoting Robert Hooke's *Micrographia* [1665], sig a2^{v}).

and conflict. Our task here is to identify the conventions by which experimental knowledge was to be produced.

In chapter 3 we discuss the state and objects of Hobbes's natural philosophy before the publication of Boyle's *New Experiments* of 1660. Our major object here is to read *Leviathan* (1651) as *natural* philosophy and as epistemology. As a treatise in civic philosophy *Leviathan* was designed to show the practices that would guarantee order in the state. That order could be, and during the Civil War was being, threatened by clerical intellectuals who arrogated to themselves a share of civic authority to which they were not entitled. Their major resources in these acts of usurpation were, according to Hobbes, a false ontology and a false epistemology. Hobbes endeavoured to show the absurdity of an ontology that posited incorporeal substances and immaterial spirits. Thus, he built a *plenist* ontology, and, in the process, erected a materialistic theory of knowledge in which the foundations of knowledge were notions of *causes*, and those causes were matter and motion. An enterprise entitled to the name of philosophy was causal in nature. It modelled itself on the demonstrative enterprises of geometry and civic philosophy. And, crucially, it produced assent through its demonstrative character. Assent was to be total and it was to be enforced.

Hobbes's philosophy, both in *Leviathan* and in *De corpore* (1655) was already in place when Boyle's experimental programme became public in the year of the Restoration. He immediately replied to Boyle's radical proposals. The analysis of Hobbes's *Dialogus physicus* forms the framework for chapter 4. In this text, Hobbes attempted to explode Boyle's experimentalism on several grounds: he argued that Boyle's air-pump lacked physical integrity (it leaked) and that, therefore, its putative matters of fact were not facts at all; he used the leakage of the pump to offer an alternative physical explanation of Boyle's findings. The pump, far from being an operational vacuum, was always full of a fraction of atmospheric air. Plenist accounts of the pump were superior to Boyle's, and Hobbes attacked Boyle as a vacuist despite the latter's professions of nescience on the vacuist-plenist debates of the past. Of greater epistemological importance was Hobbes's attack on the generation of matters of fact, the constitution of such facts into the consensual foundations of knowledge, and Boyle's segregation of facts from the physical causes that might account for them. These attacks amounted to the assertion that, whatever Boyle's experimental programme was, it was not *philosophy*. Philosophy was a causal enterprise and, as such, secured a total and irrevocable assent, not the partial assent at which Boyle aimed. Hobbes's assault identified the conventional nature of experimental facts.

In chapter 5 we show how Boyle replied to Hobbes and to two other adversaries in the 1660s: the Jesuit Franciscus Linus and the Cambridge Platonist Henry More. By examining the different nature and style of Boyle's responses, we identify that which Boyle was most concerned to protect: the air-pump as a means of generating legitimate philosophical knowledge and the integrity of the rules that were to regulate the moral life of the experimental community. Boyle treated Hobbes as a failed experimentalist rather than as someone proposing a quite different way of constructing philosophical knowledge. He used the opportunities provided by all three adversaries to exhibit how experimental controversy could be managed, without destroying the experimental enterprise itself—indeed, to show how controversy could be used to buttress the factual foundations of experimental knowledge.

In chapters 2, 4, and 5 we discuss the central role of the air-pump in the experimental programme and how critics might use imperfections in its working to attack experiment itself. In chapter 6 we attempt to do two things. First, we look at how the pump itself evolved as a material object in the 1660s, arguing that these changes embodied responses to earlier criticisms, especially those offered by Hobbes. We uncover information about the small number of pumps that were successfully built in that decade, and we show that, despite Boyle's reporting practices, no one was able to build a pump and make it operate without seeing the original. This poses problems of *replication* of greater interest than historians have previously recognized. Replication is also central to the second task of this chapter. In chapter 2 we argue that the constitution of matters of fact involved the multiplication of witnesses, and that Boyle exerted himself to encourage the reiteration of his experiments. However, shortly after the *New Experiments* appeared, another philosopher, Christiaan Huygens in the Netherlands, produced a finding (the so-called anomalous suspension of water) that seemed to invalidate one of the most important of Boyle's explanatory resources. We examine how this important anomaly was treated, and we conclude that the successful working of the air-pump was calibrated by previous commitments to whether or not such a phenomenon could exist. We analyze response to anomaly as a manifestation of the experimental form of life and of the conventions employed in the experimental community to protect itself from fatal internal discord.

Boyle's experimentalism and Hobbes's demonstrative way were both offered as solutions to the problem of order. In chapter 7 we attempt to locate solutions to this problem in the wide-ranging Restoration debate over the nature and bases of assent and order in society. This debate provided the context in which different programmes for the production and protection of order were evaluated. We seek to show here the nature of the intersection between the history of natural philosophy and the history of political thought and action. One solution (Boyle's) was to set the house of natural philosophy in order by remedying its divisions and by withdrawing it from contentious links with civic philosophy. Thus repaired, the community of natural philosophers could establish its legitimacy in Restoration culture and contribute more effectively to guaranteeing order and right religion in society. Another solution (Hobbes's) demanded that order was only to be ensured by erecting a demonstrative philosophy that allowed no boundaries between the natural, the human, and the social, and which allowed for no dissent within it.

In the concluding chapter we draw out some of the implications of this study for the history of science and the history of politics. We argue that the problem of generating and protecting knowledge is a problem in politics, and, conversely, that the problem of political order always involves solutions to the problem of knowledge.

· II ·

Seeing and Believing: The Experimental Production of Pneumatic Facts

. . . Facts are chiels that winna ding,
An' downa be disputed.
ROBERT BURNS, A Dream

ROBERT Boyle maintained that proper natural philosophical knowledge should be generated through experiment and that the foundations of such knowledge were to be constituted by experimentally produced matters of fact. Thomas Hobbes disagreed. In Hobbes's view Boyle's procedures could never yield the degree of certainty requisite in any enterprise worthy of being called philosophical. This book is about that dispute and about the issues that were seen to depend upon its resolution.

Hobbes's position has the historical appeal of the exotic. How was it possible for any rational man to deny the value of experiment and the foundational status of the matter of fact? By contrast, Boyle's programme appears to exude the banality of the self-evident. How could any rational man think otherwise? In this chapter we intend to address the problem of self-evidence by dissecting and displaying the mechanisms by which Boyle's experimental procedures were held to produce knowledge and, in particular, the variety of knowledge called "matters of fact." We will show that the experimental production of matters of fact involved an immense amount of labour, that it rested upon the acceptance of certain social and discursive conventions, and that it depended upon the production and protection of a special form of social organization. The experimental programme was, in Wittgenstein's phrases, a "language-game" and a "form of life." The acceptance or rejection of that programme amounted to the acceptance or rejection of the form of life that Boyle and his colleagues proposed. Once this point is made, neither the acceptance of the experimental programme nor the epistemological status of the matter of fact ought to appear self-evident.

In the conventions of the intellectual world we now inhabit there is no item of knowledge so solid as a matter of fact. We may revise our ways of making sense of matters of fact and we may adjust their place in our overall maps of knowledge. Our theories, hypotheses, and our metaphysical systems may be jettisoned, but matters of fact stand undeniable and permanent. We do, to be sure, reject particular matters of fact, but the manner of our doing so adds solidity to the category of the fact. A discarded theory remains a theory; there are "good" theories and "bad" theories—theories currently regarded as true by everyone and theories that no one any longer believes to be true. However, when we reject a matter of fact, we take away its entitlement to the designation: it never was a matter of fact at all.

There is nothing so given as a matter of fact. In common speech, as in the philosophy of science, the solidity and permanence of matters of fact reside in the absence of human agency in their coming to be. Human agents make theories and interpretations, and human agents therefore may unmake them. But matters of fact are regarded as the very "mirror of nature."[1] Like Stendhal's ideal novel, matters of fact are held to be the passive result of holding a mirror up to reality. What men make, men may unmake; but what nature makes no man may dispute. To identify the role of human agency in the making of an item of knowledge is to identify the possibility of its being otherwise. To shift the agency onto natural reality is to stipulate the grounds for universal and irrevocable assent.

Robert Boyle sought to secure assent by way of the experimentally generated matter of fact. Facts were certain; other items of knowledge much less so. Boyle was therefore one of the most important actors in the seventeenth-century English movement towards a probabilistic and fallibilistic conception of man's natural knowledge. Before the mid-seventeenth century, as Hacking and Shapiro have shown, the designations of "knowledge" and "science" were rigidly distinguished from the category of "opinion."[2] Of the former one could expect the absolute certainty of demonstration, exemplified by logic and geometry. The goal of physical scientists had been to model their enterprise, so far as possible, upon the

[1] For a discussion of the historical origins of the correspondence theory of knowledge and the task of philosophy, see Rorty, *Philosophy and the Mirror of Nature*, esp. pp. 129ff.

[2] Hacking, *The Emergence of Probability*, esp. chaps. 3-5; B. Shapiro, *Probability and Certainty*, esp. chap. 2.

demonstrative sciences and to attain to the kind of certainty that compelled absolute assent. By contrast, English experimentalists of the mid-seventeenth century and afterwards increasingly took the view that all that could be expected of physical knowledge was "probability," thus breaking down the radical distinction between "knowledge" and "opinion." Physical hypotheses were provisional and revisable; assent to them was not obligatory, as it was to mathematical demonstrations; and physical science was, to varying degrees, removed from the realm of the demonstrative. The probabilistic conception of physical knowledge was not regarded by its proponents as a regrettable retreat from more ambitious goals; it was celebrated as a wise rejection of a failed project. By the adoption of a probabilistic view of knowledge one could attain to an *appropriate* certainty and aim to secure *legitimate* assent to knowledge-claims. The quest for necessary and universal assent to physical propositions was seen as inappropriate and illegitimate. It belonged to a "dogmatic" enterprise, and dogmatism was seen not only as a failure but as dangerous to genuine knowledge.

If universal and necessary assent was not to be expected of explanatory constructs in science, how then was proper science to be *founded*? Boyle and the experimentalists offered the matter of fact as the foundation of proper knowledge. In the system of physical knowledge the fact was the item about which one could have the highest degree of probabilistic assurance: "moral certainty." A crucial boundary was constructed around the domain of the factual, separating matters of fact from those items that might be otherwise and about which absolute, permanent, and even "moral" certainty should not be expected. In the root metaphor of the mechanical philosophy, nature was like a clock: man could be certain of the hour shown by its hands, of natural effects, but the mechanism by which those effects were really produced, the clockwork, might be various.[3] In this chapter we shall examine the means by which the experimental matter of fact was produced.

[3] The usual form in which Boyle phrased this was that God might produce the same natural effects through very different causes. Therefore, "it is a very easy mistake for men to conclude that because an effect may be produced by such determinate causes, it must be so, or actually is so." Boyle, "Usefulness of Experimental Natural Philosophy," p. 45; see also Laudan, "The Clock Metaphor and Probabilism"; Rogers, "Descartes and the Method of English Science"; van Leeuwen, *The Problem of Certainty*, pp. 95-96; B. Shapiro, *Probability and Certainty*, pp. 44-61.

The Mechanics of Fact-Making: Three Technologies

Boyle proposed that matters of fact be established by the aggregation of individuals' *beliefs*. Members of an intellectual collective had mutually to assure themselves and others that belief in an empirical experience was warranted. Matters of fact were the outcome of the process of having an empirical experience, warranting it to oneself, and assuring others that grounds for their belief were adequate. In that process a multiplication of the witnessing experience was fundamental. An experience, even of a rigidly controlled experimental performance, that one man alone witnessed was not adequate to make a matter of fact. If that experience could be extended to many, and in principle to all men, then the result could be constituted as a matter of fact. In this way, the matter of fact is to be seen as both an epistemological and a social category. The foundational item of experimental knowledge, and of what counted as properly grounded knowledge generally, was an artifact of communication and whatever social forms were deemed necessary to sustain and enhance communication.

We will show that the establishment of matters of fact in Boyle's experimental programme utilized three *technologies*: a *material technology* embedded in the construction and operation of the air-pump; a *literary technology* by means of which the phenomena produced by the pump were made known to those who were not direct witnesses; and a *social technology* that incorporated the conventions experimental philosophers should use in dealing with each other and considering knowledge-claims.[4] Despite the utility of distinguishing the three technologies employed in fact-making, the impression should not be given that we are dealing with distinct categories: each embedded the others. As we shall see, experimental practices employing the material technology of the air-pump crystallized specific forms of social organization; these valued social forms were dramatized in the literary exposition of experimental findings; the literary reporting of air-pump performances ex-

[4] Our use of the word *technology* in reference to the "software" of literary practices and social relations may appear jarring, but it is both important and etymologically justified, as Carl Mitcham nicely shows: "Philosophy and the History of Technology," esp. pp. 172-175. Mitcham demonstrates that Plato distinguished between two types of *techne*: one that consisted mainly of physical work and another that was closely associated with speech. By using *technology* to refer to literary and social practices, as well as to machines, we wish to stress that all three are *knowledge-producing tools*.

tended an experience that was regarded as essential to the propagation of the material technology or even as a valid substitute for direct witness of experimental displays. If we wish to understand how Boyle worked to construct pneumatic facts, we must consider how each of the three technologies was used and how each bore upon the others.

The Material Technology of the Air-Pump

We start by noting the obvious: matters of fact in Boyle's new pneumatics were machine-made. His mechanical philosophy used the machine not merely as an ontological metaphor but also, crucially, as a means of intellectual production. The matters of fact that constituted the foundations of the new science were brought into being by a purpose-built scientific machine. This was the air-pump (or "pneumatical engine," or, eponymously, the *machina Boyleana*), which was constructed for Boyle by the instrument maker Greatorex and, especially, by Robert Hooke in 1658-1659. We have to describe how this machine was put together and how it worked in order to understand its role in fact-production.

Boyle intended to improve upon the design of Otto von Guericke's device, described by Caspar Schott in his *Mechanica hydraulico-pneumatica* of 1657. According to Boyle, this earlier machine (see figure 22) had several practical disadvantages: (1) it needed to be immersed in a large volume of water; (2) it was a solid vessel, such that experimental apparatus could not be inserted in it; and (3) it was extremely difficult to operate, requiring, as Boyle observed, "the continual labour of two strong men for divers hours" to evacuate it.[5] Boyle and Hooke sought to overcome these practical problems. Figure 1 is an engraving of their first successful machine, that was used to produce the forty-three experiments of *New Experiments Physico-Mechanical*.[6] The machine consisted of two main parts: a glass globe (or "receiver") and the pumping apparatus itself.

[5] Boyle, "New Experiments," pp. 6-7. (Many of Boyle's essay titles began with "New Experiments . . ."; we use this short title to refer exclusively to the "New Experiments Physico-Mechanical, touching the Spring of the Air" [1660].)

[6] This account is drawn largely from that provided by Boyle in "New Experiments," pp. 6-11. One of the best modern descriptions of this pump and its operation is Frank, *Harvey and the Oxford Physiologists*, pp. 129-130. The best overall accounts remain the nineteenth-century essays of Wilson, both his *Religio chemici*, pp. 191-219, and, especially, his "Early History of the Air-Pump."

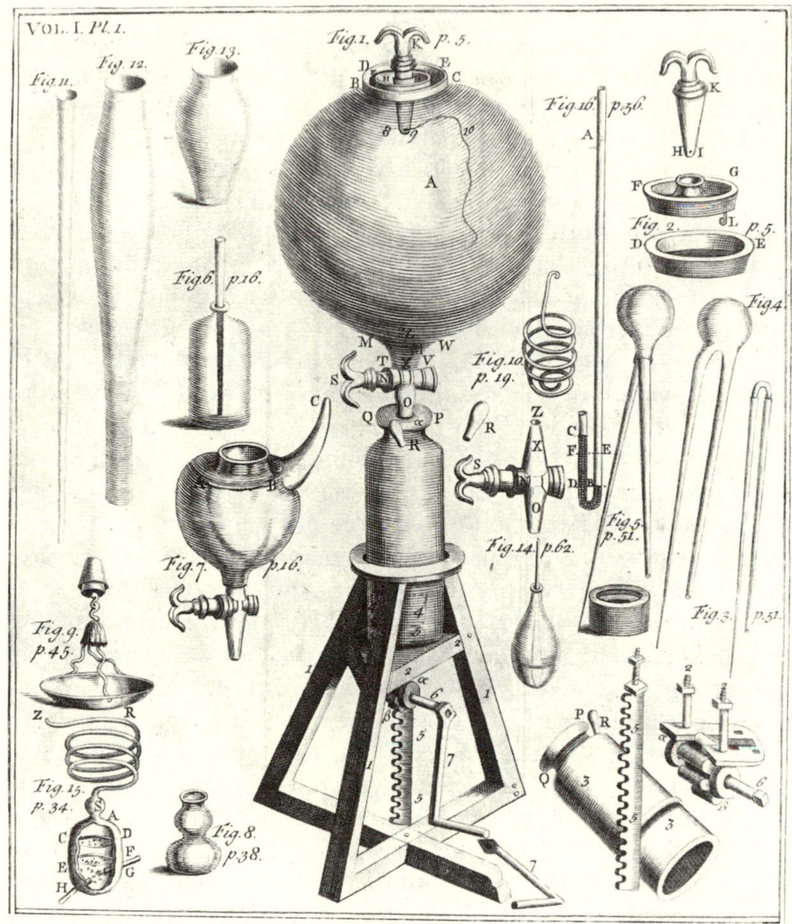

FIGURE 1

Robert Boyle's first air-pump, as it appeared in an engraving in New Experiments Physico-Mechanical *(1660). (Courtesy of Edinburgh University Library.)*

The receiver contained the space from which atmospheric air was to be removed. It was approximately thirty quarts in volume: although Boyle would, ideally, have liked a larger one, this was the limit of his "glass-men's" capabilities. In a few of his *New Experiments* Boyle used a variety of smaller receivers, some as small as one quart in volume, hoping (which proved to be untrue) that these would be easier to evacuate.[7] Experimental apparatus could be placed in

[7] Boyle, "New Experiments," p. 25.

the receiver through an aperture of about four-inch diameter at the top ("B-C"), and special arrangements could be made for instruments, like the Torricellian experiment, which were taller than even the big receiver, in which cases part of the apparatus extended through the sealed aperture above the receiver.

The receiver narrowed at its base so as to fit into a brass device ("N") containing a stopcock ("S"). This in turn was connected to a hollow brass cylinder ("3") about 14 inches long and about three inches in internal diameter. At the upper lip of the cylinder there was a small hole into which a brass valve ("R") could be inserted as required. Within the cylinder was a wooden piston (or "sucker") topped with "a good thick piece of tanned show-leather" ("4"), which provided for an exceedingly tight fit between piston and the inside of the cylinder. The piston was worked up and down by means of an iron rack ("5") and pinion ("7") device, the whole machine resting upon a wooden frame ("I").

This is how the engine worked to remove air from the receiver: with the stopcock in the closed position and the valve "R" inserted, the sucker was drawn up to the top of the cylinder; at this point there was no air between sucker and the top of the cylinder. Then the sucker was drawn down and the stopcock was opened, permitting the passage of a quantity of air from the receiver into the cylinder. The stopcock was closed, the valve was removed, and the sucker was forced up, thus expelling that quantity of air to the exterior. The process was repeated, each "exsuction" requiring progressively more force as the amount of air remaining in the receiver was diminished. (This account of how the machine worked to *remove* air, it must be noted, agrees with that provided by Boyle and modern commentators. As we shall see, Hobbes claimed that the receiver remained always full; therefore his view of how the pump operated, to be detailed in chapter 4, differed radically from Boyle's.) Later air-pumps of the 1660s and 1670s (described in chapters 5 and 6) differed from this original design in several respects: the cylinder and receiver were indirectly connected, and, after Denis Papin's innovation of 1676, there were two pumping cylinders with self-acting valves. Although we shall be almost exclusively concerned here with Boyle's air-pump as a rarefying engine, it could also be used to condense air in the receiver, simply by reversing the operations by which air was withdrawn.[8]

[8] As noted, for example, by Wilson, *Religio chemici*, pp. 197-198; and see Boyle, "New Experiments," p. 36.

The evacuation of air from the receiver of Boyle's original air-pump was an extremely difficult business, as was maintaining that exhaustion for any length of time. Among the chief difficulties was the problem of leakage. Great care had to be taken to ensure that external air did not insinuate itself back into pump or receiver through a number of possible avenues. This is not at all a trivial and merely technical point. The capacity of this machine to produce matters of fact crucially depended upon its physical integrity, or, more precisely, upon collective agreement that it was air-tight for all practical purposes. Boyle detailed the measures he had taken to seal the machine against the intrusion of external air. For example, the aperture at the top of the receiver was sealed with a special cement called *diachylon,* a mixture "which . . . would, by reason of the exquisite commixtion of its small parts, and closeness of its texture, deny all access to the external air."[9] Boyle did not provide the recipe for diachylon, but it was probably a mixture of olive oil and other vegetable juices boiled together with lead oxide. He described how the stopcock was affixed and made good so that it did not leak, using a mixture of "melted pitch, rosin, and wood-ashes." And he took special pains to recount how the leather ring around the sucker was lubricated, both to facilitate its movement in the cylinder and to "more exactly hinder the air from insinuating itself betwixt it and the sides of the cylinder": a certain quantity of "sallad oil" was poured into both receiver and into the cylinder, and more oil was used to lubricate and seal the valve "R". Boyle noted that sometimes a mixture of oil and water proved a more effective seal and lubricant.[10] In addition, the machine was liable to more spectacular assaults upon its physical integrity. Given the state of the glass-blower's art (which Boyle continually lamented), receivers were likely to crack and even to implode. Small cracks were not, in Boyle's view, necessarily fatal. The greater external pressure could act to press them together, and he provided a recipe for fixing them if required: a mixture of powdered quick-lime, cheese scrapings and water, ground up into a paste "to have a strong and stinking smell," spread onto linen plasters and applied to the crack.[11] Finally, the brass cylinder might be bent by atmospheric pressure and the force required to move the sucker: this might also affect the goodness of the seal between washer and the inside of

[9] Boyle, "New Experiments," p. 7; but see p. 35 for Boyle's surmise that even diachylon was somewhat porous to air.

[10] Ibid., p. 9.

[11] Ibid., p. 26.

the cylinder. The reasons for our detailed treatment of the physical integrity of the air-pump and the steps Boyle took to guarantee it will become clear below. For the present, we simply note three points: (1) that both the engine's integrity and its limited leakage were important resources for Boyle in validating his pneumatic findings and their proper interpretation; (2) that the physical integrity of the machine was vital to the perceived integrity of the knowledge the machine helped to produce; and (3) that the lack of its physical integrity was a strategy used by critics, particularly Hobbes, to deconstruct Boyle's claims and to substitute alternative accounts.

The Air-Pump as Emblem

Boyle's machine was a powerful emblem of a new and powerful practice. As Rupert Hall has noted:

> The air-pump was the unfailing pièce de résistance of the incipient scientific laboratory. Its wonders were inevitably displayed whenever a grandee graced a scientific assembly with his presence. After the chemist's furnace and distillation apparatus it was the first large and expensive piece of equipment to be used in experimental practice.

It was "the cyclotron of its age."[12] Similarly, Marie Boas Hall:

> . . . Boyle's air-pump together with Hooke's microscope constituted the show pieces of the [Royal] Society; when distinguished visitors were to be entertained, the chief exhibits were always experiments with the pump.[13]

As early as February 1661 the Danish ambassador "was entertained with experiments on Mr. Boyle's air-pump," and in 1667 Margaret Cavendish, Duchess of Newcastle, probably the first woman to be admitted to a meeting of the Royal Society, was treated to a similar display. According to Pepys, Margaret "was full of admiration, all

[12] A. R. Hall, *From Galileo to Newton*, p. 254, and idem, *The Revolution in Science*, p. 262; see also Price, "The Manufacture of Scientific Instruments," p. 636: the pneumatic pump "was the first large and complex machine to come into the laboratory."

[13] M. B. Hall, *Boyle and Seventeenth-Century Chemistry*, p. 185.

admiration."[14] When in 1664 the King was to be received at the Society, it was anxiously debated what successor to the pump (by then well-known to His Majesty) could so well amuse and instruct the honoured guest. As Christopher Wren wrote from Oxford,

> The solemnity of the occasion, and my solicitude for the honour of the society, make me think nothing proper, nothing remarkable enough. It is not every year will produce such a master experiment as the Torricellian, and so fruitful as that is of new experiments; and therefore the society hath deservedly spent much time upon that and its offspring.

An experimental display adequate to such circumstances ought to be both edifying and spectacular, such as those conducted with the air-pump:

> And if you have any notable experiment, that may appear to open new light into the principles of philosophy, nothing would better beseem the pretensions of the society; though possibly such would be too jejune for this purpose, in which there ought to be something of pomp. On the other side, to produce knacks only, and things to raise wonder, such as Kircher, Schottus, and even jugglers abound with, will scarce become the gravity of the occasion. It must be something between both, luciferous in philosophy, and yet whose use and advantage is obvious without a lecture; and besides, that may surprise with some unexpected effect, and be commendable for the ingenuity of the contrivance.[15]

[14] The visit of the Danish ambassador is noted in Birch, *History*, vol. I, p. 16, and that of Margaret in ibid., pp. 175, 177-178. For Pepys' remark, see Pepys, *Diary*, vol. VIII, pp. 242-243 (entry for 30 May/9 June 1667); see also Nicolson, *Pepys' 'Diary' and the New Science*, chap. 3. Margaret had recently written of her strong preference for rationalistic, rather than experimental, methods in science. Her family were Hobbes's patrons, and her anti-experimentalism reflected his sentiments closely. See Cavendish, *Observations upon Experimental Philosophy* (1666), "Further Observations," p. 4 (also sig d1): ". . . our age being more for deluding Experiments than rational arguments, which some cal a *tedious babble*, doth prefer Sense before Reason, and trusts more to the deceiving sight of their eyes, and deluding glasses, then to the perception of clear and regular Reason. . . ." Cf. R. F. Jones, *Ancients and Moderns*, p. 315n.

[15] Wren to Brouncker, 30 July/9 August 1663, in Birch, *History*, vol. I, p. 288. Preparations for the King's reception were intense, going on from April 1663 to May 1664, but we have no evidence that the royal experimental performance ever took place; see also Oldenburg to Boyle, 2/12 July 1663, in Oldenburg, *Correspondence*, vol. II, pp. 78-79. At precisely the same time that Wren wrote his letter, Boyle

No new device had taken the place of the *machina Boyleana* as an emblem of the Royal Society's experimental programme.

The powerfully emblematic status of the air-pump is manifested in its contemporary iconography. Boyle and Hooke took an active interest in the production of drawings and engravings by William Faithorne that depicted Boyle together with his pneumatic engine (see figure 16b).[16] During the mid-1660s the Somerset virtuoso John Beale was sedulously involved in celebrating the Baconian works of the Royal Society, encouraging John Evelyn to produce an appropriate iconographic drawing which, after various vicissitudes, eventually appeared as a frontispiece in some copies of Sprat's *History of the Royal Society* (1667) (see figure 2).[17] This engraving (by Wenceslaus Hollar) shows a redesigned version of Boyle's pump in the left background. (See figure 17 for an enlargement.) Through the later seventeenth and eighteenth centuries the Faithorne image was continually adapted and modified. Perhaps the richest in iconographic significance eventually appeared on the title page of the collected editions of Boyle's *Works* in 1744 and 1772 (figure 3).[18] This vignette by Hubert François Gravelot incorporated the Faithorne likenesses of Boyle and his original pump. The power of the pump is indicated by the conjunction of the Latin motto and the gesture of the classical female figure. Her left hand points to the air-pump while her right points to the heavens. The significance of the gesture is reinforced by the motto: "To know the Supreme Cause from the causes of things." It is the operation of the pneumatic engine, among all the scientific apparatus displayed in the engraving, that is going to enable the philosopher to approach God's knowledge.[19] The au-

was using similar language about "jugglers" and royal displays: "The works of God are not like the tricks of jugglers, or the pageants, that entertain princes, where concealment is requisite to wonder; but the knowledge of the works of God proportions our admiration of them." Boyle, "Usefulness of Experimental Natural Philosophy," p. 30 (1663).

[16] For a full account of seventeenth- and eighteenth-century images of Boyle, see Maddison, "The Portraiture of Boyle." For correspondence relating to the Faithorne work, see Boyle, *Works*, vol. VI, pp. 488, 490, 499, 501, 503.

[17] A detailed treatment of the circumstances attending the production of this image is in Hunter, *Science and Society*, pp. 194-197.

[18] See Maddison, "The Portraiture of Boyle," p. 158.

[19] Such a motto might have been regarded as inappropriate by many mid-seventeenth-century experimental philosophers; its apparently immodest sentiments seem to belong more to the mid-eighteenth century. Boyle agreed that one could move in understanding "from Nature up to Nature's God," yet we shall see that he set strict limits on the possibilities of causal knowledge.

FIGURE 2

Frontispiece to Sprat's History of the Royal Society *(1667). Engraving by Wenceslaus Hollar, design probably by John Evelyn for John Beale in about 1666-1667, and transferred to Sprat's book later. Boyle's revised version of the air-pump is in the centre-left background (see also figure 17). The three figures in the foreground are the president of the Royal Society, Lord Brouncker (left); the King (bust, centre, being crowned by Fame); and Francis Bacon (right). (Courtesy of the British Library.)*

FIGURE 3
Vignette by Hubert François Gravelot Bourguignon for Thomas Birch's edition of Boyle's Works *(1744 and 1772), frontispiece to vol. I. (Courtesy of Edinburgh University Library.)*

thorship of the pump is further symbolized by the line from the heaven-pointing hand to Boyle himself. Note further the spatial separation of the various items of philosophical instrumentation. On the right are instruments for experimenting on the nature of the air: the pump, a two-branch mercury barometer (leaning on the pump), and a double capillary manometer. All these are modern experimental devices, just as Boyle's pneumatics was paradigmatic of modern experimental philosophy. On the left are instruments for experimenting with fire: notably a furnace with an alembic. All these are medieval in origin, being the apparatus employed by alchemists and practitioners of the old philosophy. The female figure faces away from these, indicating not Boyle's rejection of these (since he employed them himself) but the relative value of the two programmes and their resulting intellectual products. Furthermore, those products take the form of *writings*, and the figure's feet rest upon a pile of books (the embodiment of the quest for knowledge) that belong to the assemblage of pneumatic instruments. There are no books on the left.[20] Some indication that the

[20] It is, of course, possible that our interpretation of this image is incorrect, but it is unlikely that, in its general form, it is overargued. An immense amount of

assemblage of objects and the gesture had an institutionalized status is afforded by figure 4. This is the frontispiece of a 1679 French collection of experimental essays, including a series by Boyle on tastes and smells.[21] The female figure in this case is recognizably that of Athena, goddess of wisdom. The left hand gestures to heaven, but the right holds a scroll inscribed "Nouvelles Experiences." (It is not clear whether this is a specific reference to the title of Boyle's pneumatic essays.) The female figure's feet rest on books, as they do in figure 3.

FIGURE 4
Frontispiece to anonymously edited collection of essays on natural philosophy: Recueil d'expériences et observations sur le combat, qui procède du mélange des corps *(Paris, 1679). (Courtesy of the British Library.)*

---

thought and symbolic labour went into the preparation of philosophical iconography, and such images were intended to be de-coded and reflected upon in this manner. See, for example, the treatment of frontispieces in Webster, *From Paracelsus to Newton*; also Eisenstein, *The Printing Press as an Agent of Change*, esp. pp. 258-261; C. R. Hill, "The Iconography of the Laboratory."

[21] *Recueil d'expériences et observations sur le combat qui procède du mélange des corps* (Paris, 1679). Pp. 125-220 are "Expériences curieuses de l'illustre Mr. Boyle sur les saveurs et sur les odeurs." The anonymously edited collection also included essays by Nehemiah Grew and Leeuwenhoek.

The Pump and the "Empire of the Senses"

The power of new scientific instruments, the microscope and telescope as well as the air-pump, resided in their capacity to enhance perception and to constitute new perceptual objects. The experimental philosophy, empiricist and inductivist, depended upon the generation of matters of fact that were objects of perceptual experience. Unassisted senses were limited in their ability to discern and to constitute such perceptual objects. Boyle himself reckoned "that the Informations of Sense assisted and highlighted by Instruments are usually preferrable to those of Sense alone."[22] And Hooke detailed the means by which scientific instruments *enlarged* the senses:

> . . . his design was rather to improve and increase the distinguishing faculties of the senses, not only in order to reduce these things, which are already sensible to our organs unassisted, to number, weight, and measure, but also in order to the inlarging the limits of their power, so as to be able to do the same things in regions of matter hitherto inaccessible, impenetrable, and imperceptible by the senses unassisted. Because this, as it inlarges the empire of the senses, so it besieges and straitens the recesses of nature: and the use of these, well plied, though but by the hands of the common soldier, will in short time force nature to yield even the most inaccessible fortress.[23]

In Hooke's view, the task was one of remedying the "infirmities" of the human senses "with Instruments, and, as it were, the adding of artificial Organs to the natural." The aim was the "*inlargement of the dominion*, of the Senses."[24] Among the senses, the eye was paramount, but, " 'tis not improbable, but that there may be found many *Mechanical Inventions* to improve our other Senses, of *hearing, smelling, tasting, touching*."[25]

Things would be seen that were previously invisible: the rings of Saturn, the mosaic structure of the fly's eye, spots on the sun.

[22] Westfall, "Unpublished Boyle Papers," p. 115 (quoting Boyle, "Propositions on Sense, Reason, and Authority," Royal Society, Boyle Papers, IX, f 25); see also van Leeuwen, *The Problem of Certainty*, p. 97.

[23] Birch, *History*, vol. III, pp. 364-365 (entry for 13/23 December 1677).

[24] Hooke, *Micrographia* (1665), "The Preface," sig a2^{r}; see also Bennett, "Hooke as Mechanic and Natural Philosopher," p. 44.

[25] Hooke, *Micrographia*, "The Preface," sig b2^{v}.

And other things, essentially invisible, would be given visual manifestations: the pressure of the air, aqueous and terrestrial effluvia. As Hooke said, "There is a new visible World discovered."[26] This new visible world indicated not only the potential of scientific instruments to enhance the senses; it also served as a warning that the senses were inherently fallible and required such assistance as the experimental philosopher could offer. Glanvill took the telescopic discovery of Saturn's rings as an instance of the fallibility of both unassisted sense and of the hypotheses erected upon unassisted sense:

> And perhaps the newly discovered Ring about Saturn . . . will scarce be accounted for by any systeme of things the World hath yet been acquainted with. So that little can be looked for towards the advancement of natural Theory, but from those, that are likely to mend our prospect of events and sensible appearances; the defect of which will suffer us to proceed no further towards Science, then to imperfect guesses, and timerous supposals.[27]

Scientific instruments therefore imposed both a correction and a discipline upon the senses. In this respect the discipline enforced by devices such as the microscope and the air-pump was analogous to the discipline imposed upon the senses by reason. The senses alone were inadequate to constitute proper knowledge, but the senses disciplined were far more fit to the task. Hooke described the appropriate circulation of items from the senses to the higher intellectual faculties:

> The *Understanding* is to *order* all the inferiour services of the lower Faculties; but yet it is to do this only as a *lawful Master*, and not as a *Tyrant*. . . . It must *watch* the irregularities of the Senses, but it must not go before them, or *prevent* their information. . . . [T]he true Philosophy . . . is to *begin* with the Hands and Eyes, and to *proceed* on through the Memory, to be *con-*

[26] Ibid., sig a2^{v}. There is a clear connection between these views of the role of scientific instruments and the epistemological problem of "transdiction" (inferring from the visible to the invisible) discussed by Mandelbaum, *Philosophy, Science, and Sense Perception*, chap. 2.

[27] Glanvill, *Scepsis scientifica* (1665), "To the Royal Society," sig b4^{v}; also pp. 54-55. See also B. Shapiro, *Probability and Certainty*, pp. 61-62; for an account of the observational and theoretical issues at stake in the problem of Saturn's rings, see van Helden, " 'Annulo Cingitur': The Solution of the Problem of Saturn"; idem, "Accademia del Cimento and Saturn's Ring."

> *tinued* by the Reason; nor is it to stop there, but to *come about* to the Hands and Eyes again, and so, by a *continual passage round* from one Faculty to another, it is to be maintained in life and strength, as much as the body of man is.[28]

Just as the reason disciplined the senses, and was disciplined by it, so the new scientific instruments disciplined sensory observation through their control of *access*.

Boyle's and Hooke's air-pump was, in the former's terminology, an "elaborate" device. It was also temperamental (difficult to operate properly) and very expensive: the air-pump was seventeenth-century "Big Science." To finance its construction on an individual basis it helped greatly to be a son of the Earl of Cork. Other natural philosophers, presumably as well supplied with cash as Boyle, shied away from the expense of building a pneumatic engine, and a major justification for founding scientific societies in the 1660s and afterwards was the collective financing of the instruments upon which the experimental philosophy was deemed to depend.[29] Reading histories of seventeenth-century science, one might gain the impression that air-pumps were widely distributed. They were, however, very scarce commodities. We shall present further details concerning the location and operation of air-pumps during the 1660s in chapter 6. However, the situation can be briefly summarized: Boyle's original machine was soon presented to the Royal Society of London; he had one or two redesigned machines built for him by 1662, operating mainly in Oxford; Christiaan Huygens had one made in The Hague in 1661; there was one at the Montmor Academy in Paris; there was probably one at Christ's College, Cam-

[28] Hooke, *Micrographia*, "The Preface," sig b2r. For Hooke's stress on deductions from hypotheses, which differed from Boyle's approach, see Hesse, "Hooke's Philosophical Algebra"; idem, "Hooke's Development of Bacon's Method."

[29] The only hard evidence we have found concerning the cost of this air-pump indicates that a version of the *receiver* ran to £5: Birch, *History*, vol. II, p. 184. Given the expense of machining the actual pumping apparatus, and replacement costs for broken parts (probably considerable), an estimate of £25 for the entire machine might prove conservative. Thus this pump would have cost more than the annual salary of Robert Hooke as Curator of the Royal Society, who was the London pump's chief operator. Christiaan Huygens' older brother Constantijn, much the wealthiest of the three Huygens brothers, withdrew from a pump-building project, "being afraid of the cost": Huygens, *Oeuvres*, vol. III, p. 389. Cf. van Helden, "The Birth of the Modern Scientific Instrument," pp. 64, 82n-83n; and A. R. Hall, *The Revolution in Science*, p. 263: "Everyone wanted at least to have witnessed the experiments, though few could own so costly a piece of apparatus." In chapter 6 we present some evidence on the cost of later devices.

bridge, by the mid-1660s; and Henry Power may have possessed one in Halifax from 1661. So far as can be found out, these were all the pumps that existed in the decade after their invention.

Without doubt, the intricacy of these machines and their limited availability posed a problem of access that experimental philosophers laboured to overcome. Less obviously, the control of access to the devices that were to generate genuine knowledge was a positive advantage. The space where these machines worked—the nascent laboratory—was to be a public space, but a restricted public space, as critics like Hobbes were soon to point out. If one wanted to produce authenticated experimental knowledge—matters of fact—one had to come to this space and to work in it with others. If one wanted to see the new phenomena created by these machines, one had to come to that space and see them with others. The phenomena were not on show anywhere at all. The laboratory was, therefore, a disciplined space, where experimental, discursive, and social practices were collectively controlled by competent members. In these respects, the experimental laboratory was a better space in which to generate authentic knowledge than the space outside it in which simple observations of nature could be made. To be sure, such observations were reckoned to be vital to the new philosophy and were judged vastly preferable to trust in ancient authority. Yet most observational reports were attended with problems in evaluating *testimony*. A report of an observation of a new species of animal in, for example, the East Indies, could not easily be checked by philosophers whose credibility was assured. Thus all such reports had to be inspected both for their plausibility (given existing knowledge) and for the credibility and trustworthiness of the witness.[30] Such might not be the case with experimental performances in which, ideally, the phenomena were witnessed together by philosophers of known reliability and discernment. Insofar as one insisted upon the foundational status of experimentally produced matters of fact, one ruled out of court the knowledge-claims of alchemical "secretists" and of sectarian "enthusiasts" who claimed individual and unmediated inspiration from God, or whose solitary "treading of the Book of Nature" produced unverifiable observational testimony. It is not novel to notice that the constitution of experimental knowledge was to be a public process. We stress, however, that producing matters of fact through scientific

[30] For concern with evaluating testimony in the natural history sciences, see B. Shapiro, *Probability and Certainty*, chap. 4, esp. pp. 142-143.

machines imposed a special sort of discipline upon this public. In following sections of this chapter we shall describe the nature of the discursive and social practices that Boyle recommended for the generation of the matter of fact. Before proceeding to that task we need briefly to describe what a pneumatic experiment was and how its matters of fact were said to relate to their interpretation and explanation.

Two Experiments

The text of Boyle's *New Experiments* of 1660 consisted of narratives of forty-three trials made with the new pneumatic engine. In following chapters we shall see how critics of Boyle's experimental programme managed to deconstruct the integrity of both his matters of fact and explanatory resources. These deconstructions called into question almost every aspect of Boyle's practices and findings: from the physical integrity of the air-pump to the legitimacy of making experimental matters of fact into the foundations of proper natural philosophical knowledge. For the present, however, it will be useful to describe two of Boyle's first air-pump experiments as he himself recounted them. These two experiments have not been randomly chosen. There are three reasons for concentrating upon them. First, the phenomena produced were accounted paradigmatic by advocates and critics of Boyle's philosophy. They were prizes contested between mechanical and nonmechanical natural philosophers, and between varieties of mechanical philosophers in the seventeenth century. Second, they include a contrast between an experiment which Boyle reckoned to be successful and one which he admitted to be a failure: critics such as Hobbes, as we shall see, seized upon this admission of failure as a way to undermine the whole of Boyle's experimental programme. Third, both experiments were deemed by Boyle to have a particularly intimate connection with the legitimacy of his major explanatory items in pneumatics: the pressure and the "spring" of the air. The tactical relations between experimental matters of fact and their explanation is, therefore, especially visible in these instances.

The first experiment to be described is the seventeenth of Boyle's original series. He himself referred to it as "the principal fruit I promised myself from our engine." Arguably, the air-pump was constructed chiefly with a view to performing this experiment. We shall call it the "void-in-the-void" experiment. It consisted of putting the Torricellian apparatus in the pump and then evacuating the receiver.[31] The "noble experiment" of Evangelista Torricelli was first performed in 1644. A tube of mercury, sealed at one end, was filled and then inverted in a dish of the same substance. The resultant "Torricellian space" left at the top became a celebrated phenomenon and problem for natural philosophers. For a decade after its production, the phenomenon was associated with two questions of immense cosmological importance: the real *character* of that "space" and the *cause* of the elevation of the mercury in the glass tube. The centre of interest in these questions in 1645-1651 was France, where Mersenne reported on the Italian work, and where natural philosophers such as Pascal, Petit, Roberval, and Pecquet all gave their views and experimented with the Torricellian apparatus.

Two points about the state of this problem need to be made in this connection. First, the Torricellian phenomenon was discussed in terms of long-standing debates over whether or not a vacuum could exist in nature.[32] Was this experiment decisive proof that a vacuum did exist? In practice, all possible combinations of views were held on the Torricellian space and the elevation of the mercury. Scholastic authorities maintained that the space was not void, and that the height of mercury was determined by the necessary limit to the expansion of the air left above the mercury. For Descartes, the mercury was sustained by the weight of the atmosphere, but the Torricellian space was filled by some form of subtle matter. For Descartes' inveterate opponent Roberval, the Torricellian space was indeed empty, but the height of the mercury depended upon the limit of a natural *horror vacui*. Finally, both Torricelli and Pascal held that the space was empty, and that the mercury was sustained by atmospheric weight. This experiment was therefore given various descriptions in the course of a debate which centred on the choice between plenist and vacuist theories. Given the range of views actually maintained in the 1640s and 1650s, the Torricellian problem seemed a key example of scandal in natural philosophy.[33]

Second, it seemed to participants that experimental measures

[31] Boyle, "New Experiments," p. 33. Experiment 19 used a *water* barometer.

[32] For medieval and early modern controversies over the vacuum, see Grant, *Much Ado about Nothing*, esp. chap. 4.

[33] Schmitt, "Experimental Evidence for and against a Void"; idem, "Towards an Assessment of Renaissance Aristotelianism," esp. p. 179; de Waard, *L'expérience barométrique*; Middleton, *The History of the Barometer*, chaps. 1-2; Westfall, *The Construction of Modern Science*, pp. 25-50.

offered a path away from such indecisive controversy. In his own work Blaise Pascal tried to combine experimental modesty and demonstrative compulsion to sway his opponents and critics. In treatises published in 1647-1648 Pascal described what soon became celebrated experimental variants of the Torricellian performance that he tentatively proffered as convincing evidence for his hypothesis, including a report of the Puy-de-Dôme trial of September 1648. Pascal firmly argued against men like the orthodox but Cartesian philosopher Noël for their love of theory and their premature hypothesizing. Thus the Torricellian experiment was intimately associated with the claim of experiment to settle belief about nature, to end controversy, and to generate consensus.[34]

Boyle's void-in-the-void experiment, and his interpretation of it, indicates the depth of his commitment to the role of experiment in securing assent. No less importantly, it illustrates the extent to which Boyle broke with the natural philosophical discourse in which the Torricellian experiment and its derivatives had previously been situated. The contents of the Torricellian space, whether in the receiver or outside of it, were of little concern to him. Neither was it of interest to stipulate whether or not the exhausted receiver constituted a "vacuum" within the frame of meaning of existing vacuist-plenist controversies. He would create a new discourse in which the language of vacuism and plenism was ruled out of order, or at least managed so as to minimize the scandalous disputes that, in his view, it had engendered. The receiver was a space into which one could move this paradigmatic experiment. And the discursive and social practices in which talk about this experiment was to be embedded constituted a space in which disputes might be neutralized.[35]

This is what Boyle did: he took a three-foot-long glass tube, one-

[34] Guenancia, *Du vide à Dieu*, pp. 63-100. For the French context of this work, see also Lenoble, *Mersenne*; H. Brown, *Scientific Organizations*. For the transmission of this interest to England, and, particularly, to Boyle, see Webster, "Discovery of Boyle's Law," pp. 455-457; Hartlib to Boyle, 9/19 May 1648, in Boyle, *Works*, vol. VI, pp. 77-78. For a contemporary version of the history of experimental pneumatics, see Barry, *Physical Treatises of Pascal*, pp. xv-xx.

[35] For continuing English disagreements about the nature of the Torricellian space in the 1660s: Hooke, *Micrographia*, pp. 13-14, 103-105; idem, *An Attempt for the Explication* (1661), pp. 6-50 (rewritten in *Micrographia*, pp. 11-32); Power, *Experimental Philosophy* (1664), pp. 95, 109-111; John Wallis to Oldenburg, 26 September/6 October 1672, in Oldenburg, *Correspondence*, vol. IX, pp. 258-262; see also Frank, *Harvey and the Oxford Physiologists*, chaps. 4-5, where the context of overriding interest by Oxford researchers in the *nitre* is discussed.

quarter inch in diameter, filled it with mercury, and inverted it as usual into a dish of mercury, having, as he said, taken care to remove bubbles of air from the substance. The mercury column then subsided to a height of about 29 inches above the surface of the mercury in the dish below, leaving the Torricellian space at the top. He then pasted a piece of ruled paper at the top of the tube, and, using a number of strings, lowered the apparatus into the receiver. Part of the tube extended above the aperture in the receiver's top, and Boyle carefully filled up the joints with melted diachylon. He noted that there was no change in the height of the mercury before evacuation commenced.[36] (See figure 12 for a drawing of a later version of this experimental set-up.)

Pumping now commenced. The initial suck resulted in an immediate subsidence of the mercury column; subsequent sucks caused further falls. (Boyle's primitive attempt to measure the levels reached after each suck was unsuccessful, as the mercury descended below the paper gauge.) After about a quarter-hour's pumping (how many sucks is not recorded), the mercury would fall no further. Significantly, the mercury column did not fall all the way to the level of the liquid in the dish, remaining about an inch above it. The experiment was quickly repeated in the presence of witnesses, and the same result was obtained. Boyle further observed that the fall of the mercury could be reversed by turning the stopcock to let in a little air. However, the column did not quite regain its previous height even when the apparatus was returned to initial conditions. Variants of this basic protocol were also reported: the experiment was tried with a glass mercury-containing tube sealed at the top with diachylon to test the porousness of that plaster. Boyle found that diachylon did not provide a completely tight seal. It was tried with a smaller receiver to see whether a more efficient exhaustion, and therefore a more complete fall of the mercury column, could be obtained (it could not); and it was tried in reverse (the air in the receiver was condensed by working the pump backwards) to see whether the mercury could be made to stand higher than 29 inches (it could).

So far, the account we have given has been restricted to what Boyle said was done and observed, without any of the *meanings* he attached to the experiment. For Boyle, this experiment offered an exemplar of how it was permissible to interpret matters of fact.

[36] This summary derives from the account given in Boyle, "New Experiments," pp. 33-39.

The problems were those traditionally associated with the Torricellian experiment: the elevation of the mercury and the nature of apparently void space. Boyle came to the void-in-the-void experiment with definite expectations about its outcome. The purpose of putting the Torricellian apparatus in the receiver was to imitate, and to give a visible analogy for, the impossible task of trying "the experiment beyond the atmosphere." He surmised that the normal height at which the mercury column was sustained was accounted for by "an aequilibrium with the cylinder of air supposed to reach from the adjacent mercury to the top of the atmosphere." So, "if this experiment could be tried out of the atmosphere, the quicksilver in the tube would fall down to a level with that in the vessel." This expectation was accompanied by a preformed explanatory resource: the *pressure* of the air. If the mercury descended as expected, it would be because "then there would be no pressure upon the subjacent [mercury], to resist the weight of the incumbent mercury."[37] Another, related, explanatory resource was also implicated. When Boyle initially enclosed the Torricellian apparatus in the receiver, and before he began evacuating it, he noted that the column remained at the same height as before. The reason for this, he said, must be "rather by virtue of [the] spring [of the air enclosed in the receiver] than of its weight; since its weight cannot be supposed to amount to above two or three ounces, which is inconsiderable in comparison to such a cylinder of mercury as it would keep from subsiding." When pumping began, the mercury level fell because of the diminished pressure of air in the receiver. The observation that the mercury did not in fact fall all the way down was accounted for by slight leakage:

> . . . when the receiver was considerably emptied of its air, and consequently that little that remained grown unable to resist the irruption of the external, that air would (in spight of whatever we could do) press in at some little avenue or other; and though much could not thereat get in, yet a little was sufficient to counterbalance the pressure of so small a cylinder of quicksilver, as then remained in the tube.[38]

In the next section of this chapter we examine the ways in which Boyle used the concepts of the air's weight and its spring or elasticity. But, for the present, we note that weight and spring were

[37] Ibid., p. 33.
[38] Ibid., p. 34.

the two mechanical notions that circumscribed interpretative talk about this paradigmatic experiment.

While it was permissible, even obligatory, to speak of the cause of the mercury's elevation in such terms, the treatment of the question of a void was handled in a radically different manner. This was to be made, so far as possible, into a nonquestion. Was the Torricellian space a *vacuum*? Did the exhausted receiver constitute a vacuum? The platform from which Boyle elected to address these questions was *experimental*: the way of talking appropriate to experimental philosophy was different in kind to existing natural philosophical discourse. Boyle recognized that his experiment would be deemed relevant to the traditional question posed of the Torricellian experiment, "whether or no that noble experiment infer a vacuum?" Was the exhausted receiver a space "devoid of all corporeal substance?" Boyle professed himself reluctant to enter "so nice a question" and he did not "dare" to "take upon me to determine so difficult a controversy." But settling the question of a vacuum was not what this experiment was about, nor were questions like this any part of the experimental programme. They could *not* be settled experimentally, and, because they could not, they were illegitimate questions. Plenists, those who maintained, either on mechanical or nonmechanical grounds, that there could not be a vacuum, had taken their reasons

> not from any experiments, or phaenomena of nature, that clearly and particularly prove their hypothesis, but from their notion of a body, whose nature, according to them, consisting only in extension . . . [means that] to say a space devoid of body, is, to speak in the schoolmen's phrase, a contradiction in adjecto.

But such reasons and such speech had no place in the experimental programme; they served "to make the controversy about a vacuum rather a metaphysical, than a physiological question; which therefore we shall here no longer debate. . . ."[39]

The significance of this move must be stressed. Boyle was not "a vacuist" nor did he undertake his *New Experiments* to prove a vacuum. Neither was he "a plenist," and he mobilized powerful arguments against the mechanical and nonmechanical principles adduced by those who maintained that a vacuum was impossible.[40]

[39] Ibid., pp. 37-38. The notion of body attacked here was that of Cartesian plenists.

[40] For example, ibid., pp. 37-38, 74-75; cf. C. T. Harrison, "Bacon, Hobbes, Boyle, and the Ancient Atomists," pp. 216-217 (on Boyle's "belief in the vacuum").

What he was endeavouring to create was a natural philosophical discourse in which such questions were inadmissible. The air-pump could not decide whether or not a "metaphysical" vacuum existed. This was not a failing of the pump; instead, it was one of its *strengths*. Experimental practices were to rule out of court those problems that bred dispute and divisiveness among philosophers, and they were to substitute those questions that could generate matters of fact upon which philosophers might agree. Thus Boyle allowed himself to use the term "vacuum" in relation to the contents of the evacuated receiver, while giving the term experimental meaning. By "vacuum," Boyle declared, "I understand not a space, wherein there is no body at all, but such as is either altogether, or almost totally devoid of air."[41] Boyle admitted the *possibility* that the receiver exhausted of air was replenished with "some etherial matter," "but not that it really is so."[42] As we shall see in chapter 5, during the 1660s Boyle rendered the question of an aether into an experimental programme, partly in response to plenist critics of his *New Experiments*. However, even in that research programme, the *existence* of an aether in the receiver, and therefore of a plenum, was not decided, but only whether such an aether had any experimental consequences.

Boyle's "vacuum" was a space "almost totally devoid of air": the incomplete fall of the mercury indicated to him that the pump leaked to a certain extent. The finite leakage of the pump was not, in his view, a fatal flaw but a valuable resource in accounting for experimental findings and in exemplifying the proper usage of terms like "vacuum." The "vacuum" of his exhausted receiver was thus not an experiment but a space in which to do experiments and generate matters of fact without falling into futile metaphysical dispute.[43] And it was an experimental space about which new discursive and social practices could be mobilized to generate assent.

The second of Boyle's *New Experiments* we describe can be treated more briefly. This was the thirty-first of the series, and again it dealt with a theoretically important and much debated phenome-

[41] Boyle, "New Experiments," p. 10. This was a definition apparently so novel, and so difficult to comprehend within existing philosophical discourse, that Boyle was obliged continually to repeat it in his subsequent disputes with Hobbes and Linus (see chapter 5).

[42] Ibid., p. 37.

[43] Compare the reaction of the German researchers Schott and Guericke to leakage in Boyle's pump (discussed in chapter 6). They said that their pump (in which one could not perform experiments) was therefore better than Boyle's: Schott, *Technica curiosa sive mirabilia artis* (1664), book II, pp. 75, 97-98.

non, that of *cohesion.* Two smooth bodies, such as marble or glass discs, can be made spontaneously to cohere when pressed against each other. This common phenomenon had long been a centrepiece of vacuist-plenist controversies. Lucretius used it to prove the existence of a vacuum; in the Middle Ages it was appropriated by both vacuists and plenists to support their cases; and it occupied a prominent place in Galileo's work on the problems of rigidity and cohesion. (In following chapters we shall discuss the work that Boyle did on cohesion prior to *New Experiments,* Hobbes's treatment of the phenomenon in his *De corpore* of 1655, and the continuing disputes between the two that dealt with this problem.) The fact that such surfaces displayed spontaneous cohesion was not in doubt; the proper explanation of that cohesion and of the circumstances attending their forcible separation was, however, intensely debated. It was agreed by all that it was difficult, yet possible, to separate cohered very smooth bodies by exerting a force perpendicular to the plane of their cohesion. Lucretius had argued that, since the velocity of the air rushing in from the sides to fill the space created by their separation must be finite, therefore a vacuum existed at the moment of separation. Scholastic plenists tended to stress the difficulty of separation, attributing this to the *horror vacui.* Various glosses were put upon the act of separation, all tending to establish the reality of a plenum.[44]

Boyle's idea, as with the Torricellian experiment, was to insert this phenomenon into his new experimental space. He would thus subject it to his new technical and discursive practices and use it to exemplify the effects of the air's pressure. Again, Boyle came to the experiment with an expectation of its outcome and with explanatory resources equipped to account for the outcome. If two "exquisitely polished" marble discs were laid upon each other, "they will stick so fast together, that he, that lifts up the uppermost, shall, if the undermost be not exceedingly heavy, lift up that too, and sustain it aloft in the free air." "A probable cause" of this cohesion was at hand:

> . . . the unequal pressure of the air upon the undermost stone; for the lower superfices of that stone being freely exposed to the air, is pressed upon by it, whereas the uppermost surface,

[44] See, for example, Grant, *Much Ado about Nothing,* pp. 95-100; Lucretius, *On the Nature of the Universe,* p. 12; Galileo, *Dialogues concerning Two New Sciences,* pp. 11-13; Millington, "Theories of Cohesion." Boyle used the terms "cohesion" and "adhesion" more or less interchangeably in referring to this phenomenon. As "adhesion" now suggests viscous sticking, we shall consistently use "cohesion."

> being contiguous to the superior stone, is thereby defended from the pressure of the air; which consequently pressing the lower stone against the upper, hinders it from falling.

Boyle conjectured that cohered marbles placed in the receiver that was then evacuated would fall apart as the air's pressure diminished.

This is what he did: he took marble discs 2⅓ inches in diameter and between ¼ and ½ inch thick; he then tried to make them cohere in free air. Immediately, there were problems: he could not obtain marbles ground so smooth that they would stay together for more than several minutes. Since it would take longer than that to exhaust the receiver, these were clearly unsuitable. So he moistened the interior surfaces of the pair with alcohol. This would, he reckoned, serve to smooth out residual irregularities in the marbles. Having got the marbles to cohere, he then attached a weight of four ounces to the lower stone ("to facilitate its falling off"), lowered the set by means of a string into the receiver, and commenced pumping. (For a later version of this experiment, see figure 9.) The marbles did not separate, and the experiment was accounted unsuccessful. Yet Boyle was ready with a reason why this experimental failure should not occasion the abandonment of his hypothesis: the pump leaked. That quantity of residual air, allowed in by the porousness of diachylon or by the looseness of the fit between sucker and cylinder, kept the marbles stuck together. The same leakage that permitted Boyle to offer an experimental meaning of the "vacuum" now provided a reason to hold fast to the theory of the air's pressure in the face of apparent counterevidence. In this sense, the experiment was not a failure at all.[45]

One other striking circumstance of this experiment needs to be noted. The trial was reported as a test and exemplification of the pressure of the air. In the quite brief narrative that constituted

[45] Boyle, "New Experiments," pp. 69-70. Boyle alluded here to earlier experiments on cohesion, published a year later in *The History of Fluidity and Firmness*; we discuss these in chapter 5. Readers of a realist bent, who might wish to know "what really happened" in these experiments, will necessarily be disappointed. We cannot reconstruct with any confidence what specific physical factors operated in Boyle's trials. From the point of view of modern scientific knowledge, a range of factors would have to be considered here. These include: (1) the isotropic pressure gradient on different surfaces of the marbles (as Boyle said); (2) short-range contact forces (not considered by Boyle); and (3) the phenomenon of *adhesion* due to the viscosity of the various lubricants Boyle employed (which he considered he had sufficiently allowed for).

Boyle's thirty-first experiment there was no allusion of any kind to the discursive tradition in which the phenomenon of cohesion had been paradigmatic. The phenomenon was not treated here as having any bearing upon the question of a vacuum versus a plenum. Having argued against the legitimacy of this philosophical discourse in experiment 17, Boyle now showed how one of its centrepieces could be handled as if that discourse did not exist.[46]

FACTS AND CAUSES: THE SPRING, PRESSURE, AND WEIGHT OF THE AIR

Boyle's *New Experiments* did not offer any explicit and systematic philosophy of knowledge. It did not discuss the problem of justifying inductive inference, propose formal criteria for establishing physical hypotheses, nor did it stipulate formal rules for limiting causal inquiry. What *New Experiments* did do was to *exemplify* a *working* philosophy of scientific knowledge.[47] In a concrete experimental setting it showed the new natural philosopher how he was to proceed in dealing with practical matters of induction, hypothesizing, causal theorizing, and the relating of matters of fact to their explanations. Boyle sought here to create a *picture* to accompany the experimental language-game and the experimental form of life. He did this largely by *ostension*: by showing others through his own example what it was like to work and to talk as an experimental philosopher.

Boyle's epistemological armamentarium included matters of fact, hypotheses, conjectures, doctrines, speculations, and many other locutions serving to indicate causal explanations. His overarching concern was to protect the matter of fact by separating it from various items of causal knowledge, and he repeatedly urged caution in moving from experimental matters of fact to their physical explanation. How, in practice, did Boyle manage this boundary? And

[46] We shall see that Boyle's adversaries, Hobbes and Linus, refused to allow this phenomenon to pass into the new, "nonmetaphysical" experimental discourse. Boyle's responses to them commented upon vacuist-plenist discourse and its legitimacy in this case.

[47] For an attempt to identify Boyle's "coherent and sophisticated view of scientific method," see Laudan, "The Clock Metaphor and Probabilism," pp. 81-97, esp. p. 81. We have no substantial disagreements with Laudan on Boyle's methods, but we dissent from his assessment of Boyle's philosophy as coherent and systematic. Cf. also Wiener, "The Experimental Philosophy of Boyle," and Westfall, "Unpublished Boyle Papers."

how, in practice, did he move between matters of fact and ways of accounting for them? Our best access to these questions is through an examination of Boyle's major explanatory resources in *New Experiments* and in his subsequent essays in pneumatics: the spring, pressure, and weight of the air.

The first thing to note is that the epistemological status of spring, pressure, and weight was never clearly spelt out in *New Experiments* or elsewhere. For example, in reporting the first of his *New Experiments*, the spring of the air was simply referred to as a "notion": it was "that notion, by which it seems likely, that most, if not all [his pneumatical findings] will prove explicable. . . ."[48] In other places Boyle chose to label the status of the spring an "hypothesis" or a "doctrine."[49] And, as we shall show in chapter 5, Boyle operationally treated the spring of the air as a matter of fact. In the twentieth of the *New Experiments* Boyle supposed that the fact "that the air hath a notable elastical power" has been "abundantly evinced" from his researches, "and it begins to be acknowledged by the eminentest naturalists."[50]

It would be easy to conclude, if one wanted, that Boyle was a poor formal philosopher of knowledge and a deficient formulator of scientific methodology. That is not a point we wish to make; nevertheless, there are several aspects of his procedures we need to note in this connection. First, Boyle did not detail the steps by which he moved from matters of fact to their explanation. He did not, for example, say in what ways the air's "elastical power" had been "evinced" and established; he merely announced that this had been accomplished. Second, he did not clearly discriminate between the air's spring and pressure as hypothetical causes of experimental facts and as matters of fact in their own right. Certainly, by the early 1660s (especially in his controversies with critics) Boyle was treating these explanatory items as if they were matters of fact and not hypotheses: their real existence had been *proved* by experiment, and he entertained no doubt on that score. While continuing to warn experimentalists to be circumspect in their hypothesizing and

[48] Boyle, "New Experiments," p. 11.

[49] See, for example, Boyle, "Examen of Hobbes," p. 197; idem, "Defence against Linus," pp. 119-120, 162 (and note the full title referring to the "doctrine" of the air's spring and weight). For discussion of the senses in which Boyle used the term "hypothesis," see Westfall, "Unpublished Boyle Papers," pp. 69-70: "Boyle evidently considered all generalizations in natural science to be hypotheses"; "To Boyle 'hypothesis' meant a supposition put forth to account for known facts . . ."

[50] Boyle, "New Experiments," p. 44.

to regard causal items as provisional, he treated *these* hypotheses as certainly established. And yet the criteria and rules for establishing hypotheses were not given. Third, Boyle made an unexplained distinction between the assurance we can have about the air's spring and pressure as causes and the assurance we can have about *their* causes. There was a strong boundary placed between speech about the spring as an explanation of matters of fact and speech about explanations of spring. Thus, in the first of the *New Experiments*, Boyle claimed that his "business [was] not . . . to assign the adequate cause of the spring of the air, but only to manifest, that the air hath a spring, and to relate some of its effects." Possible causes of this spring were arrayed, Boyle professing himself "not willing to declare peremptorily for either of them against the other." For instance, one might conceive of the spring as caused by the air having a real texture like that of wool fleece or sponge; or one might account for it in terms of Cartesian vortices; or one could posit that the air's corpuscles actually were "congeries of little slender springs."[51] Not only was it impossible to decide, it was, in Boyle's view, impolitic to try to decide which was the real cause. He warned against any such attempt as futile, and he never worked to specify the cause of the spring. The spring and the spring's cause were therefore treated as fundamentally different explanatory items: the former was "evinced" by the experiments; the latter was not, and, in practice, could not be. But they were both causes, and Boyle proffered no criteria for identifying in what way they were entitled to such radically different treatments. (The cause of the air's weight was, however, more straightforwardly accounted for: it was a function of the height and density of the atmospheric cylinder bearing upon any given cross-section.)

Our point may be summarized this way: the language-game that Boyle was teaching the experimental philosopher to play rested upon implicit acts of boundary-drawing. There was to be a crucial boundary between the experimental matter of fact and its ultimate physical cause and explanation. Viewed naively, or as a stranger might view it, it is unclear why the spring of the air, as the professed cause of the observed results, should be treated as a matter of fact rather than as a speculative hypothesis. Indeed, we have hinted here (and shall describe in detail in chapter 5) how the idea of the spring moved from outside to within the class of matters of fact.

[51] Ibid., pp. 11-12, 50, 54. Boyle explicitly labelled these various causal notions as "hypotheses." See also idem, "The General History of the Air," pp. 613-615.

It is also unclear upon what bases Boyle distinguished between his treatment of the spring and the cause of the spring. These are the grounds upon which one might wish to criticize Boyle as epistemologist and methodologist. However, our conclusions are not these: rather, we note that Boyle's criteria and rules for making his preferred distinctions between matters of fact and causes have the status of *conventions*. Causal talk is grounded in conventions which Boyle's reports exemplify, just as the construction of the matter of fact is conventional in nature (as we shall show in the following sections of this chapter). The ultimate justification of convention does not take the form of verbalized rules. Instead, the "justification" of convention *is* the form of life: the total pattern of activities which includes discursive practices.[52] This observation is supported by our later discussions of the ways in which Boyle's critics attempted to subvert his justifications of experimental practice and the ways in which Boyle replied.

Consider also the language Boyle used to describe his principal ontological concern: the air and its properties of spring, weight, and pressure. As we have noted, Boyle announced that the function of his pneumatic researches was "only to manifest that the air hath a spring, and to relate some of its effects."[53] Adversaries were defined by Boyle in terms of their alleged attitude to the spring of the air as a matter of fact. He argued that "the Cartesians," for example, need not grant a vacuum, nor need they abandon their notion of some form of subtle matter that could penetrate glass, but they must "add, as some of them of late have done, the spring of the air to their hypothesis." Boyle confessed in 1662 that it was more difficult to deal with adversaries, such as the Jesuit Franciscus Linus, who allowed a limited spring in the air, than it was to deal with those who denied it altogether, such as Hobbes. So in his response to Linus he claimed that "we have performed much more by the spring of the air, which we can within certain limits increase at pleasure, than we can by bare weight."[54] This comment suggests that Boyle distinguished systematically between spring and weight. He did not. Typically, he used the term "pressure" to describe

[52] This account has obvious resonances with Wittgenstein's treatment of language as secondary to patterns of activity. Language makes sense as embedded within those patterns: Wittgenstein, *Blue and Brown Books*, pp. 81-89; idem, *On Certainty*, props. 192, 204.

[53] Boyle, "New Experiments," p. 12.

[54] Boyle, "Examen of Hobbes," p. 191; idem, "Defence against Linus," pp. 121, 133.

these attributes of the air, distinguishing the specific cause of pressure only when it fitted a specific polemical purpose. In future references we shall follow Boyle in using the term "pressure" generically.

But Boyle's terminology was by no means consistent. He referred to the "pressing or sustaining force of the air," or to the "sustaining power of the air." In *New Experiments* he discussed the apparent heaviness of the cover of the receiver when evacuated, using the terms "spring of the external air," "force of the internal expanded air and that of the atmosphere," and "pressure" interchangeably. In early experiments in this text the term "protrusion" is used alongside that of "pressure."[55] These usages were no more consistent in subsequent essays on pneumatics and the air-pump trials. In the *Continuation of New Experiments* of 1669 and in later texts written against Hobbes, "pressure" referred to both weight and spring.[56] And in the central void-in-the-void experiment 17 of *New Experiments* Boyle reported that the insertion of the Torricellian apparatus in the sealed receiver did not produce a fall in the height of the mercury in the barometer. He attributed this to the "spring" of the air inside the still-unevacuated receiver, which was not affected by its removal from the "weight" of the atmosphere. Thus trials that computed the relation between the height of this mercury and the number of strokes of the sucker were interpreted as testing the relation between the air's "pressure" and its "density." "Pressure" thus embraced spring and weight.[57]

Two important moments in Boyle's exposition made this terminology highly sensitive to interpretation. First, we have introduced Boyle's experiment on the cohesion of smooth marbles *in vacuo*. This was, as we shall describe in chapter 5, a continuation of a sustained series of earlier trials in free air. In *The History of Fluidity and Firmness*, composed in 1659 and published in 1661, such cohesion was attributed to "the pressure of the atmosphere, proceeding partly from the weight of the ambient air . . . and partly from a kind of spring." This suggested that, since cohesion was due to the "pressure of the air" or "the sustaining power of the air," the removal of the air from the receiver of the air-pump would

55 Boyle, "History of Fluidity and Firmness," p. 409; idem, "New Experiments," pp. 11, 15-18, 69, 76.

56 Boyle, "Continuation of New Experiments," p. 276; idem, "Animadversions on Hobbes," p. 111.

57 Boyle, "New Experiments," pp. 33-34. Compare Webster, "Discovery of Boyle's Law," p. 470: ". . . the spring of the air, which [Boyle] now terms its pressure."

produce the separation of the cohering marbles. This trial failed, but the evidence of this failure was later used to demonstrate "the spring of the air even when rarified." In 1661 and 1662 Boyle continued to use "pressure" to embrace spring and weight in this experimental context. In *The History of Fluidity and Firmness* this usage was important, because Boyle offered an account of the cohesion of marbles that relied upon "the spring of the air" pressing upon the marbles isotropically, and *also* an account which relied upon "the pressure of the air considered as a weight." Yet Boyle used the term "pressure" for both.[58] In his response to Hobbes, Boyle still wrote that "the spring of the air may perform somewhat in the case proposed," though he emphasized that the weight of the air was more important, and continued to use the term "pressure of the fluid air" for the cause of cohesion.[59]

Second, Boyle used his term "pressure" when contesting the Scholastic argument from the *horror vacui*. Here "pressure" functioned as the *sole* alternative to an *unacceptable* mystification, whereas in the trials with marbles it functioned as a term that covered a *multiplicity* of *acceptable* explanations of a single phenomenon. In *New Experiments*, therefore, "the supposed aversation of nature to a vacuum" was presented as "accidental" and attributed to "the weight and fluidity, or at least flexility of the bodies here below; and partly, perhaps principally, of the air, whose restless endeavour to expand itself every way makes it either rush in itself or compel the interposed bodies into small spaces."[60] Finally, the spring and the weight of the air could not be easily disentangled, since one produced the other. Boyle wrote in *New Experiments* that the effects of spring were due to the release of compressed particles, and that this compression was itself due to the weight of the air. This claim was applied repeatedly in the accounts of the air-pump trials, and in each case the term "pressure" was used. In the later *Continuation* Boyle outlined the distinction between weight and pressure in a systematic fashion, *for the first time* in print. He attacked "the school-philosophers" and their use of *horror vacui*; he distinguished between the "gravity" and "the bare spring of the air," "which latter I now mention as a distinct thing from the other." Boyle acknowledged that his trials had *not* separated weight from spring, "since the weight of the upper parts of the air does, if I may so speak,

[58] Boyle, "History of Fluidity and Firmness," pp. 403-406.

[59] Boyle, "Examen of Hobbes," p. 227.

[60] Boyle, "New Experiments," p. 75.

bend the springs of the lower." Referring to the work in *New Experiments*, Boyle announced his intention of displaying the practically identical, but theoretically distinct, effects of "the pressure of all the superincumbent atmosphere acting as a weight" and "the pressure of a small portion of the air, included indeed (but without any new compression) acting as a spring." So "pressure" was to be read as an embracing term, and its ambiguities and variation of meaning were themselves a resource that Boyle used in debating the air-pump trials, notably those of the cohering marbles and of the enclosure of the mercury barometer in the receiver.[61]

Witnessing Science

We have begun to develop the idea that experimental knowledge production rested upon a set of *conventions* for generating matters of fact and for handling their explications. Taking the matter of fact as foundational to the experimental form of life, let us proceed to analyze and display how the conventions of generating the fact actually worked. In Boyle's view the capacity of experiments to yield matters of fact depended not only upon their actual performance but essentially upon the assurance of the relevant community that they had been so performed. He therefore made a vital distinction between actual experiments and what are now termed "thought experiments."[62] If knowledge was to be empiri-

[61] Ibid., pp. 13, 16; idem, "Continuation of New Experiments," pp. 176-177.

[62] See, for instance, Boyle, "Sceptical Chymist," p. 460: here Boyle suggested that many experiments reported by the alchemists "questionless they never tried." For an insinuation that Henry More may not actually have performed experiments adduced against Boyle's findings, see Boyle, "Hydrostatical Discourse," pp. 607-608. Compare the response of Boyle to Pascal's trials of the Puy-de-Dôme experiment ("New Experiments," p. 43); and by Power, Towneley, and himself ("Defence against Linus," pp. 151-155). Yet Boyle doubted the reality of Pascal's other reports of underwater trials; see "Hydrostatical Paradoxes," pp. 745-746: ". . . though the experiments [Pascal] mentions be delivered in such a manner, as is usual in mentioning matters of fact; yet I remember not, that he expressly says, that he actually tried them, and therefore he might possibly have set them down, as things that *must* happen, upon a just confidence, that he was not mistaken in his ratiocinations. . . . Whether or no Monsieur Pascal ever made these experiments himself, he does not seem to have been very desirous, that others should make them after him." For the report by Pascal that drew Boyle's censure, see Barry, *Physical Treatises of Pascal*, pp. 20-21; for the role of thought experiments in the history of science: Koyré, *Galileo Studies*, p. 97; Kuhn, "A Function for Thought Experiments"; Schmitt, "Experience and Experiment."

cally based, as Boyle and other English experimentalists insisted it should, then its experimental foundations had to be *witnessed*. Experimental performances and their products had to be attested by the testimony of eye witnesses. Many phenomena, and particularly those alleged by the alchemists, were difficult to accept by those adhering to the corpuscular and mechanical philosophies. In these cases Boyle averred "that they that have seen them can much more reasonably believe them, than they that have not."[63] The problem with eye witnessing as a criterion for assurance was one of *discipline*. How did one police the reports of witnesses so as to avoid radical individualism? Was one obliged to credit a report on the testimony of any witness whatsoever?

Boyle insisted that witnessing was to be a collective act. In natural philosophy, as in criminal law, the reliability of testimony depended upon its multiplicity:

> For, though the testimony of a single witness shall not suffice to prove the accused party guilty of murder; yet the testimony of two witnesses, though but of equal credit . . . shall ordinarily suffice to prove a man guilty; because it is thought reasonable to suppose, that, though each testimony single be but probable, yet a concurrence of such probabilities, (which ought in reason to be attributed to the truth of what they jointly tend to prove) may well amount to a moral certainty, *i.e.*, such a certainty, as may warrant the judge to proceed to the sentence of death against the indicted party.[64]

And Sprat, in defending the reliability of the Royal Society's judgments in matters of fact, inquired

> whether, seeing in all Countreys, that are govern'd by Laws, they expect no more, than the consent of two, or three witnesses, in matters of life, and estate; they will not think, they are fairly dealt withall, in what concerns their *Knowledg*, if they have the concurring Testimonies of *threescore or an hundred?*[65]

The thrust of the legal analogy should not be missed. It was not merely that one was multiplying authority by multiplying witnesses

[63] Boyle, "Unsuccessfulness of Experiments," p. 343; idem, "Sceptical Chymist," p. 486; cf. idem, "Animadversions on Hobbes," p. 110.

[64] Boyle, "Some Considerations about Reason and Religion," p. 182; see also Daston, *The Reasonable Calculus*, pp. 90-91; on testimony, see Hacking, *The Emergence of Probability*, chap. 3; on evidence in seventeenth-century English law, see B. Shapiro, *Probability and Certainty*, chap. 5.

[65] Sprat, *History*, p. 100.

(although this was part of the tactic); it was that *right action* could be taken, and seen to be taken, on the basis of these collective testimonies. The action concerned the voluntary giving of assent to matters of fact. The multiplication of witness was an indication that testimony referred to a true state of affairs in nature. Multiple witnessing was accounted an active licence rather than just a descriptive licence. Did it not force the conclusion that such and such an action was done (a specific trial), and that subsequent action (offering assent) was warranted?

In experimental practice one way of securing the multiplication of witnesses was to perform experiments in a social space. The experimental "laboratory" was contrasted to the alchemist's closet precisely in that the former was said to be a public and the latter a private space.[66] Air-pump trials, for instance, were routinely performed in the Royal Society's ordinary assembly rooms, the machine being brought there specially for the occasion. (We shall see in chapter 4 that one of the ways by which Hobbes attacked the experimental programme was to deny the Society's claim that this *was* a public place.) In reporting upon his experimental performances Boyle commonly specified that they were "many of them tried in the presence of ingenious men," or that he made them "in the

[66] The terms "laboratory" and "elaboratory" (etymologically: a place where the work is done) were very new in seventeenth-century England. The first use of the former recorded in the *Oxford English Dictionary* was in Thomas Timme's edition of DuChesne's *Practise of Chymicall and Hermeticall Physicke* (1605), part 3, sig Bb4^r (where the reference was to a place for keeping things secret); the first use of the latter was in John Evelyn's *State of France as It Stood in the IXth Year of Lewis XIII* (1652). It is plausible that the usage entered England from French and German iatrochemistry, and, thus, at least initially, that it had Paracelsian resonances. For Timme (or Tymme) as the leading ideologue of Paracelsian theory, see Debus, *The English Paracelsians*, pp. 87-97. For an exemplary use of "laboratory" to refer to a closed, private space, see Gabriel Plattes, "Caveat for Alchymists," in Hartlib, *Chymical, Medicinal and Chyrurgical Addresses* (1655; composed 1642-1643), p. 87: "A Laboratory, like to that in the City of *Venice*, where they are sure of secrecy, by reason that no man is suffered to enter in, unless he can be contented to remain there, being surely provided for, till he be brought forth to go to the Church to be buried." Compare Geoghegan, "Plattes' Caveat for Alchymists." For the "universal laboratory" developed in London by Hartlib, Clodius and Digby, see Hartlib to Boyle, 8/18 May and 15/25 May 1654, in Boyle, *Works*, vol. VI, pp. 86-89, and Clodius to Boyle, 12/22 December 1663, in Maddison, *Life of Boyle*, p. 87. For a list of the new open laboratories established in London in the 1650s and 1660s, including that of the King at Whitehall, see Gunther, *Early Science in Oxford*, vol. I, pp. 36-42; also Webster, *The Great Instauration*, pp. 48, 239, 302-303. Thomas Birch praised Boyle because "his laboratory was constantly open to the curious," while noting that Boyle suppressed his own work in poisons and on invisible or erasable ink: Boyle, *Works*, vol. I, p. cxlv.

presence of an illustrious assembly of virtuosi (who were spectators of the experiment)."[67] Boyle's collaborator Hooke codified the Royal Society's procedures for the standard recording of experiments: the register was "to be sign'd by a certain Number of the Persons present, who have been present, and Witnesses of all the said Proceedings, who, by Sub-scribing their Names, will prove undoubted Testimony."[68] And Thomas Sprat described the role of the "Assembly" in "resolv[ing] upon the matter of *Fact*" by collectively correcting individual idiosyncrasies of observation and judgment. The Society made "the whole process pass under its own eyes."[69] In reporting experiments that were particularly important or problematic, Boyle named his witnesses and stipulated their qualifications. Thus the experiment of the original air-pump trials that was "the principal fruit I promised myself from our engine" was conducted in the presence of "those excellent and deservedly famous Mathematic Professors, Dr. *Wallis*, Dr. *Ward*, and Mr. *Wren* . . . , whom I name, both as justly counting it an honour to be known to them, and as being glad of such judicious and illustrious witnesses of our experiment."[70] Another important experiment was attested to by Wallis "who will be allowed to be a very competent judge in these matters."[71] And in his censure of the alchemists Boyle generally warned natural philosophers not "to believe chymical experiments . . . unless he, that delivers that, mentions his doing it upon his own particular knowledge, or upon the relation of some credible person, avowing it upon his own experience." Alchemists were recommended to name the putative author of these experiments "upon whose credit they relate" them.[72] The credibility of witnesses followed the taken-for-granted conventions of that setting for assessing individuals' reliability and trustworthiness: Oxford professors were accounted more reliable witnesses than Oxfordshire peasants. The natural philosopher had no option but to rely for a substantial part of his knowledge on the testimony of witnesses; and, in assessing that testimony, he (no less than judge or jury) had to determine their credibility. This necessarily involved

[67] Boyle, "New Experiments," p. 1; idem, "History of Fluidity and Firmness," p. 410; idem, "Defence against Linus," p. 173.

[68] Hooke, *Philosophical Experiments and Observations*, pp. 27-28.

[69] Sprat, *History*, pp. 98-99, 84; see also B. Shapiro, *Probability and Certainty*, pp. 21-22; Glanvill, *Scepsis scientifica*, p. 54 (on experiments as a corrective to sense).

[70] Boyle, "New Experiments," pp. 33-34.

[71] Boyle, "Discovery of the Admirable Rarefaction of Air," p. 498.

[72] Boyle, "Sceptical Chymist," p. 460.

their moral constitution as well as their knowledgeability, "for the two grand requisites, of a witness [are] the knowledge he has of the things he delivers, and his faithfulness in truly delivering what he knows." Thus the giving of witness in experimental philosophy traversed the social and moral accounting systems of Restoration England.[73]

Another important way of multiplying witnesses to experimentally produced phenomena was to facilitate their *replication*. Experimental protocols could be reported in such a way as to enable readers of the reports to perform the experiments for themselves, thus ensuring distant but direct witnesses. Boyle elected to publish several of his experimental series in the form of letters to other experimentalists or potential experimentalists. The *New Experiments* of 1660 was written as a letter to his nephew, Lord Dungarvan; the various tracts of the *Certain Physiological Essays* of 1661 were written to another nephew, Richard Jones; the *History of Colours* of 1664 was originally written to an unspecified friend.[74] The purpose of this form of communication was explicitly to proselytize. The *New Experiments* was published so "that the person I addressed them to might, without mistake, and with as little trouble as possible, be able to repeat such unusual experiments. . . ."[75] The *History of Colours* was designed "not barely to relate [the experiments], but . . . to teach a young gentleman to make them."[76] Boyle wished to encourage young gentlemen to "addict" themselves to experimental pursuits and thereby to multiply both experimental philosophers and experimental facts.

In Boyle's view, replication was rarely accomplished. When he came to publish the *Continuation of New Experiments* more than eight years after the original air-pump trials, Boyle admitted that, despite his care in communicating details of the engine and his procedures, there had been few successful replications.[77] This situation had not

[73] Boyle, "The Christian Virtuoso," p. 529; also B. Shapiro, *Probability and Certainty*, chap. 5, esp. p. 179. For the role of social accounting systems in the evaluation of observation reports, see Westrum, "Science and Social Intelligence about Anomalies: The Case of Meteorites."

[74] M. B. Hall, *Boyle and Seventeenth-Century Chemistry*, pp. 40-41.

[75] Boyle, "New Experiments," p. 2.

[76] Boyle, "The Experimental History of Colours," p. 663. Certain "easy and recreative experiments, which require but little time, or charge, or trouble in the making" were recommended to be tried by ladies (p. 664).

[77] Boyle, "Continuation of New Experiments," p. 176 (dated 24 March 1667 [o.s.]; published 1669). In chapter 6 we discuss some interesting problems of replication involving Huygens' air-pump in Holland during the 1660s.

materially changed by the mid-1670s. In the seven or eight years after the *Continuation*, Boyle said that he had heard "of very few experiments made, either in the engine I used, or in any other made after the model thereof." Boyle now expressed despair that these experiments would ever be replicated. He said that he was now even more willing "to set down divers things with their minute circumstances" because "probably many of these experiments would be never either re-examined by others, or re-iterated by myself." Anyone who set about trying to replicate such experiments, Boyle said, "will find it no easy task."[78]

Prolixity and Iconography

The third way by which witnesses could be multiplied is far more important than the performance of experiments before direct witnesses or the facilitating of their replication: it is what we shall call *virtual witnessing*. The technology of virtual witnessing involves the production in a *reader's* mind of such an image of an experimental scene as obviates the necessity for either direct witness or replication.[79] Through virtual witnessing the multiplication of witnesses could be, in principle, unlimited. It was therefore the most powerful technology for constituting matters of fact. The validation of experiments, and the crediting of their outcomes as matters of fact, necessarily entailed their realization in the laboratory of the mind and the mind's eye. What was required was a technology of trust and assurance that the things had been done and done in the way claimed.

The technology of virtual witnessing was not different in kind to that used to facilitate actual replication. One could deploy the same linguistic resources in order to encourage the physical replication of experiments or to trigger in the reader's mind a naturalistic image of the experimental scene. Of course, actual replication was to be preferred, for this eliminated reliance upon testimony altogether. Yet, because of natural and legitimate sus-

[78] Boyle, "Continuation of New Experiments. The Second Part," pp. 505, 507 (1680).

[79] We prefer this term to van Leeuwen's "vicarious experience": we wish to preserve the notion that virtual witnessing is a positive action, whereas vicarious experience is commonly held not to be proper experience at all; see van Leeuwen, *The Problem of Certainty*, pp. 97-102; Hacking, *The Emergence of Probability*, chaps. 3-4.

picion among those who were neither direct witnesses nor replicators, a greater degree of assurance was required to produce assent in virtual witnesses. Boyle's literary technology was crafted to secure this assent.

In order to understand how Boyle deployed the literary technology of virtual witnessing, we have to reorient some of our common ideas about the scientific text. We usually think of an experimental report as a narration of some prior visual experience: it points to sensory experiences that lie behind the text. This is correct. However, we should also appreciate that the text itself constitutes a visual source. It is our task here to see how Boyle's texts were constructed so as to provide a source of virtual witness that was agreed to be reliable. The best way to fasten upon the notion of the text as this kind of source might be to start by looking at some of the pictures that Boyle provided alongside his prose.

Figure 1, for example, is an engraving of his original air-pump, appended to the *New Experiments*. Producing these kinds of images was an expensive business in the mid-seventeenth century and natural philosophers used them sparingly. As we see, figure 1 is not a schematized line drawing but an attempt at detailed naturalistic representation complete with the conventions of shadowing and cut-away sections of the parts. This is not a picture of the "idea" of an air-pump, but of a particular existing air-pump.[80] And the same applies to Boyle's pictorial representations of his pneumatic experiments: in one engraving we are shown a mouse lying dead in the receiver; in another, images of the experimenters. Boyle devoted great attention to the manufacture of these images, sometimes consulting directly with the engraver, sometimes by way of Hooke.[81] Their role was to be a supplement to the imaginative witness provided by the words in the text. In the *Continuation* Boyle expanded upon the relationships between the two sorts of exposition; he told his readers that "they who either were versed in such kind of studies or have any peculiar facility of imagining, would well enough conceive my meaning only by words," but others required visual assistance. He apologized for the relative poverty of the images, "being myself absent from the engraver for a good

[80] For studies of engraving and print-making in scientific texts, see Ivins, *Prints and Visual Communication*, esp. pp. 33-36; Eisenstein, *The Printing Press as an Agent of Change*, esp. pp. 262-270, 468-471. We briefly treat Hobbes's iconography in chapter 4.

[81] Hooke to Boyle, 25 August/4 September and 8/18 September 1664, in Boyle, *Works*, vol. VI, pp. 487-490, and Maddison, "The Portraiture of Boyle."

part of the time he was at work, some of the cuts were misplaced, and not graven in the plates."[82]

So visual representations, few as they necessarily were in Boyle's texts, were mimetic devices. By virtue of the density of *circumstantial detail* that could be conveyed through the engraver's laying of lines, they imitated reality and gave the viewer a vivid impression of the experimental scene. The sort of naturalistic images that Boyle favoured provided a greater density of circumstantial detail than would have been proffered by more schematic representations. The images served to announce, as it were, that "this was really done" and that "it was done in the way stipulated"; they allayed distrust and facilitated virtual witnessing. Therefore, understanding the role of pictorial representations offers a way of appreciating what Boyle was trying to achieve with his literary technology.[83]

In the introductory pages of *New Experiments*, Boyle's first published experimental findings, he directly announced his intention to be "somewhat prolix." His excuses were threefold: first, delivering things "circumstantially" would, as we have already seen, facilitate replication; second, the density of circumstantial detail was justified by the fact that these were "new" experiments, with novel conclusions drawn from them: it was therefore necessary that they be "circumstantially related, to keep the reader from distrusting them"; third, circumstantial reports such as these offered the possibility of virtual witnessing. As Boyle said, "these narratives [are to be] as standing records in our new pneumatics, and [readers] need not reiterate themselves an experiment *to have as distinct an idea of it*, as may suffice them to ground their reflexions and speculations upon."[84] If one wrote experimental reports in the correct way, the reader could take on trust that these things happened. Further, it would be as if that reader had been present at the

[82] Boyle, "Continuation of New Experiments," p. 178.

[83] Compare Alpers, *The Art of Describing*, which analyzes the purposes and conventions of realistic pictures in seventeenth-century Holland, demonstrating substantial links between English empiricist theories of knowledge and Dutch picturing. Evidently, the Dutch were trying to achieve by way of picturing what the English were attempting through the reform of prose.

[84] Boyle, "New Experiments," pp. 1-2 (emphases added). The function of circumstantial detail in the prose of Boyle and other Fellows of the Royal Society is also treated in B. Shapiro, *Probability and Certainty*, chap. 7; Lupoli, "La polemica tra Hobbes e Boyle," p. 329; Dear, "*Totius in verba*: The Rhetorical Constitution of Authority in the Early Royal Society"; and Golinski, *Language, Method and Theory in British Chemical Discourse*. We are very grateful to Dear and Golinski for allowing us to see their typescripts.

proceedings. He would be recruited as a witness and be put in a position where he could validate experimental phenomena as matters of fact.[85] Therefore, attention to the writing of experimental reports was of equal importance to doing the experiments themselves.

In the late 1650s Boyle devoted himself to laying down the rules for the literary technology of the experimental programme. Stipulations about how to write proper scientific prose were dispersed throughout his experimental reports of the 1660s, but he also composed a special tract on the subject of "experimental essays." Here Boyle offered an extended *apologia* for his "prolixity": "I have," he understated, "declined that succinct way of writing"; he had sometimes "delivered things, to make them more clear, in such a multitude of words, that I now seem even to myself to have in divers places been guilty of verbosity." Not just his "verbosity" but also Boyle's ornate sentence structure, with appositive clauses piled on top of each other, was, he said, part of a plan to convey circumstantial details and to give the impression of verisimilitude:

> . . . I have knowingly and purposely transgressed the laws of oratory in one particular, namely, in making sometimes my periods [i.e., complete sentences] or parentheses over-long: for when I could not within the compass of a regular period com-

[85] There is probably a connection between Boyle's justification of circumstantial reporting and Bacon's argument in favour of "initiative," as opposed to "magistral," methods of communication; see, for example, Hodges, "Anatomy as Science," pp. 83-84; Jardine, *Bacon: Discovery and the Art of Discourse*, pp. 174-178; Wallace, *Bacon on Communication & Rhetoric*, pp. 18-19. Bacon said that the magistral method "requires that what is told should be believed; the initiative that it should be examined." Initiative methods display the processes by which conclusions are reached; magistral methods mask those processes. Although Boyle's inspiration may, plausibly, have been Baconian, the "influence" of Bacon is sometimes exaggerated (e.g., Wallace, *Bacon on Communication & Rhetoric*, pp. 225-227). It is useful to remember that it was Boyle, not Bacon, who developed the literary forms for an actual programme of systematic experimentation; it is hard to imagine two more different forms than Bacon's aphorisms and Boyle's experimental narratives. See also a marvellously speculative paper on the *Cartesian* roots of contrasting styles of scientific exposition: Watkins, "Confession is Good for Ideas," and the better-known Medawar, "Is the Scientific Paper a Fraud?" For modern testimony to Boyle's success in winning readers' assurance, see Gillispie, *The Edge of Objectivity*, p. 103: "Truly experimental physics came into its own with Robert Boyle. He spared his reader no detail. No one could doubt that he performed all the experiments he reported . . . , bringing to his laboratory great ingenuity, incomparable patience, and that simple honesty which makes experiment really a respectful inquiry rather than an overbearing demonstration."

> prise what I thought requisite to be delivered at once, I chose rather to neglect the precepts of rhetoricians, than the mention of those things, which I thought pertinent to my subject, and useful to you, my reader.[86]

Elaborate sentences, with circumstantial details encompassed within the confines of one grammatical entity, might mimic that immediacy and simultaneity of experience afforded by pictorial representations.

Boyle was endeavouring to appear as a reliable purveyor of experimental testimony and to offer conventions by means of which others could do likewise. The provision of circumstantial details was a way of assuring readers that real experiments had yielded the findings stipulated. It was also necessary, in Boyle's view, to offer readers circumstantial accounts of *failed* experiments. This performed two functions: first, it allayed anxieties in those neophyte experimentalists whose expectations of success were not immediately fulfilled; second, it assured the reader that the relator was not wilfully suppressing inconvenient evidence, that he was in fact being faithful to reality. Complex and circumstantial accounts were to be taken as undistorted mirrors of complex experimental outcomes.[87] So, for example, it was not legitimate to hide the fact that air-pumps sometimes did not work properly or that they often leaked: ". . . I think it becomes one, that professeth himself a faithful relator of experiments not to conceal" such unfortunate contingencies.[88] It is, however, vital to keep in mind that in his circumstantial accounts Boyle proffered only a *selection* of possible contingencies. There was not, nor can there be, any such thing as a report that notes *all* circumstances that might affect an experi-

[86] Boyle, "Proëmial Essay," pp. 305-306, 316; cf. idem, "New Experiments," p. 1; Westfall, "Unpublished Boyle Papers." According to one literary historian, "though [Boyle] aims, like Dryden, to write as a cultured man would talk, his style is hurried and careless, and his sentences rattle on without form or elegance." (Horne, "Literature and Science," p. 193.)

[87] Boyle, "Unsuccessfulness of Experiments," esp. pp. 339-340, 353. Recognizing that contingencies might affect experimental outcomes was also a way of tempering inclinations to *reject* good testimony too readily: if an otherwise reliable source stipulated an outcome that was not immediately obtained, one was advised to persevere; see ibid., pp. 344-345; idem, "Continuation of New Experiments," pp. 275-276; idem, "Hydrostatical Paradoxes," p. 743; Westfall, "Unpublished Boyle Papers," pp. 72-73.

[88] Boyle, "New Experiments," p. 26; and recall Boyle's reporting of the failed experiment 31 (discussed above). In chapter 5 we return to the problem of success and failure in experiment.

ment. Circumstantial, or stylized, accounts do not, therefore, exist as pure forms but as publicly acknowledged moves towards or away from the reporting of contingencies.

The Modesty of Experimental Narrative

The ability of the reporter to multiply witnesses depended upon readers' acceptance of him as a provider of reliable testimony. It was the burden of Boyle's literary technology to assure his readers that he was such a man as should be believed. He therefore had to find the means to make visible in the text the accepted tokens of a man of good faith. One technique has just been discussed: the reporting of experimental failures. A man who recounted unsuccessful experiments was such a man whose objectivity was not distorted by his interests. Thus the literary display of a certain sort of morality was a technique in the making of matters of fact. A man whose narratives could be credited as mirrors of reality was a *modest man*; his reports ought to make that modesty visible. In treating the moral tone of experimental reporting we are therefore beginning to understand the relationship between Boyle's literary and social technologies. How experimentalists were to talk with each other was an important element in specifying the social relations that could constitute and protect experimental knowledge.

Boyle found a number of ways of displaying modesty. One of the most straightforward was the use of the *form* of the experimental essay. The essay, that is, the piecemeal reporting of experimental trials, was explicitly contrasted to the natural philosophical *system*. Those who wrote entire systems were identified as "confident" individuals, whose ambition extended beyond what was proper or possible. By contrast, those who wrote experimental essays were "sober and modest men," "diligent and judicious" philosophers, who did not "assert more than they can prove." This practice cast the experimental philosopher into the role of intellectual "underbuilder," or even that of "a drudge of greater industry than reason." This was, however, a noble character, for it was one that was freely chosen to further "the real advancement of true natural philosophy" rather than personal reputation.[89] The public display of this

[89] Boyle, "Proëmial Essay," pp. 301-307, 300; cf. idem, "Sceptical Chymist," pp. 469-470, 486, 584. Within a year, Henry Power was quoting Boyle's formulations back to him: "I beseech you to looke upon us [Yorkshire experimentalists] as Countrey-Drudges of *much greater Industry than Reason*." Power to Boyle, 10/20 November

modesty was an exhibition that concern for individual celebrity did not cloud judgment and distort the integrity of one's reports. In this connection it is absolutely crucial to remember who it was that was portraying himself as a mere "under-builder." Boyle was the son of the Earl of Cork, and everyone knew that very well. Thus, it was *plausible* that such modesty could have a noble aspect, and Boyle's presentation of self as a moral model for experimental philosophers was powerful.[90]

Another technique for showing modesty was Boyle's professedly "naked way of writing." He would eschew a "florid" style; his object was to write "rather in a philosophical than a rhetorical strain." This plain, ascetic, unadorned (yet convoluted) style was identified as *functional.* It served to display, once more, the philosopher's dedication to community service rather than to his personal reputation. Moreover, the "florid" style to be avoided was a hindrance to the clear provision of virtual witness: it was, Boyle said, like painting "the eye-glasses of a telescope."[91]

The most important literary device Boyle employed for demonstrating modesty acted to protect the fundamental epistemological category of the experimental programme: the matter of fact. There were to be appropriate moral postures, and appropriate modes of speech, for epistemological items on either side of the important boundary that separated matters of fact from the lo-

---

1662, in British Library Sloane MSS 1326 f33^{v}. For natural philosophical textbooks, see Reif, "The Textbook Tradition in Natural Philosophy."

[90] Several of the less modest personalities of seventeenth-century English science were individuals who lacked the gentle birth that routinely enhanced the credibility of testimony: for instance, Hobbes, Hooke, Wallis, and Newton. The best source for Boyle's social situation and temperament is J. Jacob, *Boyle*, chaps. 1-2.

[91] Boyle, "Proëmial Essay," pp. 318, 304. For the importance of the lens and the perceptual model of knowledge in seventeenth-century theories of knowledge, see Alpers, *The Art of Describing*, chap. 3. For Boyle, as for many other philosophers concerned with the reform of language, the goal was "plain-speaking." For the linguistic programme of the early Royal Society and its connections with experimental philosophy, see Christensen, "Wilkins and the Royal Society's Reform of Prose Style"; R. F. Jones, "Science and Language"; idem, "Science and English Prose Style"; Salmon, "Wilkins' *Essay*"; Slaughter, *Universal Languages and Scientific Taxonomy*, esp. pp. 104-186; Aarsleff, *From Locke to Saussure*, pp. 225-277; B. Shapiro, *Probability and Certainty*, pp. 227-246; Hunter, *Science and Society*, pp. 118-119; Dear, "*Totius in verba*: The Rhetorical Constitution of Authority in the Early Royal Society." For Boyle's attack on the "confused," "equivocal," and "cloudy" language of the alchemists, see "Sceptical Chymist," esp. pp. 460, 520-522, 537-539; and, for his criticisms of Hobbes's expository "obscurity," see "Examen of Hobbes," p. 227, and our discussion in chapter 5.

cutions used to account for them: theories, hypotheses, speculations, and the like. Thus, Boyle told his nephew,

> . . . in almost every one of the following essays I . . . speak so doubtingly, and use so often, *perhaps, it seems, it is not improbable*, and such other expressions, as argue a diffidence of the truth of the opinions I incline to, and that I should be so shy of laying down principles, and sometimes of so much as venturing at explications.

Since knowledge of physical causes was only "probable," this was the correct moral stance and manner of speech, but things were otherwise with matters of fact, and here a confident mode was not only permissible but necessary: ". . . I dare speak confidently and positively of very few things, except of matters of fact."[92] Boyle specifically warned readers who expected physical statements to possess "a mathematical certainty and accurateness": ". . . in physical enquiries it is often sufficient, that our determinations come very near the matter, though they fall short of a mathematical exactness."[93]

It was necessary to speak confidently of matters of fact because, as the foundations of proper philosophy, they required protection. And it was proper to speak confidently of matters of fact because they were not of one's own making: they were, in the empiricist language-game, discovered rather than invented. As Boyle told one of his adversaries, experimental facts can "make their own way," and "such as were very probable, would meet with patrons and defenders."[94] The separation of moral modes of speech and the ability of facts to make their own way were made visible on the printed page. In *New Experiments* Boyle said he intended to leave "a conspicuous interval" between his narratives of experimental findings and his occasional "discourses" on their interpretation. One might then read the experiments and the "reflexions" separately.[95] Indeed, the construction of Boyle's experimental essays

[92] Boyle, "Proëmial Essay," p. 307; on "wary and diffident expressions," see also idem, "New Experiments," p. 2. Cf. Sprat, *History*, pp. 100-101; Glanvill, *Scepsis scientifica*, pp. 170-171. For treatments of Boyle's remarks in the context of probabilist and fallibilist models of knowledge, see B. Shapiro, *Probability and Certainty*, pp. 26-27; van Leeuwen, *The Problem of Certainty*, p. 103; Daston, *The Reasonable Calculus*, pp. 164-165.

[93] Boyle, "Hydrostatical Paradoxes," p. 741. Boyle was chastising Pascal in this context.

[94] Boyle, "Hydrostatical Discourse," p. 596.

[95] Boyle, "New Experiments," p. 2.

made manifest the proper separation and balance between the two categories: *New Experiments* consisted of a sequential narrative of forty-three pneumatic experiments; *Continuation* of fifty; and the second part of *Continuation* of an even larger number of disconnected experimental observations, only sparingly larded with interpretative locutions.

The confidence with which one ought to speak about matters of fact extended to stipulations about the proper use of authorities. Citations of other writers should be employed to use them not as "judges, but as witnesses," as "certificates to attest matters of fact." If such a practice ran the risk of identifying the experimental philosopher as an ill-read philistine, it was, for all that, necessary. As Boyle said, "I could be very well content to be thought to have scarce looked upon any other book than that of nature."[96] The injunction against the ornamental citing of authorities performed a significant function in the mobilization of assent to matters of fact. It was a way of displaying that one was aware of the workings of the Baconian "idols" and was taking measures to mitigate their corrupting effects on knowledge-claims.[97] A disengagement between experimental narrative and the authority of systematists served to dramatize the author's lack of preconceived expectations and, especially, of theoretical investments in the outcome of experiments. For example, Boyle several times insisted that he was an innocent of the great theoretical systems of the seventeenth century. In order to reinforce the primacy of experimental findings, "I had purposely refrained from acquainting myself thoroughly with the intire system of either the Atomical, or the Cartesian, or any other whether new or received philosophy." And, again, he claimed that he had avoided a systematic acquaintance with the systems of Gassendi, Descartes, and even of Bacon, "that I might not be prepossessed with any theory or principles."[98]

[96] Boyle, "Proëmial Essay," pp. 313, 317.

[97] On the "idols" and fallibilism, see B. Shapiro, *Probability and Certainty*, pp. 61-62.

[98] Boyle, "Some Specimens of an Attempt to Make Chymical Experiments Useful," p. 355; idem, "Proëmial Essay," p. 302; on the corrupting effects of "preconceived hypothesis or conjecture," see idem, "New Experiments," p. 47, and, for doubts about the correctness of Boyle's professed unfamiliarity with Descartes and other systematists, see Westfall, "Unpublished Boyle Papers," p. 63; Laudan, "The Clock Metaphor and Probabilism," p. 82n; M. B. Hall, "The Establishment of the Mechanical Philosophy," pp. 460-461; idem, *Boyle and Seventeenth-Century Chemistry*, chap. 3; idem, "Boyle as a Theoretical Scientist"; idem, "Science in the Early Royal Society," pp. 72-73; Kargon, *Atomism in England*, chap. 9; Frank, *Harvey and the*

Boyle's "naked way of writing," his professions and displays of humility, and his exhibition of theoretical innocence all complemented each other in the establishment and the protection of matters of fact. They served to portray the author as a disinterested observer and his accounts as unclouded and undistorted mirrors of nature. Such an author gave the signs of a man whose testimony was reliable. Hence, his texts could be credited and the number of witnesses to his experimental narratives could be multiplied indefinitely.

Scientific Discourse and Community Boundaries

We have argued that the matter of fact was a social as well as an intellectual category, and we have shown that Boyle deployed his literary technology so as to make virtual witnessing a practical option for the validation of experimental performances. In this section we want to examine the ways in which Boyle's literary technology dramatized the social relations proper to a community of experimental philosophers. Only by establishing right rules of discourse could matters of fact be generated and defended, and only by constituting these matters of fact into the agreed foundations of knowledge could a moral community of experimentalists be created and sustained. Matters of fact were to be produced in a public space: a particular physical space in which experiments were collectively performed and directly witnessed and an abstract space constituted through virtual witnessing. The problem of producing this kind of knowledge was, therefore, the problem of maintaining a certain form of discourse and a certain mode of social solidarity.

In the late 1650s and early 1660s, when Boyle was formulating his experimental and literary practices, the English experimental community was still in its infancy. Even with the founding of the Royal Society, the crystallization of an experimental community centred on Gresham College, and the network of correspondence organized by Henry Oldenburg, the experimental programme was far from securely institutionalized. Criticisms of the experimental way of producing physical knowledge emanated from English philosophers (notably Hobbes) and from Continental writers committed to rationalist methods and to the practice of natural philosophy

Oxford Physiologists, pp. 93-97. Our concern here is not with the veracity of Boyle's professions but with the reasons he made them and the purposes they were designed to serve.

as a demonstrative discipline.[99] Experimentalists were made into figures of fun on the Restoration stage: Thomas Shadwell's *The Virtuoso* dramatized the absurdity of weighing the air, and scored many of its jokes by parodying the convoluted language of Sir Nicholas Gimcrack (Boyle). The practice of experimental philosophy, despite what numerous historians have assumed, was not overwhelmingly popular in Restoration England.[100] In order for experimental philosophy to be established as a legitimate activity, several things needed to be done. First, it required recruits: experimentalists had to be enlisted as neophytes, and converts from other forms of philosophical practice had to be obtained. Second, the social role of the experimental philosopher and the linguistic practices appropriate to an experimental community needed to be defined and publicized.[101] What was the proper nature of discourse in such a community? What were the linguistic signs of competent membership? And what uses of language could be taken as indications that an individual had transgressed the conventions of the community?

The entry fee to the experimental community was to be the communication of a candidate matter of fact. In *The Sceptical Chymist*, for instance, Boyle extended an olive branch even to the alchemists. The solid experimental findings produced by some alchemists could be sifted from the dross of their "obscure" speculations. Since the experiments of the alchemists (and the few experiments of the Aristotelians) frequently "do not evince what they are alleged to prove," the former might be accepted into the experimental philosophy by stripping away the theoretical language with which they happened to be glossed. As Carneades (Boyle's mouthpiece) said,

[99] For a major Continental critique, see R. McKeon, *Philosophy of Spinoza*, chap. 4; A. R. Hall and M. B. Hall, "Philosophy and Natural Philosophy: Boyle and Spinoza"; and, for an English attack related to Hobbes's, see J. Jacob, *Stubbe*, esp. pp. 84-108.

[100] For the extent to which experimental philosophy was "popular," see Hunter, *Science and Society*, esp. chaps. 3, 6. Shadwell's play was performed in 1676; as we shall see in chapter 4, Charles II, the Society's royal patron, was also said to have found the weighing of the air rather funny, and Petty was aware of pneumatic satire in the early 1670s: A. R. Hall, "Gunnery, Science, and the Royal Society," pp. 129-130. There is some evidence that Hooke believed *he* was Gimcrack: Westfall, "Hooke," p. 483.

[101] This is not intended as an exhaustive catalogue of the measures required for institutionalization. Clearly, patronage was necessary and alliances had to be forged with existing powerful institutions.

> your hermetic philosophers present us, together with divers substantial and noble experiments, theories, which either like peacocks feathers make a great shew, but are neither solid nor useful; or else like apes, if they have some appearance of being rational, are blemished with some absurdity or other, that, when they are attentively considered, make them appear ridiculous.[102]

Thus those alchemists who wished to be incorporated into a legitimate philosophical community were instructed what linguistic practices could secure their admission. Boyle laid down the same principles with respect to any practitioner: "Let his opinions be never so false, his experiments being true, I am not obliged to believe the former, and am left at liberty to benefit myself by the latter."[103] By arguing that there was only a contingent, not a necessary, connection between the language of theory and the language of facts, Boyle was defining the linguistic terms on which existing communities could join the experimental programme.

They were liberal terms, which might serve to maximize potential membership. Boyle's way of dealing with the Hermetics drew on the views of the Hartlib group of the late 1640s and 1650s. By contrast, there were those who rejected the findings of late alchemy (e.g., Hobbes) and those who rejected the process of assimilation (e.g., Newton). The debt to the Hartlib group is important. *The Sceptical Chymist* was drafted before summer 1658 as "Reflexions" on Peripatetic and Paracelsian chemical theory. Precedents existed for the style and tone of the dialogue in Mersenne's *Vérité des sciences* (1625), a conversation between a Christian philosopher, a sceptic, and an alchemist in which an *open* alchemical college was proposed; in Plattes' *Caveat for Alchymists* (1655), published along with Boyle's invitation to open communication in alchemy and physic, where Plattes referred to attempts to demonstrate transmutation before Parliament; and in Renaudot's *Conference concerning the Philosopher's Stone*, published in the same Hartlibian volume, in which seven men—some sceptics, some believers—publicly disputed the possibility of transmutation. Boyle distanced himself somewhat from the group in 1655-1656 when he moved to Oxford to initiate the work on air and saltpetre. But he continued his commitment to the absorption of alchemy within the rules of experimental discourse. The contrast with Newton is instructive. He behaved in an appro-

[102] Boyle, "Sceptical Chymist," pp. 468, 513, 550, 584.
[103] Boyle, "Proëmial Essay," p. 303.

priate but totally distinct manner in alchemy and in experimental philosophy, while Boyle laboured to bring alchemy into the public domain: hence Boyle's 1670s *publications* on alchemy and Newton's criticisms of Boyle's decision to publish.[104]

There were other natural philosophers Boyle despaired to recruit and to assimilate. As we shall see, Hobbes was the sort of philosopher who on no account ought to be admitted to the experimental companionship, for he denied the value of systematic and elaborate experimentation as well as the foundational status of the fact and the distinction between causal and descriptive language. The experimental and the rationalistic language-games were perceived to be radically incompatible. There could be no rapprochement between them, only a choice between the one and the other.

Manners in Dispute

Since experimental philosophers were not to be compelled to give assent to all items of knowledge, dispute and disagreement were to be expected. The task was to manage such dissensus by confining it within safe boundaries. Disagreement about causal explanations might be rendered safe insofar as it was accepted that such items were not foundational. What was neither safe nor permissible was dispute over matters of fact or over the rules of the game by which matters of fact were experimentally produced.

The problem of conducting dispute was a matter of serious practical concern in early Restoration science. During the Civil War and Interregnum "enthusiasts," hermeticists and sectaries threatened

[104] Compare Boyle, "Experimental Discourse of Quicksilver Growing Hot with Gold" (1676) and "An Historical Account of a Degradation of Gold" (1678) with Newton to Oldenburg, 26 April/6 May 1676, in Newton, *Correspondence*, vol. II, pp. 1-3. For Boyle's intention to compose "a short essay concerning chemistry," and a comment on the degradation of gold, see Hartlib to Boyle, 28 February/ 10 March 1654, in Boyle, *Works*, vol. VI, p. 79. For Boyle and the Hartlib group: O'Brien, "Hartlib's Influence on Boyle's Scientific Development"; Rowbottom, "Earliest Published Writing of Boyle"; Webster, "English Medical Reformers"; Wilkinson "The Hartlib Papers." Dobbs, *Foundations of Newton's Alchemy*, p. 72, writes that Boyle and Hartlib moved alchemy "into the area of public dialogue where assumptions underlying alchemical theory could be subjected to a critical analysis. . . . And conceptual scrutiny was being paralleled elsewhere in the group by a more open communication of empirical information." For sources of *The Sceptical Chymist*, see M. B. Hall, "An Early Version of Boyle's 'Sceptical Chymist'," which dates the "Reflexions" to 1657, and Webster, "Water as the Ultimate Principle of Nature," which gives the latest date as summer 1658.

to bring about a radical individualism in knowledge: a situation in which "private judgment" eroded any existing authority and the credibility of any existing institutionalized conventions for generating valid knowledge. Nor did the various sects of Peripatetic natural philosophers display a public image of a stable and united intellectual community. The "litigiousness" of Scholastic philosophers was commonly noted by their experimentalist critics.[105] Unless the experimental community could exhibit a broadly based harmony and consensus within its own ranks, it was unreasonable to expect it to secure the legitimacy within Restoration culture that its leaders desired. Moreover, that very consensus was vital to the establishment of matters of fact as the foundational category of the new practice.

By the early 1660s Boyle was in a position to give concrete exemplars of how disputes in natural philosophy ought to be managed. Three adversaries entered the lists, each objecting to aspects of his *New Experiments*. In chapters 4 and 5 we shall see what their objections were and how Boyle responded to each one: Hobbes, Linus, and Henry More. But even before he had been publicly engaged in dispute, Boyle laid down a set of rules for how controversies were to be handled by the experimental philosopher. For example, in *Proëmial Essay* (published 1661, composed 1657), Boyle went to great lengths to lay down the moral conventions that ought to regulate controversy. Disputes should be about findings and not about persons. It was proper to take a hard view of reports that were inaccurate but most improper to attack the character of those that rendered them, "for I love to speak of persons with civility, though of things with freedom." The *ad hominem* style must at all costs be avoided, for the risk was that of making foes out of mere dissenters. This was the key point: potential contributors of matters of fact, however misguided they might be, must be treated as possible converts to the experimental form of life. If, however, they were harshly dealt with, they would be lost to the cause and to the community whose size and consensus validated matters of fact:

> And as for the (very much too common) practice of many, who write, as if they thought railing at a man's person, or wrangling about his words, necessary to the confutation of his

[105] On Peripatetic litigiousness, see, for example, Boyle, "The Christian Virtuoso," p. 523, and Glanvill, *Scepsis scientifica*, pp. 136-137; on opposition to the sectaries' individualism, see J. Jacob, *Boyle*, chap. 3; and, for general background, see Heyd, "The Reaction to Enthusiasm in the Seventeenth Century."

> opinions; besides that I think such a quarrelsome and injurious way of writing does very much misbecome both a philosopher and a Christian, methinks it is as unwise, as it is provoking. For if I civilly endeavour to reason a man out of his opinions, I make myself but one work to do, namely, to convince his understanding; but, if in a bitter or exasperating way I oppose his errors, I increase the difficulties I would surmount, and have as well his affections against me as his judgment: and it is very uneasy to make a proselyte of him, that is not only a dissenter from us, but an enemy to us.[106]

Furthermore, even the acknowledgment that natural philosophical sects in fact existed might be impolitic. Excessive talk about sects might work to ensure their survival: "It is none of my design," Boyle said, "to engage myself with, or against, any one sect of Naturalists." The *experiments* would decide the case. The views of sects should be noticed only insofar as they were founded upon experiment. Thus it was right and politic to be severe in one's writings against those who did not contribute experimental findings, for they had nothing to offer to the constitution of matters of fact. Yet the experimental philosopher must show that there was point and purpose to legitimately conducted dispute. He should be prepared publicly to renounce positions that were shown to be erroneous. Flexibility followed from fallibilism. As Boyle wrote, "Till a man is sure he is infallible, it is not fit for him to be unalterable."[107]

The conventions for managing disputes were dramatized in the structure of *The Sceptical Chymist*. These fictional conversations (between an Aristotelian, two varieties of Hermetics, and Carneades as mouthpiece for Boyle) took the form, not of a Socratic dialogue, but of a *conference*.[108] They were a piece of theatre that exhibited how persuasion, dissensus and, ultimately, conversion to truth ought to be conducted. Several points about Boyle's theatre of persuasion can be briefly made: first, the symposiasts are imaginary, not real. This means that opinions can be confuted without exacerbating relations between real philosophers. Even Carneades, although he is manifestly "Boyle's man," is not Boyle himself: Carneades is made actually to quote "our friend Mr. *Boyle*" as a device for distancing opinions from individuals. The *author* is insulated

[106] Boyle, "Proëmial Essay," p. 312.

[107] Ibid., p. 311.

[108] See Multhauf, "Some Nonexistent Chemists."

from the text and from the opinions he may actually espouse.[109] Second, truth is not inculcated from Carneades to his interlocutors; rather it is dramatized as emerging through the conversation. Everyone is seen to have a say in the consensus which is the *dénouement*.[110] Third, the conversation is, without exception, *civil*: as Boyle said, "I am not sorry to have this opportunity of giving an example, how to manage even disputes with civility."[111] No symposiast abuses another; no ill temper is displayed; no one leaves the conversation in pique or frustration.[112] Fourth, and most important, the currency of intellectual exchange, and the means by which agreement is reached, is the experimental matter of fact. Here, as we have already indicated, matters of fact are not treated as the exclusive property of any one philosophical sect. Insofar as the alchemists have produced experimental findings, they have minted the real coins of experimental exchange. Their experiments are welcome, while their "obscure" speculations are not. Insofar as the Aristotelians produce few experiments, and insofar as they refuse to dismantle the "arch"-like "mutual coherence" of their system into facts and theories, they can make little contribution to the experimental conference.[113] In these ways, the structure and the linguistic rules of this imaginary conversation make vivid the rules for real conversations proper to experimental philosophy.

In subsequent chapters we discuss the real disputes that followed hard upon the imaginary ones of *The Sceptical Chymist.* Franciscus Linus was the adversary who experimented but who denied the power of the spring of the air; Henry More was the adversary whom Boyle wished to be an ally: More offered what he reckoned to be a more theologically appropriate account of Boyle's pneumatic

[109] Boyle, "Sceptical Chymist," p. 486. Boyle said in the preface that he would not "declare my own opinion"; he wished to be "a silent auditor of their discourses" (pp. 460, 466-467).

[110] The consensus that emerges is very like the position from which Carneades starts, but the plot of *The Sceptical Chymist* involved disguising that fact. Interestingly, the consensus is not *total* (as Jan Golinski has pointed out): Eleutherius indicates reservations about Carneades' arguments, and Philoponus (a more "hard-line" alchemist who is absent for the bulk of the proceedings) might not, in Eleutherius's opinion, have been persuaded. In later chapters we draw the contrast between the form and use of the dialogue by Boyle's anti-experimentalist adversary Hobbes.

[111] Boyle, "Sceptical Chymist," p. 462.

[112] Actually, the great bulk of the talk is between Carneades and Eleutherius. The other two participants inexplicably absent themselves during much of the symposium. This is possibly an accident of Boyle's self-confessed sloppiness with his manuscripts; see Multhauf, "Some Nonexistent Chemists," pp. 39-41.

[113] Boyle, "Sceptical Chymist," p. 469.

findings; but Hobbes was the adversary who denied the value of experiment and the foundational status of the matter of fact. Each carefully crafted response that Boyle produced was labelled as a model for how disputes should be managed by the experimental philosopher. In each response Boyle professed that his concern was not the defence of his reputation but the protection of what was vital to the collective practice of proper philosophy: the value of systematic experimentation (especially that employing "elaborate" instruments such as the air-pump), the matters of fact that experiment produced, the boundaries that separated those facts from less certain epistemological items, and the rules of social life that regulated discourse in the experimental community. The object of controversy, in Boyle's stipulation, was not fact but the interpretation of fact. And the moral tone of philosophical controversy was to be civil and liberal.

What was at stake in these controversies was the creation and the preservation of a calm space in which natural philosophers could heal their divisions, collectively agree upon the foundations of knowledge, and thereby establish their credit in Restoration culture. A calm space was essential to achieving these goals. As Boyle reminded his readers in the introdution to *New Experiments* (published in that "wonderful, pacifick year" of the Restoration), "the strange confusions of this unhappy nation, in the midst of which I have made and written these experiments, are apt to disturb that calmness of mind and undistractedness of thoughts, that are wont to be requisite to happy speculations."[114] And Sprat recalled the circumstances of the Oxford group of experimentalists that spawned the Royal Society: "Their first purpose was no more, then onely the satisfaction of breathing a freer air, and of conversing in quiet one with another, without being ingag'd in the passions, and madness of that dismal Age."[115]

Three Technologies and the Nature of Assent

We have argued that three technologies were involved in the production and validation of matters of fact: material, literary, and social. We have also stressed that the three technologies are not distinct and that the workings of each depends upon the others. We can now briefly develop that point by showing how each of

[114] Boyle, "New Experiments," p. 3. The phrase "wonderful pacifick year" is from Sprat, *History*, p. 58.

[115] Sprat, *History*, p. 53.

Boyle's technologies contributes to a common strategy for the constitution of the matter of fact. In the first section of this chapter we argued that the matter of fact can serve as the foundation of knowledge and secure assent insofar as it is not regarded as man-made. Each of Boyle's three technologies worked to achieve the appearance of matters of fact as *given* items. That is to say, each technology functioned as an *objectifying resource.*

Take, for example, the role of the air-pump in the production of matters of fact. Pneumatic facts, as we have noted, were machine-made. One of the significant features of a scientific machine is that it stands between the perceptual competences of a human being and natural reality itself. A "bad" observation taken from a machine need not be ascribed to faults in the human being, nor is a "good" observation his personal product: it is this impersonal device, the machine, that has produced the finding. In chapter 6 we shall see a striking instance of this usage. When, in the 1660s, Christiaan Huygens offered a matter of fact that appeared to conflict with one of Boyle's explanatory resources, Boyle did not impugn the perceptual or cognitive competences of his fellow experimentalist. Rather, he was able to suggest that the machine was responsible for the conflict: "[I] question not [his] Ratiocination, but only the stanchness of his pump."[116] The machine constitutes a resource that may be used to factor out human agency in the product: as if it were said "it is not I who says this; it is the machine"; "it is not your fault; it is the machine's."

The role of Boyle's literary technology was to create an experimental community, to bound its discourse internally and externally, and to provide the forms and conventions of social relations within it. The literary technology of virtual witnessing extended the public space of the laboratory in offering a valid witnessing experience to all readers of the text. The boundaries stipulated by Boyle's linguistic practices acted to keep that community from fragmenting and to protect items of knowledge to which one might expect universal assent from items of knowledge that historically generated divisiveness. Similarly, his stipulations concerning proper manners in dispute worked to guarantee that social solidarity that produced assent to matters of fact and to rule out of order those imputations that would undermine the moral integrity of the experimental form of life. The objectivity of the experimental matter of fact was an

[116] Boyle to Moray, July 1662, in Huygens, *Oeuvres*, vol. IV, p. 220. Compare Boyle's accounting for Linus's deviant findings in his attempted replication of the Puy-de-Dôme experiment: "Defence against Linus," pp. 152-153, and chapter 5 below.

artifact of certain forms of discourse and certain modes of social solidarity.

Boyle's social technology constituted an objectifying resource by making the production of knowledge visible as a collective enterprise: "It is not I who says this; it is all of us." As Sprat insisted, collective performance and collective witness served to correct the natural working of the "idols": the faultiness, the idiosyncrasy, or the bias of any individual's judgment and observational ability. The Royal Society advertised itself as a "union of eyes, and hands"; the space in which it produced its experimental knowledge was stipulated to be a *public space*. It was public in a very precisely defined and very rigorously policed sense: not everybody could come in; not everybody's testimony was of equal worth; not everybody was equally able to influence the institutional consensus. Nevertheless, what Boyle was proposing, and what the Royal Society was endorsing, was a crucially important *move towards* the public constitution and validation of knowledge. The contrast was, on the one hand, with the private work of the alchemists, and, on the other, with the individual dictates of the systematical philosopher.

In the official formulation of the Royal Society, the production of experimental knowledge commenced with individuals' acts of seeing and believing, and was completed when all individuals voluntarily agreed with one another about what had been seen and ought to be believed. This freedom to speak had to be protected by a special sort of discipline. Radical individualism—the state in which each individual set himself up as the ultimate judge of knowledge—would destroy the conventional basis of proper knowledge, while the disciplined collective social structure of the experimental form of life would create and sustain that factual basis. Thus the experimentalists were on guard against "dogmatists" and "tyrants" in philosophy, just as they abominated "secretists" who produced their knowledge-claims in a private and undisciplined space. No one man was to have the right to lay down what was to count as knowledge. Legitimate knowledge was warranted as objective insofar as it was produced by the collective, and agreed to voluntarily by those who comprised the collective. The objectification of knowledge proceeded through displays of the communal basis of its generation and evaluation. Human coercion was to have no visible place in the experimental form of life.[117]

[117] Sprat, *History*, pp. 98-99 (for the individual and the collective); ibid., p. 85, and Hooke, *Micrographia*, "The Preface," sig a2^{v} (for "eyes and hands" and "a sincere

If the obligation to assent to items of knowledge was not to come from human coercion, where did it come from? It was to be nature, not man, that enforced assent. One was to believe, and to say one believed, in matters of fact because they reflected the structure of natural reality. We have described the technologies that Boyle deployed to generate matters of fact and the conventions that regulated the knowledge-production of the ideal experimental community. Yet the transposition onto nature of experimental knowledge depended upon the routinization of these technologies and conventions. The naturalization of experimental knowledge depended upon the institutionalization of experimental conventions. It follows from this that any attack upon the validity and objectivity of experimental knowledge-production could proceed by way of a display of its conventional basis: showing the work of production involved and exhibiting the lack of obligation to credit experimental knowledge. It might also exhibit an alternative form of life by which assent might more effectively be achieved, one which would yield a superior sort of obligation to assent. In his criticisms of Boyle's programme, Hobbes endeavoured to do just this. Hobbes maintained that the experimental form of life could not produce effective assent: it was not *philosophy*.

Hand, and a faithful Eye"); Sprat, *History*, pp. 28-32 and Glanvill, *Scepsis scientifica*, p. 98 (for "tyrants" in philosophy). For the disciplining of the Royal Society's public: J. Jacob, *Boyle*, p. 156; idem, *Stubbe*, pp. 59-63; also some highly perceptive remarks in Ezrahi, "Science and the Problem of Authority in Democracy," esp. pp. 46-53.

· III ·

Seeing Double: Hobbes's Politics of Plenism before 1660

Plots have I laid, inductions dangerous . . .
SHAKESPEARE, Richard III

BOYLE'S programme for experimental philosophy was a solution to the problem of order. Natural philosophy had been in a state of scandalous dissension. Nowhere was scandal more visible than in the handling of the Torricellian phenomenon and related effects. Boyle attempted to remedy this dissension by proposing a new way of going on in natural philosophy: a new way of working, of speaking, of forming social relations among natural philosophers. To Boyle and his colleagues the experimental solution to the problem of order was possible, effective, and safe. Its practicality, potency, and innocuousness were dependent upon the erection and maintenance of a crucial boundary around the practices of the new experimental form of life. Dissension within this boundary was safe, even fertile and necessary. Dissension involving violations of this boundary, and especially involving the intrusion of rejected modes of speaking, was deemed fatal.

On one side of this boundary philosophers were enjoined to speak the language of experimental "physiology"; on the other, they spoke the traditional language of natural philosophy, now stigmatized as "metaphysics." We have seen how Boyle laboured to situate proper speech about his air-pump experiments in a new experimental discourse, one that made it unnecessary to decide upon the question of a metaphysical vacuum versus a metaphysical plenum. The "vacuum" Boyle referred to in his *New Experiments* was a new item in the vocabulary of natural philosophy: it was an operationally defined entity, reference to which was dependent upon the working of a new artificial device.

It was this usage, and the practices in which it was embedded, that Hobbes attacked in 1661. In Hobbes's view, Boyle's experimental solution to the problem of order was not possible; it was not effective; and it was dangerous. He argued, first, that the boundaries Boyle proposed to erect and maintain were guarantees of continued *disorder*, not remedies to philosophical dissension, and, second, that order could only be won and made secure by deciding upon *proper* metaphysical language, not by jettisoning that language. Specifically, Hobbes denied Boyle's right to appropriate the term "vacuum" for the new experimental discourse and he contested the legitimacy of that new discourse. Hobbes denied Boyle's claim that one could talk about a vacuum without talking metaphysics. In Hobbes's reading, Boyle was asserting that his machine had produced a metaphysical vacuum: a space devoid of all corporeal substance. Yet, as we shall show in chapter 4, Hobbes sought to demonstrate that the *machina Boyleana* had failed to achieve such a vacuum. A vacuum, thus defined, did not exist in nature and had not been produced in Boyle's experimental space.

Why did Hobbes read Boyle's texts in this way? It is not that Hobbes "misunderstood" his adversary in any simplistic way: we shall see in the next chapter that Hobbes's engagement with the details of Boyle's 1660 text was extremely close. But it was an *interested* and an historically informed reading, conditioned by the resources and the analyses Hobbes had developed prior to 1660. First, Hobbes was concerned to defend his own standing as a major natural philosopher and to defend the natural philosophical schema he had constructed and refined through the 1640s and 1650s. Second, Hobbes had developed that system as uniquely suited to securing order and achieving the proper goals of philosophy. Any other project for natural philosophy endangered order. Third, there was the heightened sensibility to the practical problem of dissension that was displayed by all English intellectuals during the making of the Restoration settlement (see chapter 7).

In this chapter we want to display the range of considerations that Hobbes brought to bear on the question of vacuism and plenism, and we want to understand what Hobbes thought was the philosophical language appropriate to discussing that question. We shall show, first, what Hobbes considered was *wrong* about vacuism, and, second, what he thought was *dangerous* about such a position. Finally, we shall point to the integrity of Hobbes's perspective: how he calculated that a proper understanding of what the natural world contained and a proper conception of philosophical practice would ensure public peace.

"Denying the Vacuum": Hobbes and Experimental Pneumatics

Hobbes began constructing a plenist natural philosophy in the 1640s. Throughout this period, he was at the very centre of the natural philosophical community in Paris. The Torricellian phenomenon and experimental pneumatics provided some of the principal concerns of this community. Hobbes was already "numbered among the philosophers" in Paris during the 1630s. In 1635-1636 he visited France when acting as tutor to the third Earl of Devonshire, William Cavendish. Hobbes corresponded with the French via Kenelm Digby, Charles Cavendish, and other members of the group connected with his employer's family.[1] Hobbes returned to Paris after his flight from England in November 1640. As a key member of the circle centred on Marin Mersenne, Hobbes now debated with Descartes, Gassendi and other natural philosophers. Mersenne published his work on optics and mechanics in 1644. Samuel Sorbière published the extended edition of Hobbes's *De cive* in 1648.[2] The texts of this period have been used to assess the sources of Hobbes's physics and to judge the character of his commitment to plenism. In particular, historians have traced his progress towards the completion of his natural philosophical work, *De corpore* (1655). The context of the composition of *De corpore* illuminates the views on the worth of experimental pneumatics which Hobbes developed before his debate with Boyle.[3]

Nothing in the argument of this book depends upon portraying Hobbes's natural philosophy as totally novel or idiosyncratic. We are concerned with the resources at Hobbes's disposal in his response to Boyle in 1661. In fact, however, contemporaries made much of the question of Hobbes's sources. His critics often wrote

[1] Halliwell, *Collection of Letters*, pp. 65-69; Tönnies, *Hobbes*, pp. 11-22; Jacquot, "Cavendish and His Learned Friends"; de Beer, "Some Letters of Hobbes," p. 196; Reik, *Golden Lands*, p. 74; Digby to Hobbes, 24 September/4 October 1637, in Tönnies, *Studien*, p. 147.

[2] Lenoble, *Mersenne*, pp. 430-436; Hobbes, "Tractatus opticus"; Köhler, "Studien," pp. 71n-72n; Hobbes to Sorbière, 6/16 May 1646, in Tönnies, *Studien*, pp. 53-54; Mersenne to Haak, 29 February/10 March 1640, in Mersenne, *Correspondance*, vol. XI, pp. 403-404.

[3] For the origins of *De corpore*: Kargon, *Atomism in England*, p. 58; Laird, *Hobbes*, pp. 115-116; Hobbes, *Critique du De Mundo*, pp. 71-88; Jacquot, "Notes on an Unpublished Work of Hobbes"; idem, "Un document inédit"; Aaron, "A Possible Draft of *De Corpore*"; Brockdorff, *Cavendish Bericht für Jungius*; Pacchi, *Convenzione*, pp. 25ff.

of groups who "have been willing to accept Mechanism upon Hobbian conditions." The danger which Hobbes posed could be defused by portraying his work as derivative. Boyle treated the views of Hobbes and Descartes together; John Wallis always pointed out Hobbes's lack of originality; in 1654 Seth Ward charged Hobbes with the theft of optical theories from another member of the Cavendish group, Walter Warner, and claimed that Hobbes's natural philosophy came from the trustier truisms of Descartes, Gassendi, and Kenelm Digby.[4] If Hobbes's natural philosophy was described as routinely mechanical, then it was possible to play down the specific challenge he offered, most obviously in his attack on Boyle. Alternatively, by treating this philosophy as utterly peculiar, it was possible to ignore Hobbes's close links with the natural philosophical community, particularly when he was in France in the 1640s. Those links provided Hobbes with resources which he used in his analysis of experiment and plenism.

Hobbes's attitude towards experimental pneumatics in this period had three salient features. First, the French experimenters and philosophers were *not* divided into two exclusive camps of plenists and vacuists. There was no consensus about the appropriate interpretation of the critical experiments produced in the 1640s. Hobbes himself used the Epicurean distinction between a microscopic array of empty spaces dispersed in matter (*vacuum disseminatum*) and a macroscopic void space produced by the absence of all body (*vacuum coacervatum*). Gassendi also used this distinction. In the texts on optics which Hobbes produced in the 1640s, he appealed to the concept of a microscopic disseminated void when describing the expansive action of the Sun. But this did not mean he was a "vacuist": he *never* accepted the reality of macroscopic empty space.[5] Second, Hobbes challenged the capacity of any set of pneumatic experiments to settle these disputes or to prove the case for an artificial vacuum. The range of rival interpretations of these experiments showed the lack of authority in current natural

[4] Glanvill, *Scepsis scientifica*, "To the Royal Society"; Ward, *Vindiciae academiarum*, p. 53; Boyle, "Examen of Hobbes"; Hobbes, "Six Lessons," p. 340; Halliwell, *Collection of Letters*, pp. 84-85.

[5] For Hobbes's use of the *vacuum disseminatum* and its context in light metaphysics, see Hobbes, "Little Treatise"; idem, *White's De Mundo Examined*, p. 101; Gargani, *Hobbes e la scienza*, pp. 98-123, 209-237; A. Shapiro, "Kinematic Optics," pp. 143-172. For the Epicurean distinction: Grant, *Much Ado about Nothing*, pp. 70-71; Webster, "Discovery of Boyle's Law," p. 443; Rochot, "Comment Gassendi interprétait l'expérience du Puy de Dôme"; Charleton, *Physiologia*, pp. 55-56.

philosophy. Finally, Hobbes identified *absurd* metaphysical language as a principal source of these difficulties in natural philosophy. He pointed out the dangerous consequences of incoherent speech about empty space, and analyzed the linguistic differences between rival natural philosophical schemes developed in the 1640s, notably that of Descartes.

Each of these features is visible in the texts Hobbes produced in France. In spring 1641 Hobbes began a heated exchange with Descartes on mechanics and optics. He rejected any Cartesian notion of an incorporeal substance and he claimed that Descartes' "subtle matter" was the same as his own model of a space-filling fluid. Descartes rejected this redefinition of terms. When Hobbes argued against plenism based on a false definition of "body," Descartes was often his prime target.[6] In the winter of 1642-1643, Hobbes composed a critique of a set of dialogues by the Catholic philosopher Thomas White. In his critique of White's *De mundo*, he pointed to the persistent divisions in the natural philosophical community on the issue of plenism. He remained sceptical of experiments on vacuism and rarefaction, "those inner mysteries of physics." He discussed two celebrated exemplars of the debate: the thermoscope, in which air heated in a bulb drove water up a tube connected to the bulb, and the wind-gun, a recently invented pneumatic device reported by Mersenne. Hobbes wrote that "the cause of the dilatation and compression of the air in the thermoscope can be sufficiently explained even if we deny the vacuum." As in his attack on Descartes, he was attentive to the connection between absurd definitions in natural philosophy and false belief: men "came to believe in the existence of innumerable daemons" just because "whatever the sight can penetrate they consider to be a vacuum." Hobbes argued that not all bodies were opaque. He had made the same point in his political tract, *Elements of Law*, written just before leaving England in 1640. Common sensations were an untrustworthy guide to the character of apparently empty space. Otherwise, he wrote, men would falsely believe that there were "insubstantial beings" or "spirits."[7]

[6] On the conflict with Descartes: Mersenne, *Correspondance*, vol. X, pp. 210-212, 426-431, 487-504, 522-534, 568-576, 588-591. Cf. Hobbes to Sir Charles Cavendish, 29 January/8 February 1641, in *English Works*, vol. VII, pp. 455-462; Hobbes, "Objectiones ad Cartesii Meditationes"; idem, *Critique du De Mundo*, pp. 16-20; Tönnies, *Studien*, p. 115; Hervey, "Hobbes and Descartes"; Brandt, *Hobbes' Mechanical Conception*, pp. 138-142.

[7] Hobbes, *White's De Mundo Examined*, pp. 46-48, 54; for White's natural philos-

The danger of such beliefs and the character of experiment preoccupied Hobbes when he composed his discussion of plenism in *De corpore*. Hobbes began writing this work after his optical treatise had been published by Mersenne in 1644. A definitive version of *De corpore* may have existed by 1648. During this period, Hobbes continued his discussion of the interstitial void and his assault on Cartesian metaphysics. His colleagues and patrons, Charles Cavendish and his brother the Earl of Newcastle, reached Paris in 1645, and their correspondence reveals something of the contents of Hobbes's project. He told Newcastle in 1646 that a disseminated vacuum might be formed by the "dilatation" of the Sun, and attacked Cartesian plenism: "For who knowes not that Extension is one thing and the thing extended another."[8] In February 1648, Hobbes repeated these views for Mersenne, arguing specifically against the ideas of the Jesuit philosopher Noël, who he claimed had made an illegitimate use of Cartesian subtle fluids in discussing the behaviour of the thermoscope and the thermometer.[9] At the same time, however, Hobbes also considered the new trials produced in experimental pneumatics and challenged their role in settling the divisions of natural philosophy.

Mersenne published the reports of the Torricellian phenomenon when he returned from Florence to Paris in spring 1645. Between autumn 1646 and autumn 1648 experimenters such as Pascal and Roberval developed a range of critical phenomena including the void-in-the-void trial and the Puy-de-Dôme experiment.[10] Hobbes may have witnessed some of these trials in Paris and the reports were transmitted to England from spring 1648. Boyle now first learnt of these accomplishments in experimental pneumatics through the correspondence of Cavendish, Haak, and Hartlib.[11] For Hobbes, such reports were by no means decisive. In February

---

ophy, see Henry, "Atomism and Eschatology"; for "spirits," compare Hobbes, "Human Nature," pp. 60-62.

[8] Pacchi, *Convenzione*, p. 28; Köhler, "Studien," p. 72n; Kargon, *Atomism in England*, p. 57.

[9] Hobbes to Mersenne, 7/17 February 1648, in Tönnies, *Studien*, pp. 132-134. For Noël, see Fanton d'Andon, *L'horreur du vide*, pp. 47-57; Noël, *Le plein du vide.*

[10] de Waard, *L'expérience barométrique*, pp. 117-123; Fanton d'Andon, *L'horreur du vide*, pp. 1-41; Sadoun-Goupil, "L'oeuvre de Pascal," pp. 249-277; Middleton, *History of the Barometer*, pp. 3-32; Auger, *Roberval*, pp. 117-133. For these trials, see also Pascal, *Oeuvres*, pp. 195-198, 221-225.

[11] Hartlib to Boyle, May 1648, in Boyle, *Works*, vol. VI, pp. 77-78; Cavendish to Petty, April 1648, in Webster, "Discovery of Boyle's Law," p. 456; Haak to Mersenne, July 1648, in H. Brown, *Scientific Organizations*, p. 271.

1648 he invited Mersenne and his colleagues to try experiments about the transmission of sound and light through the Torricellian space. Experimenters in Paris and London announced that "they do not yet wish to declare that it is a true vacuum in the glass above the mercury."[12] When Mersenne discussed these trials with Hobbes, however, he expressed his doubts whether any such experiments were conclusive. In May 1648 Hobbes told Mersenne that "all the experiments made by you and by others with mercury do not conclude that there is a void, because the subtle matter in the air being pressed upon will pass through the mercury and through all other fluid bodies, however molten they are. As smoke passes through water."[13]

Hobbes commented here on trials performed by Mersenne's colleagues, including Roberval. Roberval's trial with a carp's bladder which expanded in the Torricellian space was first performed in the same month. It did much to damage any faith in the experimental production of a macroscopic void. The experiments Roberval showed to Mersenne and his colleagues in Paris, including Hobbes, suggested that "nothing really certain can be established about that space *which seems void*, through which light and colours pass." In spring 1648 Mersenne wrote about Noël's attack on Pascal and the challenge posed by Roberval's work: "We are beginning to believe here that it is not a vacuum." In May he confessed that the carp's bladder experiment in the Torricellian space was "an insoluble business." For Roberval, this trial showed that the air was elastic and that the Torricellian space was not totally void. Thus the doctrines of a Cartesian plenum and of a macroscopic space devoid of body were both called into question. Rival experimenters proffered this range of exemplary phenomena to sustain entirely opposed natural philosophies and to evince the effects of different forms of subtle fluid.[14]

Hobbes's attitude towards these standard pneumatic experiments was informed by the French debates of the 1640s. Pascal's trials were neither decisive nor unambiguous. Hobbes used the dissem-

[12] Hobbes to Mersenne, 7/17 February 1648, in Tönnies, *Studien*, pp. 132-134; Haak to Mersenne, 24 March/3 April 1648, in H. Brown, *Scientific Organizations*, p. 58.

[13] Hobbes to Mersenne, 15/25 May 1648, in H. Brown, "Mersenne Correspondence"; Pacchi, *Convenzione*, p. 238.

[14] For Roberval's trials: Auger, *Roberval*, pp. 128-130, and Webster, "Discovery of Boyle's Law," pp. 449-450, 496-497. Compare Mersenne to Huygens, 7/17 March and 22 April/2 May 1648, in Huygens, *Oeuvres*, vol. 1, pp. 84, 91. For Noël's attack, October 1647 to April 1648, see the correspondence between Noël and Pascal, in Pascal, *Oeuvres*, pp. 199-221.

inated vacuum in his optical treatises of 1644-1646, but this did not commit him to the vacuist interpretation of experimental pneumatics. Instead, in his exchanges with Mersenne, Gassendi, Cavendish, and Descartes, he argued that the problem in pneumatics was an issue of *right language* in philosophy. No isolated experimental trial could decide on the character of the subtle fluids which filled space. These fluids mediated all natural action; their investigation depended upon a *prior* analysis of the character of body, space, and motion, and did not solely derive from uncertain trials and untrustworthy sense-data. In 1657 Hobbes explained to Sorbière, a disciple of Gassendi, that even "Epicurus's opinions do not seem absurd to me, in the sense in which it seems to me he understands the vacuum. For I believe that what he calls the *vacuum*, Descartes calls *subtle matter*, and I call the *purest aetherial substance*."[15] This discrimination between insecure experience and secure philosophical language is equally apparent in the completed version of *De corpore*, published in 1655 after Hobbes's return to England.[16]

In the first three parts of *De corpore*, Hobbes established the philosophical basis of his scheme of natural knowledge. His detailed confrontation with experimental pneumatics is confined to Part IV, which deals with that branch of natural philosophy termed "Physics."[17] Elsewhere, Hobbes illustrated the definitional exercise that led to proper language. In his definitions of "place" and "congruity," Hobbes appeared to admit the possibility of empty space:

> . . . can any man that has his natural senses, think that two bodies must therefore necessarily touch one another, because no other body is between them? Or that there can be no *vacuum*,

[15] Hobbes to Sorbière, 27 January/6 February 1657, in Tönnies, *Studien*, pp. 71-73. Compare Descartes to Mersenne, 22 February/4 March 1641, in Mersenne, *Correspondance*, vol. x, p. 524, who cited Hobbes's statement that "by *spirit* I understand a *subtle fluid body*, therefore it is the same thing as his [i.e., Descartes'] *subtle matter*."

[16] The Latin was published in 1655; the English, "Concerning Body," in the following year. There are significant differences between the two versions, but these do not involve the points we wish to make and we cite the English version. For the most detailed account of the natural philosophy of *De corpore*: Brandt, *Hobbes' Mechanical Conception*; see also Watkins, *Hobbes's System*, chaps. 3-4; Kargon, *Atomism in England*, chap. 6; Gargani, *Hobbes e la scienza*, chap. 4.

[17] The first three parts of *De corpore* are concerned with the production of knowledge of effects from the right definition of a cause; the section on "Physics" is devoted to the inferior method of ascertaining possible causes from knowledge of effects or appearances. (We discuss the status of these methods in chapter 4.) For Hobbes's critique of sensory data and of "single and particular . . . propositions," see "Concerning Body," chap. 25, esp. p. 388.

> because *vacuum* is nothing, or as they call it, *non ens*? Which is as childish, as if one should reason thus: no man can fast, because to fast is to eat nothing; but nothing cannot be eaten.[18]

In fact, Hobbes did not argue here that a vacuum *does* exist: he showed against the Scholastics that its existence or nonexistence could not be established through absurd speech and improper use of words. The proper analysis of mechanical motion and continuity of matter demanded some "fluid medium which hath no vacuity."[19] In Hobbes's physics, this space-filling medium implied that motion would be performed in closed curves. This *simple circular motion* was fundamental to Hobbes's physical explanations. So much was established through the definitional knowledge of causes. In the final section on "Physics," especially in the chapter "Of the World and of the Stars," Hobbes now addressed himself directly to the physical problem of the vacuum and the claims of experimental pneumatics. His answer was unambiguous. In his survey of pneumatics all the arguments for the vacuum and all the phenomena said to support its existence were rejected. Hobbes provided a dress rehearsal of his criticisms of Boyle's experimental findings in 1661.

He "that would take away vacuum, should without vacuum show us such causes of these phenomena, as should be at least of equal, if not greater probability."[20] This Hobbes accomplished, systematically disposing of the major phenomena purporting to demonstrate a vacuum in nature and in experimental systems. Part, but only part, of Hobbes's physical arguments against the vacuum involved the invocation of a fluid aether. He supposed that the world contains the aggregate of visible bodies, such as the Earth and the stars; invisible bodies, such as "the small atoms which are disseminated through the whole space between the earth and the stars"; and "lastly, that most fluid ether, which so fills all the rest of the universe, as that it leaves in it no empty place at all."[21] For our purposes the most important parts of Hobbes's arguments against

[18] Hobbes, "Concerning Body," pp. 107, 109; cf. p. 124. For a similar use of Hobbes against Descartes on this issue, see Barrow, *Usefulness of Mathematical Learning*, p. 140.

[19] Hobbes, "Concerning Body," pp. 321-322, 332, 341-342.

[20] Ibid., p. 425.

[21] Ibid., p. 426. For other mentions of a space-filling fluid aether, see pp. 448, 474, 480-481, 504, 519. We shall show in the following chapter that the aether was not the only resource Hobbes used to argue against the existence of a vacuum in experimental systems.

the vacuum dealt not with the general constitution of the world but with the paradigmatic experiments in the Torricellian tradition.

Here Hobbes offered what was for him a fairly detailed account of an experimental protocol. Fill with mercury a glass cylinder "of sufficient length," closed at one end; stop the open end with your finger and insert this end in a dish of the same liquid; then remove your finger. The mercury level in the glass tube subsides, leaving a space at the top end of the glass. "From whence they [vacuists] conclude that the cavity of the cylinder above the quicksilver remains empty of all body." "But," Hobbes remarked, "in this experiment I find no necessity at all of vacuum." His alternative physical explanation of what is in the Torricellian space and how it gets there set the pattern for his later controversies with Boyle and the "Greshamites." It was predicated upon the existence of a plenum and circulatory movement of matter in the plenum. Consider, Hobbes said, what happens when the mercury descends: the level of the mercury in the dish supporting the cylinder must rise; and, as it rises, so much of the contiguous air "must be thrust away as may make place for the quicksilver which is descended." That air must go somewhere, and, as it moves, it pushes away the air next to it, "and so successively, till there be a return to the place where the propulsion first began":

> And there, the last air thus thrust on will press the quicksilver in the vessel with the same force with which the first air was thrust away; and if the force with which the quicksilver descends be great enough, . . . it will make the air penetrate the quicksilver in the vessel, and go up into the cylinder to fill the place which they [vacuists] thought was left empty.

So the Torricellian space, for some philosophers crucial proof of the existence of a vacuum, is actually *full*, and it is full of atmospheric air. Hobbes realized that he must also provide an explanation of why the mercury stops at some level rather than empties itself completely. His answer pointed to the relationship between the height from which the column of mercury descends, the force that this descent imparts to the contiguous air, and thus the force with which the circulated air returns to the surface of the mercury in the dish below. At about 26 inches, he claimed, an equilibrium is attained between the "endeavour" of the mercury downwards and its resistance to being penetrated by the circulated air.[22]

[22] Ibid., pp. 420-422. This general explanation of the Torricellian experiment is

Hobbes used the same resources of circulatory motion and the plenum to account for several other standard "experiments" in medieval and early modern natural philosophy. For example, the normal operation of the gardener's watering-pot was adduced to demonstrate the nonexistence of a vacuum. The lower surface of the typical canister is pierced by numerous small holes; the narrow mouth can be stopped with one's finger. When this is done, the water no longer flows out. Why is this? The water cannot flow out of the holes in the gardener's pot when the mouth is stopped because, Hobbes said, there is no ultimate place for the air below to go. The flow may be restored by removing one's finger from the mouth, in which case the contiguous air, "by continual endeavour," may proceed to the mouth "and succeed into the place of the water that floweth out." Or, even if one continued to stop the mouth, the flow may be ensured if the holes in the bottom of the pot are sufficiently large that the water passing out "can by its own weight force the air at the same time to ascend into the vessel by the same holes."[23] In this and in several other classic experiments Hobbes discussed, one of the explanatory tactics he brought to bear was the ability of air, when sufficiently strongly impelled, to penetrate water or mercury, particularly at the edges of the liquid-containing vessel. This air powerfully penetrated the tightest of the physical boundaries the vacuists deployed to keep it out.[24]

One experiment of special interest in the present context concerned the cohesion of smooth marbles or glasses. In the preceding chapter we briefly discussed Boyle's trials on cohesion *in vacuo*, the thirty-first of his *New Experiments.* We noted the way in which Boyle attempted to appropriate this paradigmatic phenomenon for the discursive practices of the new experimental programme. In the *De corpore* written five years before, Hobbes addressed cohesion in order to demonstrate that its correct physical explanation was incompatible with vacuism. As a mechanical philosopher, Hobbes had no use for the *horror vacui* to support his plenism. Instead, he

---

also to be found in Hobbes, "Seven Philosophical Problems," pp. 23-24; "Decameron physiologicum," pp. 92-93; and in "Dialogus physicus," pp. 256-257. We cannot discuss here Hobbes's concept of "endeavour" (or "*conatus*"), which he defined ("Concerning Body," p. 206) as "motion made in less space and time than can be given . . .; that is, motion made through the length of a point, and in an instant or point of time." On this subject, see Brandt, *Hobbes' Mechanical Conception*, pp. 300-315; Watkins, *Hobbes's System*, pp. 123-134; Bernstein, "*Conatus*, Hobbes and the Young Leibniz"; Sacksteder, "Speaking about Mind."

[23] Hobbes, "Concerning Body," pp. 414-415.

[24] Ibid., pp. 420, 423-424.

assimilated the phenomenon to his theory of hardness. If, in fact, two cohered bodies were perfectly hard and perfectly smooth, it would, he argued, be impossible to pull them apart by exerting a perpendicular force. This was because the separation would entail an infinite velocity for the inrushing air. For Hobbes, infinite velocity as well as infinite hardness was just as impossible as the notion of an infinite world. So the physical explanation of separation that Hobbes offered invoked finite velocity and finite hardness: if one applied very great force to the cohered bodies, they would flex and allow a successive flowing-in of air.[25] The difference between Hobbes and Boyle on cohesion was not, therefore, a difference in mechanism or a difference in their attitudes towards the *horror vacui*: both embraced the former and abominated the latter. It was a difference in conceptions of proper speech about such phenomena, and, therefore, a difference in exemplifying how the natural philosopher was to go on.

There were, in Hobbes's view, no decisive arguments and no crucial experiments to support the idea that a vacuum existed in nature or could be made by experimenters. In *De corpore* Hobbes had "taken away vacuum" on physical grounds. In his plenist physics he had supplied arguments of "at least equal, if not greater probability" to account for the phenomena adduced in favour of vacuism. He did not claim to have done the experiments himself, although he was familiar enough with several of them and possibly witnessed some of the most important performances in the Torricellian programme. We now ask about the purposes served by the definitive *plenism* which Hobbes adopted in the wake of this experience and which he later brought to bear upon Boyle's *New Experiments.* There is one highly relevant text from this period that is rarely exploited by historians of science, for it is typically regarded as the province of historians of politics, namely *Leviathan* of 1651. In this book Hobbes took away vacuum on definitional, historical, and, ultimately, on political grounds. The vacuism Hobbes attacked was not merely absurd and wrong, as it was in his physical texts; it was *dangerous.* Speech of a vacuum was associated with cultural resources that had been illegitimately used to subvert proper authority in the state.[26]

[25] Ibid., pp. 418-419; and see similar accounts in Hobbes, "Seven Philosophical Problems," pp. 17-19, and "Decameron physiologicum," pp. 90-91.

[26] On the religious and moral significance of the vacuist-plenist controversies of the Middle Ages and early modern period, see Grant, *Much Ado about Nothing*, chaps. 5-7.

Leviathan's Political Ontology

We want to read *Leviathan* as natural philosophy. In *Leviathan,* and particularly in the section entitled "Of the Kingdom of Darkness," Hobbes drew a picture of the natural world and the sorts of things it contained. This ontology was to be condemned because it was created and sustained for ideological purposes. Certain groups of intellectuals had deployed this ontology, not for proper philosophical purposes, but to serve their social interests and to buttress their illegitimate claims to authority. This was a corrupt and a corrupting philosophy of nature. Its dissemination had been a source of social disaster and it would continue to exert a corrosive effect on social order unless and until its illegitimacy was exposed. Proper philosophy was to be assessed by its contribution to public peace.

In this illegitimate philosophy of nature one of the most important notions was that of *incorporeal substance.* To Hobbes such talk of incorporeal substance was at once an absurdity of language-use, an impossibility in right philosophy, and one of the key ideological resources used in priestcraft. Nor were the priests without powerful allies in such usages. For Peripatetic philosophers the idea of incorporeal substance underpinned their use of "substantial forms" and "separate essences"; just as for priests incorporeal substance was fundamental to their conceptions of "spirit," "soul," and the eschatological use of these items.

For Hobbes, perhaps even more than for Boyle, right philosophy was predicated upon the proper use of language. We shall see that one route to proper philosophical language lay through a definitional exercise. However, in arguing with priests, it was apposite to inspect the Scriptural warrant for language-use and the meaning of terms. Was there any Scriptural justification for conceiving of the soul as an incorporeal substance? Or even for regarding a soul as the unique property of man? In Hobbes's reading there was no such warrant. In the Bible there was indeed much speech of souls, of angels, and of spirits. But there was no speech that definitively indicated that these were incorporeal. Soul might be attributed to any living creature, and it had no existence apart from the body. Angels, good and evil, were spoken of, but nowhere did Scripture say that they were incorporeal.[27] Nor was there Biblical warrant

[27] Hobbes, "Leviathan," pp. 615, 644; on Hobbes and the soul, see Willey, *Seventeenth Century Background,* pp. 100-106; Sacksteder, "Speaking about Mind"; Watkins, *Hobbes's System,* chap. 6.

for referring even to *God* as an incorporeal entity.[28] References to spirit abound in Holy Writ, where the term may signify wind, or God's gift of grace, or zeal, or a dream, or vital strength; but *not* incorporeal substance. In the Bible, spirit, said Hobbes, was uniformly used to refer "either properly [to] a real substance, or metaphorically, [to] some extraordinary *ability* or affection of the mind, or of the body."[29] If we want to refer to the Deity as spiritual, we can do so legitimately: as a way of expressing our desire to pay Him honour. Yet, even here, there was a risk of taking terms of respect to be ontological terms, of transforming "attributes of honour" into "attributes of nature."[30]

If priests spoke nonsense, so did their professional bedfellows, the Scholastic philosophers. Again, "absurd speech" and improper language-use were at the root of the problem. Scholastic notions of "substantial forms," "abstract essences" and "separate essences" were meaningless. Worse, they were mystifying and pernicious to the quest for genuine philosophical explanation: "Which insignificancy of language, though I cannot note it for false philosophy; yet it hath a quality, not only to hide the truth, but also to make men think they have it, and desist from further search."[31] What counted as physical explanation in Aristotelianism, depending as it did upon the doctrine of substantial forms, was nonsense. For example, bodies were said to sink because they were "heavy": "But if you ask what they mean by *heaviness*, they will define it to be an endeavour to go to the centre of the earth. So that the cause why things sink downward, is an endeavour to be below: which is as much to say, that bodies descend, or ascend, because they do." Aristotelian teleological explanation of physical phenomena was funny; it was "as if stones and metals had a desire, or could discern the place they would be at, as man does; or loved rest, as man does not; or that a piece of glass were less safe in the window, than falling into the street."[32] But bodies do not move themselves; they

[28] Hobbes, "Leviathan," pp. 92, 96-97. While Hobbes professed his belief in such a God, a conception of the Deity as a corporeal being, stripped of Providential power, was unusable for most clerics, and Hobbes was widely identified as an atheist. For detailed treatments: K. Brown, "Hobbes's Grounds for Belief"; Glover, "God and Hobbes"; Damrosch, "Hobbes as Reformation Theologian"; Hunter, *Science and Society*, chap. 7; Mintz, *Hunting of Leviathan*; Klaaren, *Religious Origins of Modern Science*, pp. 99-100.

[29] Hobbes, "Leviathan," pp. 383-387.

[30] Ibid., pp. 672, 680.

[31] Ibid., pp. 670-672, 686.

[32] Ibid., p. 678; see Willey, *Seventeenth Century Background*, pp. 99-100.

have no essences, so to speak, poured into them and separable from their corporeal nature; their brute and corporeal nature *is* their nature. Man and his will were to be conceived in a structurally parallel manner. Man is subject to "appetites" of desire and aversion, analogous to the physical forces acting upon stones. "Deliberation" consists of the alternate action of these appetites, and in deliberation the last appetite "immediately adhering to the action . . . is what we call the *will*."[33] In general, therefore, neither man nor inanimate objects were to be thought of as having a dual nature. Neither is compounded of matter *plus* a separable and incorporeal spiritual essence, form, or will. To speak otherwise of existents in the natural world was to speak absurdly; no one who spoke in this way could be an authentic philosopher.

Why did priests and their allies speak absurdities? Hobbes said that it was because they conceived it to be in their interests to do so: because if such notions were disseminated and credited, priests and their allies would benefit. Hobbes proposed to discredit priestly absurdities and bad philosophy by telling an interest-story. He would put and answer the question "*Cui bono*?"[34] Priests and Scholastics had sought to prosper at the expense of peace and good order in the polity. Aristotelian doctrine of separate essences had historically been deployed in an illegitimate strategy of social control. It had been used to obtain for priests a share of that authority that belonged solely to the civil sovereign. The *Politics* of Aristotle, Hobbes said, was "repugnant to government": this "doctrine of *separated essences*, built on the vain philosophy of Aristotle, would fright [men] from obeying the laws of their country, with empty names; as men fright birds from the corn with an empty doublet, a hat, and a crooked stick."[35] Priestcraft had made up a scarecrow philosophy; it had traded upon man's natural anxiety about the future and man's natural tendency to construct causal explanations out of whatever resources were at hand. If there was no visible and

[33] Hobbes, "Leviathan," pp. 48-49, 679. For a fuller account of Hobbes and the will, see Watkins, *Hobbes's System*, chap. 7. For discussions of the disputes over free will between Hobbes and John Bramhall, Bishop of Derry, in the 1650s, see Mintz, *Hunting of Leviathan*, chap. 6; Damrosch, "Hobbes as Reformation Theologian." Damrosch usefully suggests that Bramhall recognized in Hobbes's writings some of the central theological resources of English Calvinism and responded accordingly.

[34] Hobbes, "Leviathan," p. 688.

[35] Ibid., pp. 669, 674. On Aristotle as anti-monarchist, see Hobbes, "Behemoth," p. 362; and, for the view of Hobbes's sometime ally Henry Stubbe that "the politics of Aristotle suit admirably with our monarchy," see J. Jacob, *Stubbe*, p. 87 (quoting Stubbe, *Campanella Revived* [London, 1670], pp. 12-13).

known cause of events, then some invisible power or agent was widely supposed to be at work.[36] Priestcraft had encouraged these natural dispositions, and it had used man's ignorance, nervousness and insecurity to buttress the *independent* moral and political authority of the Church. To this end, priests had propagated the bad philosophy of incorporeal substances and immaterial spirits.

Hobbes's indictment of priestcraft and its use of bad philosophy was highly detailed. It took in the major points at which such priestly resources were brought into play in usurping power. *Leviathan* offered an elaborate analysis of the conceptual resources deployed in religious rituals. Consider, for example, the ritual of Holy Communion whose supposed efficacy provided the priest with his access to divine power. Holy Communion was "most gross idolatry" if it was understood that Christ was literally present in the consecrated bread. The bread was to *signify* Christ, no more. What priestcraft did was to translate "consecration into conjuration"; it was the trade of the juggler and the magician. Likewise, we construed baptism correctly if we understood it symbolically: the use of "enchanted oil and water" as "things of efficacy to drive away phantasms, and imaginary spirits" was, however, a pious and pernicious fraud. There were no such spirits, nor did the consecrated oil possess any spiritual properties apart from its material constitution. Moreover, there was no such phenomenon as *possession* by spirits. The ritual of exorcism "purges" a man of that which in reality can never infect him.[37] And the same may be said of the idea that man has a soul that lives after his death and is separated from his body, and of the notion that such disembodied spirits fly to a local heaven or hell. Such a conception proceeded directly from the doctrine of separate essences. But in truth the soul has no "existence separated from the body"; and it is a nonsense to say that it survives apart from the body. There was Scriptural support for Hobbes's idea of a corporeal soul: "*Eat not the blood, for the blood is the soul; that is, the life.*"[38] Hobbes offered a striking gloss upon what was meant in the

[36] Hobbes, "Leviathan," pp. 93, 95, 98.

[37] Ibid., pp. 610-613, 644. On conflicting seventeenth-century views of "possession," see Walker, *Unclean Spirits*.

[38] Hobbes, "Leviathan," p. 615 (quoting Deuteronomy xii, 23). There are clear and significant similarities between Hobbes's view of the soul and that of many radical sectaries during the Civil War and Interregnum. The heresy of "mortalism" was prevalent among the sects (and among a number of English men of science). Mortalists maintained that the soul either died with the body or that it slept until the general resurrection, and many mortalists identified the soul with the blood. Hobbes did not, of course, share political aims with the sectaries. Nevertheless, the

Bible by references to "*life eternal* and *torment eternal*": these terms denoted the avoidance or not of "the calamities of confusion and civil war." The "Kingdom of God" was and will be on Earth, on His coming again. Hell and heaven were not places; they were states of mind or conditions of social disorder and order.[39] Neither did the device of the Holy Trinity provide any support for conceiving of God, or any of His manifestations and incarnations, as incorporeal substance. God was referred to as triune in nature because He had been *represented* thrice: by Moses (the Father), by Jesus (the Son), and by the Apostles and their successors (the Holy Spirit).[40]

All this absurd speech of incorporeal substance and its uses had been deployed as a tool of interested groups of professionals. Who, Hobbes asked, "will not obey a priest, that can make God, rather than his sovereign, nay than God himself? Or who, that is in fear of ghosts, will not bear great respect to those that can make the holy water, that drives them from him?" Priestcraft disseminated this bad religion founded on bad philosophy in order to usurp power: "By their demonology, and the use of exorcism, and other things appertaining thereto, they keep, or think they keep, the people in awe of their power." They meant "to lessen the dependance of subjects on the sovereign power of their country."[41] And this was visible in the most concrete act of usurpation: priests' contention that kings ruled by gift of bishops. To say that the authority, power, and legitimacy of the King proceeded *Dei gratia* was not to say that they derived from a Pope or from bishops, for the ultimate religious authority was the civil sovereign; he defined what religion consisted of; he was himself a mortal God.[42] "For whatsoever power ecclesiastics take upon themselves, (in any place where they are subject to the state), in their own right, though they

---

similarity in conceptual resources may proceed from a shared analysis of the role of the Established Church and of the uses to which the Church put its notions of the soul, the afterlife, moral accountability, and the like. In this connection, see C. Hill, *World Turned Upside Down*, pp. 387-394; idem, "Harvey and the Idea of Monarchy"; idem, *Milton and the English Revolution*, chap. 25.

[39] Hobbes, "Leviathan," pp. 437, 444-445, 455; cf. Walker, *Decline of Hell.* Again, there are striking similarities with radical sectaries' denial of local hell and heaven: C. Hill, *World Turned Upside Down*, chap. 8.

[40] Hobbes, "Leviathan," pp. 486ff. See also Warner, "Hobbes's Interpretation of the Trinity" and, on the political significance of anti-Trinitarianism, see Leach, "Melchisedech and the Emperor."

[41] Hobbes, "Leviathan," pp. 675, 693.

[42] Ibid., pp. 607-608.

call it God's right, is but usurpation."[43] From that usurpation priests profited. They took tithes in the name of God and by their supposed administration of an invisible spirit world, so that the people owed a "double tribute, one to the state, another to the clergy." Monks claimed privilege; priests claimed civil exemptions; pastors made themselves into governors of the polity. And in this there was little difference between the "Roman, and the presbyterian clergy."[44]

Double tribute ended in civil war and confusion. This was what would inevitably happen if one allowed authority and power in the state to be fragmented and dispersed among professional groups each claiming its share. All professional groups had been at fault, but none more so than the clergy. In 1656 Hobbes explained to one of the hunters of *Leviathan* how he came to write that book: it "was the considerations of what the ministers before, and in the beginning of, the civil war, by their preaching and writing did contribute thereunto."[45] And in *Behemoth* of 1668 Hobbes offered a particular historical analysis of the Civil War just concluded. Among those most responsible for these calamities were the "seducers" of the people, and among these seducers none had been more reprehensible than those ecclesiastics who claimed delegated power from God that bypassed the civil authority and those, whether ordained or not, who claimed private inspiration directly from the Deity. He condemned them all for what they had done to cause confusion and war: "Ministers, as they called themselves, of Christ; and sometimes, in their sermons to the people, God's ambassadors; pretending to have a right from God to govern every one in his parish, and their assembly the whole nation." Papists were, of course, particularly odious, as they claimed not just a share of power with the civil authority, but the role of delegating absolute spiritual power to the state through their intermediacy. But religious groups whose beliefs and practices appeared most opposed to Rome and Canterbury were equally pernicious: the Protestant sects. Independents, Anabaptists, Fifth-Monarchy Men, Quakers, and Adamites: "These were the enemies which arose against his

[43] Ibid., p. 688. For brief accounts of Hobbes and Erastianism: Clark, *Seventeenth Century*, pp. 218-222; Peters, *Hobbes*, pp. 239-244; Goldsmith, *Hobbes's Science of Politics*, pp. 214ff.; Strauss, *Political Philosophy of Hobbes*, chap. 5. Hobbes was anti-professional; he took a parallel view of lawyers' usurpation of power; see Hobbes, "Dialogue between a Philosopher and a Student," p. 5, and chapter 7 below.

[44] Hobbes, "Leviathan," pp. 608, 689-691, 609-610; also idem, "Behemoth," pp. 215-216.

[45] Hobbes, "Six Lessons," p. 335.

Majesty from the private interpretation of the Scripture, exposed to every man's scanning in his mother-tongue."[46] Private judgment and personal interpretation were the ultimate threats to social order. After the Bible had been translated into English,

> every man, every boy and wench, that could read English, thought they spoke with God Almighty, and understood what he said. . . . The reverence and obedience due to the Reformed Church here, and to the bishops and pastors therein, was cast off, and every man became a judge of religion, and an interpreter of the Scriptures to himself.[47]

The problem was one of divided loyalties arising from a divided vision of reality. Thus, "*Temporal* and *spiritual* government, are but two words, brought into the world, to make men see double, and mistake their *lawful sovereign*." The remedy was to resolve this division. There was, he said,

> no other government in this life, neither of state, nor religion, but temporal; nor teaching of any doctrine, lawful to any subject, which the governor both of the state, and of religion forbiddeth to be taught. . . . And that governor must be one; or else there must needs follow faction and civil war in the commonwealth, between the *Church* and *State*; between *spiritualists* and *temporalists*; between the *sword of justice*, and the *shield of faith*. . . .[48]

This "seeing double" could be remedied by collapsing the hierarchical division between matter and spirit; and the triumph of the civil sovereign could be assured by collapsing that hierarchy in favour of matter. It was to that end that *Leviathan* proffered a materialist and monist natural philosophy. The universe, being "the aggregate of all bodies, there is no real part thereof that is not also *body* . . . and therefore *substance incorporeal* are words, which when they are joined together, destroy one another, as if a man should say, an *incorporeal body*."[49] The world is like this:

> The world, (I mean not the earth only . . . but the *universe*, that is the whole mass of all things that are), is corporeal, that

[46] Hobbes, "Behemoth," pp. 167, 171. On *Behemoth*, see MacGillivray, "Hobbes's History of the Civil War"; on Hobbes versus the sects, see Pocock, "Time, History and Eschatology," esp. pp. 180-187.

[47] Hobbes, "Behemoth," p. 190.

[48] Hobbes, "Leviathan," pp. 460-461.

[49] Ibid., p. 381.

> is to say, body; and hath the dimensions of magnitude, namely, length, breadth, and depth: also every part of body, is likewise body, and hath the like dimensions; and consequently every part of the universe, is body, and that which is not body, is not part of the universe: and because the universe is all, that which is no part of it, is *nothing*; and consequently *no where*.[50]

The world is full of body; that which is not body does not exist. And there can be no vacuum. The argument proving this was not developed within the discourse of natural philosophy that we described earlier in this chapter. Instead, the argument against vacuum was presented within a political context of use. In the cause of securing public peace Hobbes elaborated and deployed an ontology which left no space for that which was not matter, whether this was a vacuum or incorporeal substance. He recommended his materialist monism because it would assist in ensuring social order. He condemned dualism and spiritualism because they had in fact been used to subvert order. As we will see in chapter 5, the political purpose behind Hobbes's taking away of vacuum was not missed by his critics, including Robert Boyle.

Leviathan's Political Epistemology

Let us now read *Leviathan* as epistemology. Within Hobbes's overall attempt to show men the nature of obligation and the foundations of secure social order, he developed a theory of knowledge. He displayed how knowledge was generated, its relations with the physiology of the human organism, and he showed men how to go about achieving the highest and most useful form of knowledge: *philosophy*. The connections between the epistemological enterprise in *Leviathan*, on the one hand, and the ontological and political exercises were substantial and clear. First, a proper theory of what existed in the world and a proper way of producing knowledge proceeded from the same starting point: an agreement to settle the definitions of words and their uses so that absurdities were avoided; then an agreement to use correct method to move from these definitions to their consequences. Second, Hobbes's theory of how man perceived and produced knowledge used the monist and materialist ontology: his psychology was one of matter and motion,

[50] Ibid., p. 672. On Hobbes's usage of "body," "substance," and "matter," see Sacksteder, "Speaking about Mind," p. 68; cf. Watkins, *Hobbes's System*, esp. pp. 125-132.

leaving no space for notions of an incorporeal soul. Finally, both ontology and epistemology were of equal importance in attaining and securing public peace. Disorder and civil war were as likely to be produced from an incorrect appreciation of how knowledge was generated and what its nature was, as they were from incorrect ideas of what sorts of things existed. Show men what knowledge is and you will show them the grounds of assent and social order.[51]

There was already in existence a model of properly grounded and properly generated knowledge, a model that could be imitated by all intellectuals genuinely aiming to secure assent and good order. This was "geometry, which is the only science that it hath pleased God hitherto to bestow on mankind." When its methods had been rightly followed, geometry yielded irrefutable and incontestable knowledge. One *may* make a mistake in geometry, but one will not continue in error when the mistake has been revealed: "All men by nature reason alike, and well, when they have good principles. For who is so stupid, as both to mistake in geometry, and also to persist in it, when another detects his error to him?" Such knowledge was "indisputable." And it was ideally situated to provide a model of intellectual assent, since the geometrical course was not the preserve of professionals but was open to anyone with natural reason.[52] To end with agreement, one must start with agreement. Again, geometry showed how this must be done, for it started with definitions, "settling the significations of . . . words" and "plac[ing] them at the beginning of [the] reckoning." The settling of a definition was a social act, to be contrasted with private intellectual activity: "When a man's discourse beginneth not at definitions, it beginneth either at some other contemplation of his own, and then it is still called opinion; or it beginneth at some saying of another, of whose ability to know the truth, and of whose honesty in not deceiving, he doubteth not."[53] In chapter 4 we will see in

[51] For a good account of Hobbes's theory of knowledge in relation to his political philosophy, see Watkins, *Hobbes's System*, chaps. 4, 8. Verdon, "On the Laws of Physical and Human Nature," treats the connection as "analogical." Our stress here is not on analogy, nor on priority of development, but on common context of use. Here, and in other connections, see the excellent book by Gideon Freudenthal, *Atom und Individuum*, esp. chaps. 5, 9.

[52] Hobbes, "Leviathan," pp. 23-24, 35; cf. "Six Lessons," pp. 211-212. On mathematics and agreement as a sign of "science," see Missner, "Skepticism and Hobbes's Political Philosophy," pp. 410-411. For Hobbes's views on the equal natural rationality of all men, see "Leviathan," pp. 30-35, 110-111.

[53] Hobbes, "Leviathan," pp. 24, 53-54; cf. idem, "Concerning Body," p. 84, where one of the properties of definition is "that it takes away equivocation, as also all that

what way geometrical knowledge was grounded in social acts and in what the force of geometrical inference consisted. For the present it is enough to note that the assigning of definitions was the way to start on an intellectual enterprise aimed at securing universal assent, and that such knowledge was contrasted to the belief, opinion or judgment of any one individual.

The social production of knowledge was carried on from its definitional starting-point by the use of "right reason." There was no professional mystique about this "reason"; it was "reckoning," that is, "adding and subtracting, of the consequences of general names agreed upon for the *marking* and *signifying* of our thoughts."[54] Such reasoning must be rigorously continued throughout a chain of thought or else the result would not be that certainty at which one was aiming: "For there can be no certainty of the last conclusion, without a certainty of all those affirmations and negations, on which it was grounded and inferred." There must be no weak links in the chain; if reason was wanting at any step, one would "not know anything, but only believeth."[55] Belief was thus rigidly distinguished from knowledge and from "science." The methods used in generating knowledge ensured that it was not private belief. Such private belief could never underwrite the universal assent at which philosophy was aimed.[56]

In the preceding chapter we examined Boyle's conception of the matter of fact, showing how the fact was made into the foundation of proper experimental knowledge. We displayed the social mechanisms Boyle and his colleagues mobilized to constitute the matter of fact. What, then, was the status of factual knowledge in Hobbes's scheme? Interestingly, *Leviathan* radically downgraded the standing of factual knowledge, distinguished it from "science" and "philosophy" and assimilated it to the experiences of individuals. To Hobbes, knowledge of fact, "as when we see a fact doing, or remember it done," was "nothing else, but sense and memory."[57] The "nothing else" derived from Hobbes's theory of sensory impression. These impressions were caused by the motions of matter impinging on man's sensory organs, and carried on to the brain and heart.

---

multitude of distinctions, which are used by such as think they may learn philosophy by disputation."

54 Hobbes, "Leviathan," p. 30. On Hobbes's usage of "reckoning," see Sacksteder, "Some Ways of Doing Language Philosophy," p. 477.

55 Hobbes, "Leviathan," pp. 31-32.

56 Ibid., pp. 35, 52-53.

57 Ibid., p. 71.

Therefore, our sense that such impressions correspond to the external objects themselves was, to Hobbes, but "seeming, or fancy." The same impressions could be obtained dreaming or waking, by the motions of matter in a real external object or by rubbing the eye.[58] Thus, in Hobbes's view, factual knowledge, based on sensory impressions, did not have an epistemologically privileged position. It did not matter how one proposed socially to process such factual knowledge, the limitations remained. Factual knowledge, it was true, had a valuable role to play in constituting our overall knowledge, but it was not of the sort to secure certainty and universal assent. Indeed, Hobbes wished to call the body of factual knowledge by a different name, to distinguish it from "philosophy" or "science." The "register of the *knowledge of fact*" Hobbes called "history," "natural history" being the catalogue of "such facts, or effects of nature, as have no dependence of man's *will*." Thus the fundamental distinction Hobbes made between factual knowledge (or history) and philosophy involved the exercise of man's agency. Man had no control over the effects of nature, but he did over settling definitions and agreeing notions of intelligible cause. Philosophy and science were constituted by the knowledge of consequences and causes, and, again, the model was provided by geometry: "As when we know, that, if the figure shown be a circle, then any straight line through the centre shall divide it into two equal parts." "And this," Hobbes said, "is the knowledge required in a philosopher."[59]

We have shown that for Boyle and the early Royal Society there were two major threats to the social forms of experimental philosophy: the private judgment of "secretists" and enthusiasts and the tyranny of "modern dogmatists." For Hobbes in 1651 and later, only private judgment counted as a potentially fatal threat to good philosophy and to good order. If the aim was certain knowledge and irrevocable assent, then the way towards it could not traverse anything as private and unreachable as individual states of belief. Knowledge, science and philosophy were set on one side; belief and opinion on the other. The former were certain, hard and indisputable; the latter were provisional, variable and inherently contentious, affected by man's shifting passions and special inter-

[58] Ibid., pp. 1-2. Hobbes's sensory theories are more extensively developed in "Human Nature," esp. pp. 1-19. See also Barnouw, "Hobbes's Causal Account of Sensation" and Sacksteder, "Hobbes: Man the Maker," pp. 86-87.

[59] Hobbes, "Leviathan," p. 71; see pp. 72-73 for Hobbes's table of the taxonomy of the sciences. On conceptions of "history" and its relationship to causal accounting, see B. Shapiro, *Probability and Certainty*, chap. 4.

ests. The consequences of mistaking belief for knowledge, and of attempting to ground order in that which makes for disorder, were disastrous. This was why Hobbes's prescription for making proper knowledge belonged in *Leviathan*: it was a prescription for avoiding civil war.

We have already alluded to two sources of improper knowledge which Hobbes identified as contributors to civil war. First, there was priestly and Scholastic absurd speech, reasoning upon which produced "contention and sedition, or contempt."[60] Second, there was the private judgment of the radical Protestant sects, specifically condemned in *Behemoth*.[61] Their doctrine of private judgment in religious matters was a particularly virulent form of treason. The claim of each individual to decide upon religious truths and principles was the ultimate fragmentation of authority. There must, in Hobbes's view, be no "private measure of good," as this would be "pernicious to the public state." There must be no right to interpret Scriptures personally; that right belonged solely to the civil sovereign who properly had the authority to decide upon the meaning of Scripture and religious doctrine.[62] There were no legitimate grounds for sectaries' claims to divine inspiration, or, at least, to credit such claims. To say that a man "speaks by supernatural inspiration, is to say that he finds an ardent desire to speak, or some strong opinion of himself, for which he can allege no natural and sufficient reason." Any individual making such a claim may, to be sure, be telling the truth; yet he is but a man and "may err, and, which is more, may lie." How may we know a genuine prophet, a man truly speaking God's message? Such a man does miracles and preaches nothing but the doctrines established by the civil authority. There could be, therefore, no authentic word from God which preached rebellion. Moreover, "miracles now cease, [and] we have no sign left, whereby to acknowledge the pretended revelations or inspirations of any private man; nor obligation to give ear to any

[60] Hobbes, "Leviathan," p. 37.

[61] There are important links between Hobbes's discussion of private judgment in religion and his treatment of the "intellectual virtues" in poetry and philosophy; see "Leviathan," pp. 56-70. "Fancy" is private judgment and is the source of poetry; "acquired wit" is secured by "method and instruction" and equals "reason": this is the basis of public assent and is the source of philosophy. For Hobbes's literary theories: Selden, "Hobbes and Late Metaphysical Poetry"; Thorpe, *Aesthetic Theory of Hobbes*, esp. pp. 79-117; James, *The Life of Reason*, pp. 34-49.

[62] Hobbes, "Leviathan," pp. 680-681, 685.

doctrine, farther than it is conformable to the Holy Scriptures . . . without enthusiasm or supernatural inspiration."[63]

Hobbes, then, held no brief for the priestcraft by which the Established Church had taken for itself power that belonged to the civil sovereign; neither did he support the conduct of the sectarian opponents of the Church. They were deluded and dangerous; and through their doctrine of private judgment they had a major share of responsibility for the Civil War. Any society that countenanced individualistic claims to knowledge would inevitably fall into chaos. Yet in attacking the conceptual resources of the Church and the bases of the Church's independent authority Hobbes provided ammunition for all the Church's enemies.[64] With one hand Hobbes took away the legitimacy of private judgment; with the other he took away the right of religious authorities to exercise control of individuals' private beliefs. As agent of the civil sovereign, the Church had the right to control *behaviour* and verbal *profession* of belief. It did not, however, have the right to attempt to extend its control to men's minds. So claims to have had direct communication with God could not be credited because they could not be validated by others; yet, if the person who made such a claim "be my sovereign, he may oblige me to obedience, so, as not by act or word to declare I believe him not; but not to think otherwise than my reason persuades me."[65] Any other agent had no such right to exact obedience in such matters, and no agent had the right to exact belief.

There is no more fundamental contrast between Hobbes's and Boyle's strategies for knowledge-production than that which concerns states of belief. We have seen that Boyle's experimental matters of fact were founded in states of belief: individuals were freely

[63] Ibid., pp. 362-365. On private judgment among the Protestant sects, see C. Hill, *World Turned Upside Down*, chap. 6.

[64] This suggests a possible answer to a problem widely noted in Hobbes scholarship: Hobbes argued in favour of absolute deference to duly constituted authority, yet the "Hobbists" attacked in seventeenth-century polemic included notorious "scoffers" at that civil and religious authority. Hobbes argued in support of a social and political order the conceptual resources to justify which he had removed: incorporeal spirit, heaven and hell, free will, the efficacy of ritual, etc. On Hobbist "atheists": Aylmer, "Unbelief in Seventeenth-Century England," esp. pp. 36-45; on Hobbes's political followers: Skinner, "Hobbes and His Disciples"; idem, "History and Ideology"; idem, "Ideological Context of Hobbes's Political Thought"; Macpherson, "Introduction" (to *Leviathan*), esp. pp. 23-24; Warrender, "Editor's Introduction" (to *De cive*), pp. 16-26.

[65] Hobbes, "Leviathan," p. 361.

to witness and then freely to say what it was they believed to be the case. Knowledge was constituted when all believed alike. Likewise for Boyle's clerical allies, religion was a matter of belief and giving witness to that belief. There was to be no disjunction between belief and professions. This transit of belief is what Hobbes wished to expel from both religion and natural philosophy. His strategy was one of behavioural control, not one of internal moral control. It was not that the control of belief was wrong; it was that such control was impractical and an inadequate surety for order. Belief and opinion were items pertaining to individuals, and, as such, could not be manipulated into bases for public order. There were several serious problems for efforts to found order in belief. Individual states of belief were in principle uncontrollable because they were in practice unreachable. I cannot know what you believe; I can only know what you say you believe; you may be lying. I can force you to make a profession, but I cannot guarantee that this profession corresponds to your state of belief. And because belief and opinion belonged to individual men and were subject to their passions and interests, they constituted too shifting a ground on which to erect the frameworks of social order.

On these bases Hobbes contrasted belief to behaviour and reason. Both behaviour and reason were in the public domain: behaviour because it was visible to all; reason because all men had it and had it in equal measure. Actions could be successfully controlled, if necessary by coercion. A strategy aimed at regulating verbal professions of belief while leaving private states of belief intact therefore made practical sense. It takes, Hobbes said, "till perhaps a little after a civil war" to realize "that it is men, and arms, not words and promises, that make the force and power of the laws." "What man," he asked, "that has his natural senses . . . believes the law can hurt him; that is, words and paper, without the hands and swords of men?" Thus the sovereign power and its spiritual arm had no brief "to extend the power of the law, which is the rule of actions only, to the very thoughts and consciences of men, by examination, and *inquisition* of what they hold, notwithstanding the conformity of their speech and actions."[66] Hobbes did not shrink from the most extreme consequences of his strategy of behavioural control. What if the sovereign commanded one to deny Christ? Then one must make the commanded verbal profession, for "Profession with the tongue is but an external thing, and no more than any other gesture

[66] Ibid., pp. 683-684.

whereby we signify our obedience."[67] Coercion, therefore, had its place in the maintenance of order. So too did reason. A man who made an error in reasoning could be put right by pointing out to him his offence against the rules of reasoning. One did not, as a matter of practice, identify these rules as the property of any one man or group of men; one could not ask of them "*Cui bono?*" Thus the application of irresistible physical force and the application of reason were similar exercises, having comparable results. Both were means by which assent might effectively be secured. Strategies that depended upon belief could not work.

Nevertheless, any strategy designed to secure assent could be subverted by ignorant or privately interested men. Hobbes was concerned, just as Boyle was, with the problem of "manners" in philosophical disputation. Consensus could not be generated unless philosophers were prepared to conduct themselves properly; incivility was an invitation to dissensus: "For evil words by all men of understanding are taken for a defiance and a challenge to open war." In his vituperative geometrical exchanges with the "Egregious Professors" at Oxford (John Wallis and Seth Ward) that immediately followed the publication of *De corpore*, Hobbes read his adversaries a lesson on the connection between proper manners and proper knowledge:

> It cannot be expected that there should be much science of any kind in a man that wanteth judgment; nor judgment in a man that knoweth not the manners due to a public disputation in writing; wherein the scope of either party ought to be no other than the examination and manifestation of the truth.[68]

He condemned the use of "contumelious language," intemperate expressions, wilful misrepresentations, and *ad hominem* argumentation. Like Boyle, Hobbes considered that philosophers should deal with one another as Christian gentlemen. He did not, however, practise the turning of the other cheek; rather, he recommended "Vespasian's Law": "it is uncivil to give ill language first, but civil and lawful to return it."[69] And when Hobbes invoked Vespasian's Law, as he did with Wallis and Ward, all similarities with Boyle's language disappeared: "So go your ways, you *Uncivil Ecclesiastics,*

[67] Ibid., p. 493.

[68] Hobbes, "Six Lessons," pp. 331-332. Cf. the view of "manners" in idem, "The Art of Rhetoric," esp. pp. 466-472.

[69] Hobbes, "Six Lessons," pp. 331-332, 356; see also idem, "Stigmai," p. 386, and "Leviathan," chap. 11.

Inhuman Divines, Dedoctors of morality, Unasinous Colleagues, Egregious pair of Issachars, most wretched Vindices and Indices Academiarum. . . ."[70] And this summing-up of one of Wallis's attacks: ". . . all error and railing, that is, stinking wind; such as a jade lets fly, when he is too hard girt upon a full belly."[71]

The Ends of Philosophy

Boyle aimed to achieve peace and to terminate scandal in natural philosophy by securing a space within which a specified kind of dissent was manageable and safe. In the experimental form of life it was legitimate for philosophers to disagree about the causes of natural effects: causal knowledge was removed from the domain of the certain or even the morally certain. For Hobbes there was no philosophical space within which dissent was safe or permissible. Dissent over physical causes was a sign that one had not begun to do philosophy or that the enterprise in question was not philosophy. Philosophy was defined as a constitutively causal enterprise; causal knowledge was one of its starting points. Philosophy was "such knowledge of effects or appearances, as we acquire by true ratiocination from the knowledge we have first of their causes or generation: And again, of such causes or generations as may be from knowing first their effects."[72] In the next chapter we shall discuss the unequal status of these two methods, but, for the present, it is enough to note that any programme which attempted to erect a procedural boundary between speech of matters of fact and speech of their physical causes was not, on this basis, philosophical. The aim of philosophy was the highest degree of certainty that could be obtained. Philosophy was contrasted to other intellectual enterprises precisely on the grounds of the degree of certainty one could expect of each. Authentic natural philosophy, founded upon proper method, was new, no older than the revolution made by

[70] Hobbes, "Six Lessons," p. 356. Hobbes particularly disliked Wallis's puns on his plebeian name: *hob* is Old English for a rustic ghost or spectre (as *hob-goblin*). Wallis referred to Hobbes as the *Empusa* (the goblin sent by Hecate), which jest, said Hobbes, "is lost to them beyond sea" (ibid., p. 355); see also Laird, *Hobbes*, pp. 19-20. For indications how disturbed Hobbes was by this slur, see letters from Henry Stubbe to Hobbes dating from March to April 1657, in Nicastro, *Lettere di Stubbe a Hobbes*, pp. 16-17, 26-28.

[71] Hobbes, "Considerations on the Reputation of Hobbes," p. 440.

[72] Hobbes, "Concerning Body," p. 3; cf. pp. 65-66, 387 for variants on this definition.

Galileo, Harvey, and not least by Hobbes himself. Before this, Hobbes said, "There was nothing certain in natural philosophy but every man's experiments to himself, and the natural histories, if they may be called certain, that are no certainer than civil histories." Natural history yielded an inferior grade of certainty, and was excluded from the ambit of philosophy, "because such knowledge is but experience, or authority, and not ratiocination." Sensory knowledge could not be formed into the foundations of philosophy: sense and memory, which were "common to man and all living creatures," constituted knowledge, but, because they were not given by reason, "they are not philosophy." Experience "is nothing but memory."[73]

It is vital that we understand what our ends are when we do philosophy. The production of certainty would terminate disputes and secure total assent. Philosophy was one of the most profoundly useful of the arts of peace:

> But what the *utility* of philosophy is, especially of natural philosophy and geometry, will be best understood by reckoning up the chief commodities of which mankind is capable, and by comparing the manner of life of such as enjoy them, with that of others which want the same.

These commodities included the technological benefits of genuine knowledge, but they also included the fruits of moral and civic philosophy, whose methods overlapped with that of natural philosophy. The utility of these

> . . . is to be estimated, not so much by the commodities we have by knowing these sciences, as by the calamities we receive from not knowing them. Now, all such calamities as may be avoided by human industry, arise from war, but chiefly from civil war; for from this proceed slaughter, solitude, and the want of all things.[74]

In Hobbes's view the elimination of vacuum was a contribution to the avoidance of civil war. The dualist ontology deployed by priests spoke of existents which were not matter: this made men "see double" and resulted in the fragmentation of authority which led inexorably to chaos and civil war. Aristotelians spoke of separated essences which were poured into corporeal entities; vacuists

[73] Ibid., pp. vii-lx, 11, 3; cf. idem, "Human Nature," p. 29.

[74] Hobbes, "Concerning Body," p. 8.

populated the spaces they prohibited to matter with immaterial spirits. These were the ontological resources of the enemies of order. Moreover, a dualist ontology created absurdities in our notion of physical cause. By understanding what sorts of things existed in the natural world, we would understand what sort of thing could be considered as a cause. For Hobbes, there was only one cause of the movement of material body: the movement of a contiguous body. Thus the language employed in speaking of physical causation was the language of materialist monism.[75] So there was no sense in which causal speech could be deemed less certain and more divisive than speech of matters of fact. Causal language as well as ontological language emerged from the same exercise in settling the right definitions and uses of words. Both arose from, and were dependent upon, agreement: they could not be the *source* of disagreement. For Hobbes, the rejection of vacuum was the elimination of a space within which dissension could take place.

[75] Ibid., p. 124: "There can be no cause of motion, except in a body contiguous and moved." Cf. p. 390.

· IV ·

The Trouble with Experiment: Hobbes Versus Boyle

... the laws of inference can be said to compel us; in the same sense, that is to say, as other laws in human society.
WITTGENSTEIN, Remarks on the Foundations of Mathematics.

ROBERT Boyle's *New Experiments Physico-Mechanical* was published in the summer of 1660. Following the Restoration of the King in May 1660 and the gathering of "many Worthy Men" in London during the summer of that year, the Royal Society received a formal constitution at Gresham College in November 1660.[1] Hobbes was now faced, not with experiment merely as a useful adjunct to the pursuit of natural philosophy, but with a fully developed experimental programme for natural philosophy. Publications on trials of the air-pump and on other related experiments were shortly to come from Henry Power, Robert Hooke, John Wallis, and, of course, from Boyle himself, who began to produce a profusion of tracts on the new experimental philosophy.[2] Boyle and his colleagues now argued that no philosophy of nature could hope to establish a solid foundation for assent unless it was grounded in experimental practices: the procedures set forth in *New Experiments* and the essays that quickly followed its publication. In December

[1] The phrase "many Worthy Men" is Thomas Sprat's description of the returning royalist exiles. He continues: they "began now to imagine some greater thing; and to bring out experimental knowledge, from the *retreats*, in which it had long hid it self, to take its part in the *Triumphs* of that universal Jubilee. And indeed Philosophy did very well deserve that Reward: having been always Loyal in the worst of times." Sprat, *History*, pp. 58-59. Hobbes, however, reckoned himself ("the first of all that fled") to be at least as loyal as the experimental philosophers: Hobbes, "Considerations on the Reputation of Hobbes," p. 414.

[2] These tracts include: Cowley, *Proposition for the Advancement of Experimental Philosophy* (1661); Hooke, *Attempt for the Explication of the Phaenomena* (1661); idem, *Micrographia* (1665; commissioned March 1663); Power, *Experimental Philosophy* (1664; written by August 1661); Wallis, *Hobbius heauton-timorumenos* (1662); plus all the writings that Boyle published in 1660-1662. See also M. B. Hall, "Salomon's House Emergent," esp. pp. 180-182.

1660 the Society meeting at Gresham announced that it would limit its membership to fifty-five, plus those of the rank of baron and above, and that it had received royal approval from Charles II.[3] Hobbes immediately responded to Boyle and these changed circumstances: the *Dialogus physicus de natura aeris* was published in August 1661.[4]

Hobbes's criticisms of Boyle's work and of the experimental programme took several major forms:

- He was sceptical about the allegedly public and witnessed character of experimental performances, and, therefore, of their capacity to generate consensus, even within experimental rules of the game.
- He regarded the experimental programme as otiose. It was pointless to perform systematic series of experiments, since if one could, in fact, discern causes from natural effects, then a single experiment should suffice.
- He denied the status of "philosophy" to the outcome of the experimental programme. "Philosophy" for Hobbes was the practice of demonstrating how effects followed from causes or of inferring causes from effects. The experimental programme failed to satisfy this definition.
- He systematically refused to credit experimentalists' claims that one could establish a procedural boundary between observing the positive regularities produced by experiment (facts) and identifying the physical cause that accounts for them (theories).
- He persistently treated experimentalists' "hypotheses" and "conjectures" as statements about real causes.
- He contended that, whatever hypothetical cause or state of nature Boyle adduced to explain his experimentally produced phenomena, an alternative and superior explanation could be proffered and was, in fact, already available. In particular, Hobbes stipulated that Boyle's explanations invoked vacuism. Hobbes's alternatives proceeded from plenism.
- He asserted the inherently defeasible character of experimental

[3] Three months later, on 20/30 March 1661, it "was resolved, that the number of the members of the society be enlarged," and by 20/30 May 1663 there were 115 Fellows: Birch, *History*, vol. 1, pp. 5, 19, 239-240.

[4] Hobbes's two other specifically anti-experimentalist treatises were *Problemata physica* (1662) and *Decameron physiologicum* (1678). The former appeared only in Latin in Hobbes's lifetime, and was republished in Amsterdam in 1668. *Problemata physica* was translated and published in English as *Seven Philosophical Problems* (1682). We quote from the English Molesworth edition.

systems and therefore of the knowledge experimental practices produced. Hobbes noted that all experiments carry with them a set of theoretical assumptions embedded in the actual construction and functioning of the apparatus and that, both in principle and in practice, those assumptions could always be challenged.[5]

Experimental Spaces

In his dedication of the *Dialogus physicus* to Samuel Sorbière Hobbes identified his opponents as a collectivity and the air-pump as their emblematic device:[6]

> Those Fellows of Gresham who are most believed, and are as masters of the rest, dispute with me about physics. They display new machines, to show their vacuum and trifling wonders, in the way that they behave who deal in exotic animals, which are not to be seen without payment. All of them are my enemies.[7]

Who were these Fellows at Gresham, how many were they, and how did they come to be in this place? The "experimentalist" interlocutor in Hobbes's dialogue replied. They were

[5] The resonance with the "Duhem-Quine" thesis is intentional. We shall see that Hobbes's particular objections to Boyle's experimental systems provide a concrete exemplar of this "modern" thesis concerning the impossibility of crucial experiments; see Duhem, *Aim and Structure of Physical Theory*, chap. 6; Quine, *From a Logical Point of View*, esp. pp. 42-44.

[6] Samuel Sorbière (1615-1670) was a French physician who had been involved in the founding of the Montmor Academy and, later, of the Académie Royale des Sciences. He had translated some of Hobbes's work and had corresponded with Huygens, with whom he was elected to the Royal Society in June 1663. On his return to France Sorbière wrote *Relation d'un Voyage en Angleterre* (Paris, 1664), which angered the Royal Society by describing it as divided into sects. The Society considered cancelling Sorbière's membership, and Sprat replied with *Observations on Monsieur de Sorbière's Voyage* (London, 1665). On Sorbière and these episodes, see Cope and Jones, "Introduction [to Sprat, *History*]," pp. xviii; Sorbière to Oldenburg, 5/15 December 1663, in Oldenburg, *Correspondence*, vol. II, pp. 133-136, esp. p. 135n; Birch, *History*, vol. II, pp. 456-459; Guilloton, *Autour de la 'Relation' du Voyage de Sorbière*; "Memoirs for the Life of Sorbière," in Sorbière, *Voyage to England*, pp. i-xix.

[7] Hobbes, "Dialogus physicus," pp. 236-237. Note that the sentence "All of them are my enemies" was not in the original 1661 text, but was added for the 1668 Amsterdam edition of Hobbes's *Opera philosophica*. This indicates that Hobbes's view of the Royal Society and the experimental programme had, if anything, hardened as a result of his exchanges with Boyle.

> About fifty men of philosophy, most conspicuous in learning and ingenuity [*ingenio*], [who] have decided among themselves to meet each week at Gresham College for the promotion of natural philosophy. When one of them has experiences [*experientiae*] or methods [*artis*] or instruments for this matter, then he contributes them. With these things new phenomena are revealed and the causes of natural things are found more easily.

Hobbes proceeded directly to the question of whether this new experimental space was in fact open and public. He asked, why just fifty men? "Cannot anyone who wishes come, since, as I suppose, they meet in a public place, and give his opinion on the experiments [*experimenta*] which are seen, as well as they?" Interlocutor answered: "Not at all." Hobbes persisted: "By what law would they prevent it? Is this Society not constituted by public privilege?" He forced out of interlocutor the telling admission that "the place where they meet is not public." And Hobbes concluded, therefore, that its experiments were not, in practice, available to be witnessed by everyone, but only by a self-selected few: "If it pleased the master of the place, they could make one hundred men from the fifty."[8]

This was a damning judgment for two reasons: first, Hobbes showed that the experimentalists did not, as they appeared to claim, occupy a public space. Access was in fact restricted, and, because of that, the witnessing of experiments, upon which the making of matters of fact depended, was a private and, possibly, a partial affair. How do we know that these *are* authentic matters of fact if they are generated within a private space? Second, Hobbes insisted that the space occupied by the Gresham experimentalists had a "master." It had a master who decided who could come in and who could not. And it also had "masters of the rest": those "who are most believed." Hobbes had a vivid image of what it might mean to be a "master" of a philosophical place. He recalled his personal experiences of the meetings in Paris in the 1640s "at the convent of the Minims." Father Mersenne presided, and "whoever might have demonstrated a problem, would produce it for him to be

[8] Hobbes, "Dialogus physicus," p. 240. We shall discuss Hobbes's use of the dialogue form below. For the present, note that it is the experimentalist interlocutor (B) who describes the Greshamites as "men of philosophy," not Hobbes (A). In view of the date at which the *Dialogus* was probably composed, Hobbes was correct about the Society's limit of "about fifty"; see note 3 above.

examined by him and by others." "I think," Hobbes said to his interlocutor, "you also do the same."[9] Thus Hobbes disputed the social character of the space the Greshamites said they had created. He said they had a "master" who exercised his authority in the constitution of their knowledge; they said they were free and equal men, whose matters of fact mirrored the structure of reality.

In denying the Society's stipulations about the nature of its organization and the audience it provided for experimental displays, Hobbes was undermining the justifications which the Greshamites offered for the integrity of experimental findings. These findings, Hobbes claimed, were not witnessed by all; because of the nature of the social space the experimentalists elected to occupy, they were not even available for public witness. Even so, there were immense problems for the very notion of witnessing. Suppose that the experimentalists made their space a truly public one, into which everyone could enter. What would be seen by each man witnessing the experiments? The problems of witnessing experiments, Hobbes suggested, were not different in kind to those involved in evaluating testimony in natural history. It is right, Hobbes agreed, "not to believe in histories blindly." "But are not those phenomena which can be seen daily by each of you suspect, unless all of you see them simultaneously?" Were the experimental displays, in fact, witnessed simultaneously? Were they witnessed together by all members of the experimental collective? If they were not witnessed simultaneously and together, then in what ways was the evaluation of experimental testimony different from the evaluation of testimony generally? Hobbes strongly implied that there was no substantive difference, and, therefore, that the experimental form of life had not discovered a royal road to the making of objective knowledge. Hobbes then briefly treated the *necessity* of the programme dedicated to reiterated series of experimental performances. Why, after all, was it required to produce a great number of these displays instead of just one? Why were the *artificial* phenomena generated by experiment deemed superior to the experience each man has to himself? "[A]re there not enough [experiments], do you not think, shown by the high heavens and the seas and the broad Earth?"[10] Hobbes judged that the experimental programme was otiose. We shall see below that his reasons for this conclusion concerned the validity of inductive inference from effects to causes.

[9] Ibid., pp. 241-242. Interlocutor denied this.

[10] Ibid., p. 241.

But his position on the necessity of repeated artificial productions was clear: "[Experimentalists] fall back on this one thing, that they procure new phenomena, when from the experience of one phenomenon alone the causes are known by reasoning about motion."[11]

Experimental Airs

Hobbes did not rest his criticisms of Boyle and the Greshamites solely on abstract programmatic grounds. The *Dialogus physicus* offered a detailed critical account of how the air-pump worked, or rather how it did not work in the manner claimed for it. The air-pump, Hobbes decided, was not a reliable philosophical instrument. It did not operate in the way that Boyle said it did; the physical integrity of the machine was massively violated, and, therefore, the claim that it produced a vacuum in the receiver (a space devoid of air) was without foundation. In this demonstration it mattered little to Hobbes whether "vacuum" was construed as total or partial (as in Boyle's qualified operational definition). Hobbes attempted to show that all the pump's phenomena were best accounted for by supposing that the receiver was always *full*.

First, Hobbes found it essential to sort out correct ideas about the constitution of the air. According to Hobbes, it was impossible to understand the air-pump experiments "unless the nature of the air is known first."[12] In the *Dialogus* the air's constitution mattered for three reasons: first, because Hobbes's stipulation about the ultimate fluidity of the air ruled out the possibility of an absolutely impermeable seal, and thus the possibility of a secure vacuum; second, because Hobbes's description of the air as a mixture of different fluids enabled him to offer explanations of the phenomena displayed by the pump; finally, because Hobbes claimed that Boyle's unwillingness to offer a certain cause of the spring of the air, and satisfaction with showing that the air had a spring, were marks of Boyle's inadequate conception of natural philosophy. Hobbes assumed a set of basic hypotheses about the structure of the air. Since the air contained a purer and subtler part, no air-pump *could* be impermeable, and, since the air contained a grosser and more earthy part, there was an easily identifiable mechanical

[11] Hobbes, "Mathematicae hodiernae," p. 228. This was published in July 1660, at about the same time that the Society began meeting at Gresham, but it contains material that parallels the later *Dialogus physicus*.

[12] Hobbes, "Dialogus physicus," pp. 243-244.

cause of the spring. From this basis Hobbes proceeded to show why the Greshamites were wrong to claim that their machine produced a vacuum.

In the *Dialogus* both Hobbes and his interlocutor excused themselves for not having a picture of Boyle's air-pump.[13] Nevertheless, the description of the pump and its operation was, on the whole, both highly detailed and accurate. The experimentalist maintained that a vacuum was produced when the sucker was pulled down and the valves appropriately arranged. In Hobbes's view this basic supposition was in error, and, therefore, none of the physical explanations Boyle offered had any validity. Hobbes's demonstration paralleled the more general discussions of the Torricellian experiment and the working of the gardener's pot in *De corpore*, but in this context he was particularly concerned to identify specific and necessary faults in the machine's seals. Put simply, the air-pump leaked. It was no good trying to patch it up, because it was always *bound to* leak. Here is how Hobbes reckoned the air-pump actually worked: when the sucker is pulled down, that much less space is left in the plenum outside, and, by the pulsion of contiguous volumes of air,

> . . . of necessity the air is forced into the place left by the sucker and enters between the convex surface of the sucker and the concave surface of the cylinder. For supposing the parts of the air are infinitely subtle, it is impossible but that they insinuate themselves by this path left by the sucker.

There were several routes by which air might invade the supposedly evacuated machine. The contact between the leather washer (figure 1, "4") and the inner walls of the brass cylinder "cannot be perfect at all points" and there must be space left for the passage of "pure air." Second, since the force with which the sucker is drawn down is very great, this "distends the cavity of the cylinder a little bit" and another path for the passage of air is formed. Finally, "if any hard atoms get in between the edges of the two surfaces, pure air gets in that way." So the retraction of the sucker in an allegedly closed system in fact produces no vacuum at all. Moreover, the "pure air" in the cylinder and receiver is forced, by its manner of entry, to move in a circuit. And, as Hobbes said, "there is nothing . . . which could weaken its [circulatory] motion," since "there can be nothing which can impart motion to itself or diminish it."[14]

[13] Ibid., pp. 235, 242.

[14] Ibid., pp. 245-246.

The imperfection of the pump's seals and the resulting violent passage of the air through these pathways remained a standard component of Hobbes's later tracts against Boyle and experimental practices. In the *Problemata physica* of 1662 Hobbes developed his earlier notions concerning the goodness of the seal between washer and cylinder:

> Truly I think it close enough to keep out straw and feathers, but not to keep out air, nor yet matter. For suppose they were not so exactly close but that there were round about a difference for a small hair to lie between; then will the pulling back of the cylinder of wood [i.e., the sucker] force so much air in, as in retiring it forces back, and that without any sensible difficulty. And the air will so much more swiftly enter as the passage is left more narrow. Or if they touch, and the contact be in some points and not in all, the air will enter as before, in case the force be augmented accordingly. Lastly, though they touch exactly, if either the leather [i.e., the washer] yield, or the brass [i.e., the cylinder], which it may do, to the force of a strong screw, the air will again enter. . . . The effect therefore of their pumping is nothing else but a vehement wind, a very vehement wind.[15]

And in the *Decameron physiologicum* of 1678 Hobbes again allowed "no such exact contiguity [between leather and brass], nor such fastness of the leather: for I never yet had any that in a storm would keep out either air or water." Once air passed through these imperfect seals the violent circulation thus set up could account for "all the alterations that have appeared in the engine."[16] Just as Hobbes showed that Boyle's device was a physically open, not a closed, system, so he showed that its alleged findings were open to reformulation, that they were not necessarily the phenomena they were claimed to be.

Hobbes did not attempt to provide alternative physical accounts of all the phenomena of the *machina Boyleana*: he focused on just a few of Boyle's series, some of which he glossed in several lines and some of which he dealt with in greater detail, evidently regarding them as vitally important. We discuss some of these "crucial" phenomena below. First, however, we must examine the tactics of the *Dialogus* in stipulating proper meanings for the terms "air," "aether," and "vacuum," since these stipulations inform the rest of

[15] Hobbes, "Seven Philosophical Problems," pp. 20-21.

[16] Hobbes, "Decameron physiologicum," pp. 94-95.

Hobbes's arguments against his adversary. The salient features of this analysis are that common air consists of a mixture of earthy and aqueous effluvia with pure air [*aer purus*], this latter pure component sometimes called the "aether"; that such fluids are indefinitely fluid (their fluidity is not due to some minimal nonfluid particle but to the nature of a fluid medium); and, finally, that the term "vacuum" must properly denote places utterly devoid of all matter. Thus, "I suppose the air is a fluid . . . easily divisible into parts which are always still fluid and still air, such that all divisible quantities are there in any quantity. Nor do I suppose as much, but I also believe that we only understand an air purified from all effluvia of earth and water, such as may be considered as aether." Hobbes went on to argue that the Royal Society was wrong in its notion of fluidity: "You make me despair of fruit from your meeting by saying that they think that air, water and other fluids consist of nonfluids. . . . If such is to be said, then there is nothing that is not fluid."[17]

Hobbes laid down the proper use of the word "vacuum." He argued that this word must mean truly empty space, and, therefore, that the pump could not produce a vacuum. His tactic here was analyzed by John Wallis in the *Hobbius heauton-timorumenos* of 1662:

> For Mr. Hobs is very dexterous in confuting others by putting a new sense on their words rehearsed by himself: different from what the words signifie with other Men. And therefore if you [Boyle] shall have occasion to speak of Chalk, He'll tell you that by Chalk he means Cheese: and then if he can prove that what you say of Chalk is not true of Cheese, he reckons himself to have gotten a great victory. And in like manner, when that Heterogeneous Mixture (whatever it be) wherein we breath is commonly known by the name of Air and this Air wherein we live abounds, you say, with parts of such a nature; he tells you that, by air, he understands such an aether as is among the stars, and that in this air there be no such particles.[18]

In the 1640s Hobbes had confronted the same difficulty in his argument with Descartes, who, as we noted in chapter 3, had complained of the identification of his own subtle matter with that of Hobbes.[19] In 1657 Hobbes told Samuel Sorbière that in the context

[17] Hobbes, "Dialogus physicus," pp. 244-245.

[18] Wallis, *Hobbius heauton-timorumenos*, p. 154.

[19] Descartes to Mersenne, 22 February/4 March 1641, in Mersenne, *Correspondance*, vol. x, p. 524.

of the experiment of the gardener's pot, the Epicureans "call a vacuum what Descartes calls subtle matter and what I call the purest aethereal substance; of which no part is an Atom, but which can be divided . . . into parts which are always divisible."[20] Now, in 1661, Hobbes repeated his view on what constituted a true void. His interlocutor in the *Dialogus* reported the view of the Royal Society:

> [O]thers, no less authoritative among us, are of the opinion that it would not be very repugnant if by vacuum were understood a place empty of all corporeal substances. For supposing air to be made up of particles which cannot be put together without interstices, they see that it is necessary that these interstices be full of corporeal substance, or (as I will say more openly) of bodies. But they do not believe what the *plenists* understand of such a vacuum, especially recently.[21]

Hobbes responded violently to these views: he denied that any "plenist" limited real ontology to visible substance, and cited Democritus, Epicurus, and Lucretius as authorities (albeit vacuists) for the definition of vacuum as the absence of body, visible and invisible: "None of those whom you call *plenists* understands the vacuum as anything but a place in which there is no corporeal substance at all. If someone speaking negligently were to say, 'in which there is no visible body or air,' then he would be saying that he understands by air all that body which fills all the space left by the Earth and the stars."[22] Having defined "vacuum" this way, Hobbes now set himself the task of showing that some substance, even if invisible, was always present in the receiver of the air-pump.

With this stipulation, Hobbes placed a very firm boundary-condition on the results of the air-pump trials. They *could not* show that a vacuum existed. Very few of Boyle's colleagues ever claimed that the trials had shown such a thing. John Wallis told Boyle in 1662 that "I do not remember that you have therein anywhere declared your Opinion whether there be or be not, a Vacuum, but onely related matter of fact," though Wallis and Boyle had stated that "much of what we call Air is drawn out of the Recipient," which Wallis claimed "Mr. Hobs doth not Deny."[23] In Boyle's own response to Hobbes, he wrote that "the atmosphere or fluid body, that sur-

[20] Hobbes to Sorbière, 27 January/6 February 1657, in Tönnies, *Studien*, p. 72.

[21] Hobbes, "Dialogus physicus," p. 275.

[22] Ibid., p. 276.

[23] Wallis, *Hobbius heauton-timorumenos*, p. 152; cf. A. R. Hall, *The Scientific Revolution*, p. 212.

rounds the terraqueous globe, may, besides the grosser and more solid corpuscles wherewith it abounds, consist of a thinner matter, which for distinction-sake I also now and then call ethereal."[24] Similarly, Henry Power, one of Boyle's collaborators, said of the Torricellian apparatus that "in the superior part of the tube there is no absolute vacuity," and was careful to distinguish between the views of those like himself who maintained that a subtle fluid was present above the mercury and those disciples of Gassendi, true vacuists, who "will admit of no aether or forrain substances to enter the pores thereof."[25] Finally, as we shall see in chapter 6, the development of the phenomenon of anomalous suspension in the air-pump prompted successive writers to theorize ever more complex mixtures of subtle fluids present in the receiver, and authorities such as Wallis, Huygens and Hooke all wrote of such fluids and their important functions and effects. In September 1672 Wallis explained that what "Wee mean by 'Air' " was a mixture of a purer subtle matter (to be identified with Cartesian "*materia subtilis*" or Hobbesian "*aer purus*") and a grosser group of terrestrial effluvia (identified with Huygens' "air"): "[T]herefore where I speak of 'Vacuity' . . . I do expressly caution . . . not to be understood as affirming absolute Vacuity (which whether or no there be, or can be, in nature, I list not to dispute)."[26] In this context Hobbes's stipulations about the meaning of "air" and "void" were telling: they called for, and obtained, responses from the experimentalists.

How, then, did Hobbes describe the air? In the *Dialogus*, he said that his "hypotheses" about the air's constitution were twofold: "first, that many earthy particles are interspersed in the air, to whose nature simple circular motion is congenital; second, the quantity of these particles is greater in the air near the Earth than in the air further from the Earth."[27] Hobbes mobilized the tripartite typology of visible matter, invisible matter and a fluid space-filling aether which he had outlined in *De corpore*.[28] The invocation of an ultimately fluid body in which grosser substances were mixed was typical of Hobbes's physical armamentarium; he used it, for ex-

[24] Boyle, "Examen of Hobbes," p. 196; we discuss Boyle's aether experiments in chapter 5.

[25] Power, *Experimental Philosophy*, pp. 132-140, 101-103.

[26] Wallis to Oldenburg, 26 September/6 October 1672, in Oldenburg, *Correspondence*, vol. IX, p. 259; cf. Hooke, *Micrographia*, pp. 12-16, 103-105; Huygens to J. Gallois, July 1672, in Huygens, *Oeuvres*, vol. VII, pp. 204-206.

[27] Hobbes, "Dialogus physicus," p. 253.

[28] Hobbes, "Concerning Body," p. 426.

ample, in his miasmal theory of the plague.[29] In the *Problemata physica* of 1662 Hobbes repeated his view on ultimate fluidity and said that "it is that internal motion which distinguisheth all natural bodies one from another."[30] Furthermore, the argument that fluidity and firmness were due to the motion of particles of body and that the air was a mixture of varying gross particles had itself been developed against Cartesian opposition in the 1640s. Where Descartes explained fluidity by the motion of particles, Hobbes explained *firmness* by such motion, as Cavendish told Jungius in 1645 and as Hobbes himself told Mersenne: "Those who wish their matter to be Body and their subtle to be Subtle must necessarily wish the same as that which is signified by different names."[31] In each of the phenomena Hobbes now examined in the *Dialogus*, he used the contrasting fluidity of pure air and earthy effluvia to explain the observed effects. In so doing, Hobbes showed how it was always possible to generate such explanations from his two hypotheses of fluidity and firmness; he also showed that the invocation of absolute vacuity was both unnecessary and unphilosophical. Once again Hobbes picked out a central problem of the air-pump research programme. Boyle laboured to establish the air's spring as a matter of fact, eschewing any systematic attempt to explain the spring or to prove the vacuum. Hobbes's polemic disproved the vacuum and offered a physical explanation of the apparent spring. Hobbes's interlocutor agreed that "Your hypothesis pleases me more than that of the elastic force of the air. For I see that the truth of the vacuum or of the plenum depends upon the former's truth, whereas from the truth of the latter, nothing follows for either part of the question."[32] It was Boyle's nescience, and his *recommendation* of nescience as an appropriate philosophical stance, that Hobbes found objectionable: hence his effort in the *Dialogus* to show how *easily* his two hypotheses could explain all the phenomena whose cause Boyle said he was unable to find.

In dealing with phenomena which did not obviously involve the air's spring, Hobbes's task was straightforward, given his stipulation of the meaning of the term "vacuum." For example, he agreed with

[29] Hobbes, "Decameron physiologicum," p. 129; cf. p. 136.

[30] Hobbes, "Seven Philosophical Problems," p. 12.

[31] Hobbes to Mersenne, 20/30 March 1641, in Tönnies, *Studien*, p. 115; Sorbière to Mersenne, May 1647, in ibid., pp. 64-65; Hobbes to Mersenne, 28 January/7 February, 1641, in Hobbes, *Latin Works*, vol. v, p. 284; Brockdorff, *Cavendish Bericht für Jungius*, p. 3; Gargani, *Hobbes e la scienza*, p. 217.

[32] Hobbes, "Dialogus physicus," p. 262.

Boyle that animals died in the "evacuated" receiver; they were, however, literally blown to death by a violent circulatory wind, not deprived of vital air. (This was a phenomenon that Hobbes especially commended to the attention of his medical friend, Samuel Sorbière.) Candles went out for the same reason. It was difficult, Hobbes assented, to lift up the cover on the top of the "exhausted" receiver, but this was because of the nature of the plenum and the vehement circulation within the glass.[33] However, in phenomena that were explicitly said to demonstrate the presence of a "spring" in the air, Hobbes took the opportunity to develop a thoroughgoing mechanical account, using his hypothesis of the structure and fluidity of the air and its effluvia. Hobbes used the term *antitupia* for "spring," a word which Henry More had used in 1647 and claimed to have found in Sextus Empiricus.[34] Again and again, the phenomena that Boyle prized as clear examples of the effects of the spring were appropriated as exemplars of the effects of uneven mixtures of earthy and subtle particles moving with a simple circular motion. The cause of the rapid ascent of the sucker when released after an exsuction was not the difference in air pressure but the difference in the number of earthy particles: on pulling back the sucker, pure air could leak into the receiver but earthy particles could not, so a greater proportion of these latter remained outside the sucker and they pressed the sucker up very rapidly.[35] The ascent of water in a hydroscope was explained in the same terms:

> The air, with which the spherical glass was full in the beginning, being moved by those earthy particles in the simple circular motion which we described a little earlier, being compressed by the force of injection, that of it which is pure leaves by penetrating the injected water for the outer air, leaving a place for the water. So it follows that those earthy particles are left less space in which to exercise their natural motion. Thus impinging on one another, they force the water to leave: and, in leaving, the external air (since the universe is supposed to be full) penetrates it, and successively takes up the place of the

[33] Ibid., pp. 235, 253-254, 257-258, 260, 263-264. In the case of animals' death within the receiver, Hobbes offered a choice of non-Boylean explanations: either the violent wind or some form of suction that interrupted respiration.

[34] Ibid., p. 271; idem, "Decameron physiologicum," p. 108; More, *Philosophicall Poems*, "Interpretation Generall," p. 423.

[35] Hobbes, "Dialogus physicus," p. 253.

> air that leaves, until the same quantity of air being replaced, the particles regain the liberty natural to their motion.[36]

Hobbes's account of these phenomena appealed to the space necessary for the earthy particles to "exercise their natural motion": this motion itself produced the rigidity of bodies, as he had indicated in *De corpore*, and this rigidity accounted for resistance to compression and for the force of motion of bodies that contained the earthier air. The consequences of his initial definition of air and vacuum were, therefore, considerable: the subtler part of the air was just that part which *rendered a vacuum impossible*, while the grosser parts were those which performed *the effects Boyle interpreted as "spring."* These two types of matter often combined in their effects. For example, in Boyle's experiment in which a moderately inflated bladder was inserted in the receiver and the glass exhausted, the bladder was observed to inflate and finally to burst. Roberval had developed a similar experiment as evidence for the presence of a *subtle fluid* in the Torricellian space; Boyle used it as evidence for the air's *elasticity*. Hobbes's explanation again adduced the vehement circulation of the *purer part* of the air:

> [E]very skin is made up of small threads, which because of their shapes cannot touch accurately in all points. The bladder, being a skin, must therefore be pervious not only to air but to water, such as sweat. Therefore, there is the same compression of the air compressed inside the bladder by force as there is outside, whose endeavour, its motion following paths that intersect everywhere, tends in every direction towards the concave surface of the bladder. Whence it is necessary that it swells in every direction and, the strength of the endeavour increasing, it is at last torn open.[37]

These emphases on the porosity of the materials in the pump, the ability of the purer air to penetrate all such materials, and the evident power of the simple motion of these particles all characterized the accounts Hobbes offered in the *Dialogus*. As we have seen, these resources were explicitly developed here to contest the vacuum and the spring as Hobbes now defined them.

The experiment to which Hobbes devoted the greatest attention was the thirty-first of Boyle's *New Experiments*, the one in which cohered marbles were placed in the receiver in the (unfulfilled)

[36] Ibid., pp. 274-275.
[37] Ibid., pp. 266-267.

expectation that they would fall apart upon evacuation. Hobbes took up the challenge of explaining this problematic experiment and of reducing its apparent troubles to Hobbesian order. In the *Dialogus*, Hobbes's interlocutor put Boyle's case. He sketched the general form of Boyle's account of cohesion, and claimed that

> . . . if marbles thus cohering were transferred into the receiver and suspended therein, the air being sucked out, were the lower marble to cease sticking to the upper, it would not be possible to doubt that the assigned cause was true. They were moved into the receiver, but without the success expected. For by no further means would they cease to cohere, unless it happened that they were not joined together well enough.

In reply Hobbes suggested that "there was nothing in this [experiment] which should be done by the weight of the atmosphere," and that "No stronger or more evident argument could be devised against those who assert the vacuum than this experiment."[38] Part of the explanation followed Hobbes's treatment of cohesion and of gravity developed in *De corpore*.[39] This was extended and refined in response to Boyle's intervention. Hobbes pointed out that in a plenum the separation of the marbles would either require an instantaneous motion or else moving two bodies simultaneously to the same space, "to say either of which is absurd." He then examined the two possible accounts which Boyle could offer of the phenomenon. One involved the concept of atmospheric weight. Hobbes first defined weight ("fully acknowledged by them, as by everyone else") as "an endeavour along straight lines from all places to the centre of the Earth," which therefore operates in a pyramid whose vertex is in the Earth's centre. The upper marble, Hobbes argued, acted, so to speak, as a "weight shadow" for the lower marble, which received no reflected endeavour from the surface of the Earth: "So nothing arises as a result of atmospheric endeavour sustained by the lower marble to prevent its separation from contact with the upper one."[40]

The other possible explanation that Boyle might offer involved the spring of the air. Hobbes's interlocutor asked: ". . . cannot that elastic force which they say is in the air contribute anything to sustaining the marble?" By no means, Hobbes answered,

[38] Ibid., pp. 267-268.

[39] Hobbes, "Concerning Body," pp. 419, 511-513.

[40] Hobbes, "Dialogus physicus," pp. 268-269.

> . . . the endeavour of the air is no greater towards the centre of the Earth than to any other point in the universe. Since all heavy things tend from the edge of the atmosphere to the centre of the Earth, and thence again to the edge of the atmosphere by the same lines of reflection, the endeavour upwards would be equal to the endeavour downwards, and thence by mutually annihilating each other they would endeavour neither way.[41]

Hobbes could easily assert that his plenist account was the correct one because he had already produced the claim that the receiver was full, and he now condemned Boyle's explanations as "dreams": ". . . if I should deny it to be possible by human art to make the surfaces of two hard bodies touch so accurately that not the least creatable particle could be let through, then I do not see how their hypothesis could be rightly sustained, nor how our negation could be rightly argued to be unproven."[42] Hobbes evidently regarded the point as made: he had seized upon an experimentally produced phenomenon that was both central and troublesome to Boyle's programme; he had provided a physical accounting compatible with his own natural philosophy. Moreover, in his confidence, Hobbes had, as it were, "laid a bet" on future trials of this experiment: if Boyle made the experiment into a success (that is, if the marbles separated in the receiver), then, according to his interlocutor, "it would not be possible to doubt" that Boyle's accounting was superior. In the next chapter we shall see how Boyle responded to this challenge.

The Engines of Philosophy

Some historians have dismissed Hobbes's criticisms from consideration on the grounds that he did not perform experiments himself, or at least that he did not repeat those of Boyle's experiments to which he took exception: indeed, as we shall see in the next chapter, this was one of Boyle's own tactics for rejecting Hobbes's views. For this reason we need to give especially careful attention to Hobbes's opinions on the role and value of experimental procedures in natural philosophy. Let us start by confronting claims that Hobbes actually *approved of* experiment and accorded it a cen-

[41] Ibid., p. 269.
[42] Ibid., p. 271.

tral place in properly constituted philosophy. For example, J.W.N. Watkins' excellent book on Hobbes's philosophy attempts to refute the "popular idea that Hobbes despised experiments. . . . He only despised haphazard experimenting."[43] In evidence Watkins cites Hobbes's injunction in the *Decameron physiologicum* that in examining physical hypotheses "you must furnish yourself with as many experiments . . . as you can."[44] Watkins also cites, but does not quote, remarks Hobbes made in reply to one of Wallis's attacks. The passage starts:

> Every man that hath spare money, can get furnaces, and buy coals. Every man that hath spare money, can be at the charge of making great moulds, and hiring workmen to grind their glasses; and so may have the best and greatest telescopes. They can get engines made, and apply them to the stars; recipients made, and try conclusions. . . .[45]

(We shall pick up the remainder of this passage shortly.)

In addition, there are intriguing remarks on the subject of experiment in Hobbes's *Six Lessons* of 1656. Here he attempted to exonerate himself from Wallis's charge that he had denigrated the experimental work of Hobbes's friend William Harvey. The story concerned a visit made to Harvey by the Flemish Jesuit Moranus. According to Hobbes, the Jesuit, a man of "but common and childish learning," refused to be instructed by the learned physiologist, but merely vented his own meretricious opinions. In so doing, said Hobbes, "He took occasion, writing against me, to be revenged of Dr. Harvey, by slighting his learning publicly; he tells me that his learning was only experiments; which he says I say have no more certainty than civil histories. Which is false." Hobbes then quoted his remarks on this subject in the "Epistle Dedicatory" of *De corpore*: "Before these [Galileo and Harvey], there was nothing certain in natural philosophy but every man's experiments to himself, and the natural histories, if they may be called certain, that are no certainer than civil histories." Hobbes pointed out that "I except expressly from uncertainty the experiments that every man maketh to himself," and that there was no slur on Harvey.[46]

[43] Watkins, *Hobbes's System*, p. 70n; cf. Laird, *Hobbes*, p. 116: "Hobbes did not despise experimental investigations."

[44] Hobbes, "Decameron physiologicum," p. 88.

[45] Hobbes, "Considerations on the Reputation of Hobbes," p. 436. Watkins does not quote anything from the *Dialogus physicus* in *Hobbes's System*.

[46] Hobbes, "Six Lessons," pp. 338-339. Hobbes gave the Latin version of the

Hobbes's sensitivity about the appropriation of Harvey's reputation by the experimentalists of the Royal Society was even more evident in the introduction to the *Dialogus physicus*. Hobbes wanted to show that all sense experience is a consequence of an external motion. His interlocutor reported that he was dazed by the brightness of the sun. Hobbes invited him to sit down "until that excessive motion of the organ of vision settles down." Interlocutor replied:

> You advise well. Truly, I am of the opinion that lassitude of this kind due to solar heat has the habit of increasing mental cloudiness a little. But I do not see enough of the way in which either light or heat produces such effects. Since the time you first demonstrated it to us, I have no longer doubted that not only all feeling but also all change is some motion in the feeling body and in the moving body, and that this motion is generated by some external mover. For previously almost everyone denied it; for whether standing, sitting, or lying down, they nevertheless understood well enough that they were feeling.

Our own feelings appear to be within us, and, therefore, if such appearances were to be the grounds of knowledge, they would lead us to erroneous conclusions. For instance, Hobbes continued, such feelings led people to doubt "whether their own blood moved; for no one feels the motion of their blood unless it pours forth." Interlocutor agreed: "Indeed everyone doubted it before Harvey. Now, however, the same people both confess that Harvey's opinion is true and they are also beginning to accept your beliefs about the motion by which vision is produced. For in our Society there are few who feel otherwise."[47] Hobbes's contention was that it was Harvey, using correct philosophical method, who convinced men of the motion of the blood, not personal experience, and he assimilated the status of his own optical theory to that of Harvey's views of circulation. If, Hobbes argued, Harvey was to be a hero to the Greshamites, then so should Hobbes. Rightly understood, he said, Harvey and Hobbes were methodological allies, both denying the foundational nature of personal experience.

The "experiments that every man maketh to himself" are *experience*. Being, as Hobbes had said, but "sense and memory," they

---

passage from *De corpore*; we provide the English, from "Concerning Body," pp. viii-ix. For Hobbes's views regarding Galileo, Harvey and the methods of the Padua School, see Watkins, *Hobbes's System*, pp. 55-65.

47 Hobbes, "Dialogus physicus," pp. 239-240. The same example of the Sun and personal feeling was also used in "Decameron physiologicum," pp. 117-118.

generated certainty in him who has the experience; they could not, however, produce the collective certainty which was the prerogative of *philosophy*. Hobbes's views on the role of experimental practices in natural philosophy were, in any case, spelt out clearly elsewhere. In *Decameron physiologicum* Hobbes explicitly devalued the role of formal experimental procedures compared to those experiences of natural phenomena any man can have: "As for mean and common experiments, I think them a great deal better witnesses of nature, than those that are forced by fire, and known but to a very few."[48] And, after enjoining his interlocutor to "furnish yourself with as many experiments (which they call phenomenon) as you can," the agreeable interlocutor assented: "What I want of experiments you may supply out of your own store, or such natural history as you know to be true; though I can be well content with the knowledge of causes of those things which everybody sees commonly produced."[49] Of course, the best evidence of Hobbes's opinion of experimentation in natural philosophy is contained in the *Dialogus physicus*, where it is worked out in the concrete context of his reaction to the Greshamites' programme. However, let us continue with the passage which starts with Hobbes's account of what "Every man that hath spare money" can do with "furnaces," "telescopes," and "engines." This is how Hobbes concluded:

> They can get engines made, and apply them to the stars; recipients made, and try conclusions; *but they are never the more philosophers for all this*. It is laudable, I confess, to bestow money upon curious or useful delights; but that is none of the praisis of a philosopher. And yet, because the multitude cannot judge, they will pass with the unskilful, for skilful in all parts of natural philosophy . . . So also of all other arts; not every one that brings from beyond seas a new gin, or other jaunty device, *is therefore a philosopher*. For if you reckon that way, not only apothecaries and gardeners, but many other sorts of workmen, will put in for, and get the prize.[50]

And again: "If the sciences were said to be experiments of natural things, then the best of all physicists are quacks [*pharmacopoei*]."[51]

[48] Hobbes, "Decameron physiologicum," p. 117.

[49] Ibid., p. 88; cf. p. 143. Watkins substitutes an ellipsis for the round-bracketed phrase; see *Hobbes's System*, p. 70.

[50] Hobbes, "Considerations on the Reputation of Hobbes," pp. 436-437 (emphases added).

[51] Hobbes, "Mathematicae hodiernae," p. 229. Others have translated *pharmacopoei*

"Ingenuity," Dogmatism, and the Experimental Community

The point to be made is not that Hobbes "despised" experiment, nor that he argued that experiments ought not to be performed, nor even that experiments had no significant place in a properly constituted philosophy of nature. What Hobbes was claiming, however, was that the systematic doing of experiments was not to be equated with philosophy: going on in the way Boyle recommended for experimentalists was not the same thing as philosophical practice. It was not the case that one could ground philosophy in experimentally generated matters of fact. This experimental way and the philosophical way were fundamentally different: they differed in their capacity to secure assent among intellectuals and peace in the polity. The distinction that Hobbes wanted to make involved four considerations that were regarded as intimately related in mid-seventeenth-century schemes: the status of the philosopher's role, his social and moral character, the thought processes involved in doing intellectual work, and the nature of the knowledge that was the outcome of this work. By claiming that adopting an experimental form of life changed proper physicists into "quacks," Hobbes was saying something highly derogatory about the experimentalist's role, character, and practice. Machine-minders were not, in Hobbes's view, to be accounted philosophers. Philosophers should not be identified with mechanical tricksters who produced "various spectacles of an amusing nature."[52]

The modes of thought associated with the philosopher and the mechanic were different. In the *Dialogus physicus* Hobbes insisted upon that contrast: "Ingenuity is one thing and method [*ars*] is another. Here method is needed."[53] The repeated juxtaposition in Hobbes's critiques of method or philosophy, on the one hand, and ingenuity, on the other, is significant. It is plausible that Hobbes was making a substantive point about the experimental mentality by way of etymological punning. The Latin *ingenium* denotes "natural ability, cleverness, inventiveness." In Latin *ingenio* also means a kind of mill, and, from this root, are derived the Old French

as "pharmacists"; in this context "quacks" clearly renders Hobbes's meaning more accurately.

[52] Hobbes, "Dialogus physicus," p. 235. Hobbes specified that it was "a man well-known in breeding and ingenuity" who had produced these trivial spectacles; the juxtaposition was presumably meant to jar.

[53] Ibid., p. 236.

engin and the Middle English *gin*. Thus Hobbes's identification of ingenuity with, as it were, "engine philosophy" is precisely right for the evaluation he wanted to be placed upon the experimental programme and its products: it relied upon the intellectual processes of artificers and mechanics and, therefore, it yielded an inferior grade of knowledge.[54] That is why Hobbes contrasted "workmen," "apothecaries" and "gardeners" with "philosophers" and why he insisted that not every procurer of "jaunty devices" was a "philosopher." The philosopher was not *banausic*.[55]

Hobbes and Boyle had two things in common in this connection: first, they both gauged the worth of knowledge by taking into consideration the moral constitution and known probity of its producers. This was taken for granted in mid-seventeenth-century calculations, and the problem of assessing testimony made these calculations important, as we have discussed in chapter 2. Second, both Hobbes and Boyle reckoned that the philosopher should be seen as *noble*. Yet their characterizations of the philosopher's role and practice were diametrically opposed. Whose version of the philosopher was, indeed, noble? We have seen that Boyle and his colleagues liked to describe the experimental philosopher as "humble," "modest," an "under-builder," and a "drudge," while specifying that this was a noble character. Boyle and his associates in the Royal Society wanted, for specified purposes, to use the language of the craftsman and to put on the guise of the humble artisan. Hobbes was trying to insinuate that, through their celebration of ingenuity, the Greshamites *really were* making philosophy ignoble. This could have been a seriously damaging imputation in early Restoration society. Hobbes and Boyle agreed that worthy knowledge was produced by worthy men. Yet to Boyle and his friends ingenuity was to be celebrated and the knowledge produced by machines was to be accounted valuable. No stigma was said to be attached to machine minding, no odium to its intellectual products, and no contrast was made between experimental manipulations with machines and philosophy. The Greshamites enjoyed addressing each other as ingenious: the ingenious Mr. Boyle, the ingenious Mr. Wren. But this was an ingenuity made noble by the participation in experimental labour of noble, honest, and trust-

[54] Ibid., p. 278, where the role of mechanic is explicitly contrasted to that of the philosopher; and see Bennett, "Hooke as Mechanic and Natural Philosopher." Sources for the etymology are the *Oxford English Dictionary* and Partridge, *Origins*, under "General."

[55] For the notion of a banausic intellect, see Shapin and Barnes, "Head and Hand."

worthy men. That is one reason why, as Robert Greene has perceptively noted, much mid-seventeenth-century usage freely interchanged "ingenuity" and "ingenuousness."[56] Still, despite the public pronouncements, it is not the case that the Fellows of the Royal Society treated mechanics and gardeners *as* philosophers, or that they regarded the testimony of artisans as on a par with that of gentlemen. And it is well to remember that the "ingenious Mr. Boyle" possibly never tended his air-pump himself: the work was done under his supervision by "strong workmen" and skilled instrument-makers.

Now that we understand aspects of Hobbes's condemnation of experimental practice we can parenthetically discuss his relations with the Royal Society as a corporate body. Why was Hobbes not a Fellow? Was he "excluded," and, if so, on what grounds? On the face of it this is not a matter we need to treat in any detail: ours is not a study of the Royal Society itself but of conflicting strategies for generating natural knowledge in mid-seventeenth-century England. Nevertheless, since some recent work has addressed the question of Hobbes's nonmembership in the Royal Society, we ought to point out in what ways our material bears upon that issue. Quentin Skinner has argued against the view that Hobbes was kept out of the Society either because of his religious heterodoxy or because of his opinions on experimentalism and natural philosophy in general. In Skinner's account, within "the broad strategy of mid-seventeenth-century science Hobbes and the Royal Society stand on the same 'side' throughout." He concludes: "The exclusion of Hobbes is then readily explained: no one wants to encourage a club bore."[57] More recently, Hunter has echoed Skinner's judgment, pointing out that Hobbes had friends in the Royal Society, notably Sir John Hoskyns and John Aubrey.[58]

It is, in fact, Aubrey who provides the best apparent support for the claim that there was mutual respect and good will between Hobbes and leading lights in the Royal Society. According to Aubrey, Hobbes "had a high esteeme for the Royall Societie . . . , and

[56] Greene, "Whichcote, Wilkins, 'Ingenuity,' and the Reasonableness of Christianity," esp. pp. 227-229. In the religious context that Greene examines, "ingenuity" often referred to cleverness in exegesis or to the use of reason in theology: "ingenuity" was contrasted with "grace."

[57] Skinner, "Hobbes and the Early Royal Society," pp. 231, 238. Cf. C. Hill, *Some Intellectual Consequences*, pp. 63-64.

[58] Hunter, *Science and Society*, pp. 178-179; idem, "The Debate over Science," pp. 189-190; cf. idem, *The Royal Society and Its Fellows*, p. 6.

the Royall Societie (generally) had the like for him: and he would long since have been ascribed a member there, but for the sake of one or two persons whom he tooke to be his enemies." These "enemies" were, Aubrey noted, "Dr. Wallis (surely their Mercuries are in opposition), and Mr. Boyle. I might add Sir Paul Neile, who disobliges everybody." Furthermore, Aubrey quoted Hobbes's remark in *Behemoth* that " 'Naturall Philosophy was removed from the Universities to Gresham Colledge,' meaning the Royall Societie that meets there." Aubrey pointed out that Hobbes's portrait hung in the Society's meeting-place, and noted that Hobbes fell out with Henry Stubbe "for that he wrote against the lord chancellor Bacon, and the Royall Society."[59]

Aubrey's remarks bear inspection. His *Life of Hobbes*, from which they come, is a partial account, written after Hobbes's death, by a man who was a friend, a Fellow, and a friend of several of Hobbes's bitter enemies.[60] It was an exercise in posthumous reconciliation, and it played down the series of bitter controversies between Hobbes and leading Fellows. Aubrey offered no extended account of the *Dialogus physicus*, in which Hobbes declared all the "Greshamites" to be his "enemies."[61] The remark he quoted from Hobbes's *Behemoth* is taken out of context. It is not praise of the Royal Society, nor, indeed of the professors of Gresham College, but part of another extended indictment of the universities and the clergy for their divisive role in society, and it is not even absolutely clear that "Gresham College" is meant to refer to the Royal Society.[62] Even Aubrey's claim about Hobbes's portrait bears closer examination. The picture in question was painted by J. B. Caspars in 1663, and was commissioned by Aubrey himself, who presented it seven years later to the Royal Society. In notes to the manuscript version of his *Life of Hobbes* Aubrey asked himself whether it would be "improper for me to mention my owne guift?" and ultimately decided not to.[63] Still, there is no reason to doubt that Hobbes did have his friends and admirers among the Fellowship. In addition to Aubrey and Hoskyns, one might include John Evelyn, Sir William Petty, Sir Kenelm Digby, and, of course, his patron William Cavendish, 3rd Earl of Devonshire. Moreover, Hobbes had been the aman-

59 Aubrey, "Life of Hobbes," pp. 371-372.

60 On the writing of Aubrey's *Life of Hobbes*, see Hunter, *Aubrey and the Realm of Learning*, pp. 78-80.

61 Hobbes, "Dialogus physicus," p. 237.

62 Hobbes, "Behemoth," p. 348.

63 Aubrey, "Life of Hobbes," p. 354; cf. Powell, *Aubrey and His Friends*, p. 102.

uensis of the great Bacon and the friend of William Harvey, two of the Society's heroes.[64]

Of greater interest in this connection is Hobbes's relationship with the new monarch Charles II, the Society's "founder" and patron. After the Restoration Charles continued to receive his former mathematics tutor at Court. While their relationship had something of a "joking" character to it, it appears to have been publicly affectionate: the King liked to refer to the old philosopher as "the beare," and, because, as Aubrey said, the "wittes at Court were wont to bayte him," Charles would greet his approach by crying, " 'Here comes the beare to be bayted!' "[65] The connection was solid enough for the King to grant Hobbes a substantial (albeit irregularly paid) pension, and Hobbes dedicated his *Problemata physica* of 1662 to the King, using the occasion to apologize for any offence that *Leviathan* may have given. The King also possessed a portrait of Hobbes (by Samuel Cooper) which, according to Aubrey, he "conserves as one of his great rarities in his closet at Whitehall."[66] There is also

[64] A list of Hobbes's friends, with comments, is in Aubrey, "Life of Hobbes," pp. 365-371. Aubrey says that Robert Hooke "loved him," but adds that Hooke "was never but once in his company" (p. 371). In the event, there are records that Hooke met Hobbes at least twice, once in July 1663 and once in June 1674 at Aubrey's house; see Hooke to Boyle, 3/13 July 1663, in Boyle, *Works*, vol. VI, pp. 486-487 (cf. Gunther, *Early Science in Oxford*, vol. VI, pp. 139-141, where Hooke's references to Hobbes are unflattering; and, for the 1674 meeting, Hooke, *Diary*, p. 108). Of Hobbes's relationship with Bacon, Aubrey wrote: "The Lord Chancellor Bacon loved to converse with him. . . . His lordship would often say that he better liked Mr. Hobbes's taking his thoughts, then any of the other, because he understood what he wrote": "Life of Hobbes," p. 331. Hobbes's friend Sorbière also remarked upon the relationship with Bacon: Hobbes "is upon the Matter the very Remains of *Bacon*, to whom he was Amanuensis in his Youth": Sorbière, *Voyage to England*, p. 40. Sprat, protecting the Royal Society's hero, took violent exception to the claim that there was any similarity between Hobbes and Bacon, and maintained that Sorbière did not understand Hobbes's philosophy: "Of this I will give an unanswerable Testimony, and that is *the Resemblance that he makes of him to the Lord* Verulam, between whom there is no more likeness than there was between St. *George* and the Waggoner. . . . I scarce know Two Men in the World that have more different Colours of Speech than these Two Great Wits": Sprat, *Observations on Sorbière's Voyage*, p. 163.

[65] Aubrey, "Life of Hobbes," p. 340. Sorbière also reported, in reference to the English clergy and the Oxford mathematicians, that Hobbes "was like a Bear, whom they baited with Dogs to try him": Sorbière, *Voyage to England*, p. 40.

[66] According to Sorbière, the King's pension of £100 per year showed how far Charles was "from laying any stress upon Dr. *Wallis*'s Arguments" that Hobbes's politics were anti-royalist: Sorbière, *Voyage to England*, p. 39. On Hobbes's royal pension and other finances, see Laird, *Hobbes*, pp. 20-21; Hobbes to the King, 1663?, in Hobbes, *English Works*, vol. VII, pp. 471-472; Hobbes to Aubrey, 7/17 September 1663, in Tönnies, *Studien*, p. 108. For Hobbes's remarks to the King concerning

the intriguing suggestion that Charles would not have been unhappy to see Hobbes elected to the Society. Sorbière related an interview with the King in which it was "agreed on all Hands, that if Mr. *Hobbs* were not so very Dogmatical, he would be very Useful and Necessary to the Royal-Society; for there are few People that can see farther into things than he, or have applied themselves so long to the Study of Natural Philosophy." Early in the Society's career the King signalled his opinion of Hobbes's worth as a mathematician when he forwarded to the Royal Society one of Hobbes's geometrical demonstrations.[67] As several historians have suggested, the closeness of the King's association with the great dogmatist must have constituted a considerable threat to the experimentalists of the Royal Society. The King, on whom rested the Society's hopes of material support, was a patron of the new science, but there is little evidence that he discriminated markedly between the rationalist and the experimentalist programmes. Indeed, as Pepys reported, he was known to jest at those very experimental activities which the Society treated as emblematic: "Gresham College he mightily laughed at, for spending time only in weighing of ayre, and doing nothing else since they sat." Nor is it clear what the Royal Society made of the King's sport of placing bets on the outcome of the Society's pneumatic experiments.[68] The stream of fulsome praise directed from the Society to the King was closely connected to its expectations of royal patronage—expectations that were rarely satisfied. Meanwhile, anxious eyes were turned towards "Hobbist" morality at Court. And, as we noted in chapter 3, Hobbes's philosophical standing on the Continent was substantial,

---

Leviathan, see "Seven Philosophical Problems," pp. 3-6. On Hobbes's portraits, see Sorbière, *Voyage to England*, pp. 39-40; Laird, *Hobbes*, p. 25n; Aubrey, "Life of Hobbes," p. 338. There was also an engraving made in 1664 by William Faithorne (who engraved Boyle's portrait as well) in a series devoted to "distinguished royalists."

[67] Sorbière, *Voyage to England*, p. 40; cf. Laird, *Hobbes*, p. 21. For the King's present to the Royal Society, see Birch, *History*, vol. I, p. 42: the gift was made on 4/14 September 1661, just weeks after Hobbes published his *Dialogus physicus*. The only other communication from Hobbes that was placed in the Royal Society's *Letter-Book* was a letter of 10/20 December 1668 concerning a fasting woman, and given to the Society by Daniel Colwall: Birch, *History*, vol. II, pp. 333-334; also in Hobbes, *English Works*, vol. VII, pp. 463-464.

[68] Pepys, *Diary*, vol. V, pp. 32-33 (entry for 1/11 February 1664). On 12/22 January 1671 Sir Robert Moray reported to the Royal Society that "the King has laid a wager of fifty pounds to five for the compression of air by water; and that it was acknowledged, that his Majesty had won the wager": Birch, *History*, vol. II, p. 463. The King also referred to the experimental philosophers as "court jesters"; see Middleton, "What did Charles II Call the Fellows of the Royal Society?"

just as his political reputation and influence were feared to be in England. "Baiting the bear" was, therefore, an important tactic in policing the boundaries of the new experimental philosophy and in displaying publicly what counted as proper scientific activity and what did not. As de Beer remarks, "It may be said that Hobbes did influence the early policy of the Royal Society, for he set for all time the standard of the sort of man who must not be elected into the Fellowship."[69]

Hobbes continued to engage the Royal Society in controversy through the 1670s. There is no evidence that he assimilated, or responded to, Boyle's post-1660 researches in pneumatics, but the geometrical disputes with John Wallis, begun in the 1650s, flared up repeatedly. In 1671 and 1672 Hobbes attacked Wallis in his *Rosetum geometricum* and *Lux mathematica*, and in a pamphlet addressed *To the Right Honourable and Others, the Members of the Royal Society*.[70] Wallis used the *Philosophical Transactions* to reply, but Hobbes was not allowed access to its pages.[71] Increasingly irritated by this slight, Hobbes wrote to Oldenburg in November 1672 requesting that "if hereafter I shall send you any paper tending to the advancement of Physiques or Mathematiques, and not too long, you will cause it to be printed by him that is Printer to the Society, as you have done often for Dr. Wallis. It will save me some charges."[72] Oldenburg consulted with Wallis about the wisdom of acceding to Hobbes's wish, but the professor, while expressing cool disinterestedness in the matter, judged that Hobbes had nothing to contribute to geometry.[73] Thus armed with advice, Oldenburg wrote to Hobbes in a mollifying tone:

> . . . I have no mind to repeat [Wallis's] Answer, being far more inclined, if I were capable, to make you friends, than set you further asunder. Neither is ye R. Society willing to enter into ye decision of the disputes betwixt you, having regard to yr

[69] de Beer, "Some Letters of Hobbes," p. 197; also Bredvold, "Dryden, Hobbes, and the Royal Society," pp. 422-423; Laird, *Hobbes*, pp. 20-21.

[70] The latter appears in *English Works*, vol. VII, pp. 429-448, under the title "Three Papers Presented to the Royal Society against Dr. Wallis."

[71] See, for example, *Philosophical Transactions*, no. 72 (19/29 June 1671), pp. 2185-2186; no. 75 (18/28 September 1671), pp. 2241-2250; no. 87 (14/24 October 1672), pp. 5067-5073.

[72] Hobbes to Oldenburg, 26 November/6 December 1672, in Oldenburg, *Correspondence*, vol. IX, pp. 329-330; also in Hobbes, *English Works*, vol. VII, pp. 465-466, and in Aubrey, "Life of Hobbes," pp. 362-363.

[73] Wallis to Oldenburg, 26 December 1672/5 January 1673, in Oldenburg, *Correspondence*, vol. IX, p. 372.

> age, and esteeming yr parts, but doubting you doe mistake in these controversies. However, I am ready to comply with your desires in yt particular, wch concerns ye publication of such papers, you shall send me tending to ye advancement of Physiques and Mathematiques, as are not too long, nor interwoven with personal reflexions; in a word, yt shall be licensed by ye Council of ye R.S.[74]

Apart from another mathematical tract of 1674, Hobbes retired from the fray.[75] He did not reply to Oldenburg, and he never had anything printed by the Royal Society.

The central issue was indeed Hobbes's "dogmatism." But we would miss the point if we separated claims that Hobbes was personally dogmatic from the dogmatism he was seen to recommend in natural philosophical practice. From what we know of his character, there is little doubt that Hobbes could be a difficult person, set in his ways, and not relishing contradiction. In contrast to the "modest" and "humble" experimental philosophers of the Royal Society, Hobbes confidently claimed that he had developed a complete and self-sufficient system of philosophy: they were to come to him to remedy defects in their thought. On the other hand, his friends liked him, and said so. He had a good line in humour, which he delivered in a mild West Country accent. He sang "pricksongs" (badly), got drunk only rarely, played tennis once a week (at the age of 78), gave alms freely, and was regarded as extraordinarily handsome (see figure 5). There was, however, no Mrs. Hobbes, although there was talk of an illegitimate daughter whom he supported. He swore a bit more than was considered strictly proper, but there was no hard evidence of personal libertinism, and he received the sacraments on his deathbed.[76] On these bases it is difficult to recognize in Hobbes the archetypal "club bore." He was, in the event, much more clubbable than one of his major antagonists, John Wallis, F.R.S., who, according to Aubrey, "makes it his Trade to be a comon spye, steals from every ingeniose persons discourse, and prints it. . . . He is a most ill-natured man, an egregious lyar and back-biter, a flatterer and fawner."[77] Boyle himself

[74] Oldenburg to Hobbes, 30 December 1672/9 January 1673, ibid., pp. 374-375.

[75] Hobbes, *Principia et problemata aliquot geometrica* . . . (1674).

[76] This is according to Aubrey, "Life of Hobbes," pp. 340, 347-353; see also Sorbière, *Voyage to England*, p. 27.

[77] Aubrey to Hobbes, 24 June/4 July 1675, as quoted in Hunter, *Aubrey and the Realm of Learning*, p. 224; cf. Aubrey, *Brief Lives*, vol. II, pp. 280-283. In fact, Wallis did act as a code breaker for the New Model Army in the 1640s.

FIGURE 5
Portrait of Thomas Hobbes at age 81 (1669) by J. M. Wright (in the National Portrait Gallery, London, and reproduced by its permission).

was a renowned valetudinarian; he was long-winded but jealous of his privacy: in later years he went so far as to put up visiting hours on his door. At times, his company could be dreary. (For portraits of Boyle, see figure 16.)

When the leading lights of the Royal Society censured Hobbes's dogmatism, they tended to conjoin comments on his personal qualities with judgments upon his philosophical programme. Neither could be tolerated; both personal and programmatic dogmatism were anathema to the practice of experimental philosophy. Sprat's attack on the "Modern Dogmatists," while characteristically not mentioning Hobbes by name, could only have been composed with Hobbes in view. Sprat made his point with a telling analogy between philosophy and politics. Having rejected the dogmatic "tyranny" of the ancients, modern dogmatists directly

> . . . fell to form and impose new Theories on Mens Reason, with an usurpation, as great as that of the others: An action, which we that live in this Age, may resemble to some things that we have seen acted on the Stage of the World: For we also have beheld the Pretenders to publick Liberty, turn the greatest *Tyrants* themselves.

"Methinks," Sprat said, "there is an agreement, between the growth of *Learning*, and of *Civil Government*." Tyranny in each was to be combatted, and there was no reason to prefer the tyranny of any modern philosopher to that of the ancients. No man could rightly claim to have produced a complete and satisfactory system of philosophy. All such dogmatic systems contained the seeds of dissent and, therefore, of their own destruction:

> It is probable, that he, who first discover'd, that all things were order'd in *Nature* by *Motion* [i.e. Hobbes]; went upon a better ground then any before him. But now if he will onely manage this, by nicely disputing about the Nature, and Causes of *Motion* in general; and not prosecute it through all particular Bodies: to what will he at last arrive, but onely to a better sort of *Metaphysicks*? And it may be, his Followers, some Ages hence, will divide his Doctrine into as many distinctions, as the *Schole-men* did that of *Matter*, and *Form*: and so the whole life of it, will also vanish away, into air, and words, as that of theirs has already done.

Sprat discerned a causal connection between philosophical dogmatism and the social relationships this engendered. Dogmatism inclined men to become "imperious," to be unshakable in their convictions, and to be "impatient of contradiction." It produced egotism and individualism, which is a "Temper of mind, of all others the most pernicious." It was pernicious because it disrupted the social relationships which could alone produce and sustain factual natural knowledge. By contrast, the experimental philosophers of the Royal Society were "modest, humble, and friendly"; they were tolerant of differing opinions and worked collectively towards attainable and solid goals.[78]

We now have an answer to the question, "Why was Hobbes excluded from the Royal Society?" It is an answer that does not attempt to distinguish assessments of Hobbes's personality from judgments of his philosophical programme. The connections among personal characteristics, social relationships, and philosophical practices were perceived, as Sprat's polemic shows, to be substantial and vital. The modest and humble Boyle was juxtaposed to the intolerant and confident Hobbes, just as the modest and humble experimental programme was contrasted to Hobbes's overweening rationalism. Each philosophical programme was predicated upon its distinctive social relationships, and each valued a characteristic philosophical persona. The social order implicated in the rationalistic production of knowledge threatened that involved in the Royal Society's experimentalism. Thus our excursion into Hobbes's relations with the Royal Society is not, in fact, peripheral to our major concern with conflicting knowledge-generating strategies. Hobbes's anti-experimentalism, as expressed in the *Dialogus physicus* and elsewhere, gave grounds for his exclusion.[79]

Experiments and Causes

In chapter 2 we discussed Boyle's proposal to erect a procedural boundary between speech of matters of fact (as experimentally produced and manifested) and speech of the physical causes of these facts. In this practice one recognized that God might produce the same effect by a number of different causes, and one professed the appropriately nescient attitude towards the search for real

[78] Sprat, *History*, pp. 28-34. Paul Wood differs from earlier writers on the reliability of Sprat's *History* as an "official" and authoritative account of the Society's activities, while crediting it as sanctioned apologetic: "Methodology and Apologetics." For similar views of dogmatism, see Glanvill, *Vanity of Dogmatizing* (1661); idem, *Scepsis scientifica* (1665).

[79] We therefore find ourselves in some agreement with writers less recent than Skinner and Hunter. However, we do not equate anti-experimentalism with "anti-science," nor need we accept that Hobbes "did not understand" or "did not appreciate" what experimentalism entailed: see Laird, *Hobbes*, p. 24; de Beer, "Some Letters of Hobbes," p. 197; R. F. Jones, *Ancients and Moderns*, p. 128; Bredvold, "Dryden, Hobbes, and the Royal Society," pp. 424-425.

causes.[80] Such real causes *might* ultimately be unveiled by experimental philosophy, but it was wisest and safest to treat causal inquiry with modest caution. Knowledge of causes was at best conjectural and the quest for real causes ought to be carefully segregated from the factual enterprise that laid down the foundations of experimental philosophy. It was this boundary, and the epistemological hierarchy it manifested between knowledge of effects and knowledge of causes, that Hobbes attacked as unphilosophical. In Hobbes's view, in order to be counted as philosophy an intellectual practice could not affect nescience concerning the causes of things. Indeed, philosophy could proceed *from* correct knowledge of causes to knowledge of effects. These programmatic differences between the enterprises proposed by Boyle and Hobbes were concretely expressed in the *Dialogus physicus*.

Hobbes's experimentalist interlocutor had provided an account of the spring of the air and of how this spring might be conceived in terms of wool fleece. Hobbes interrupted: ". . . I ask you, is this not the rule for all hypotheses, that all things which are supposed must be of a possible, that is, conceivable, nature?" The experimentalist concurred and suggested that the elastical hypothesis was supported by the restorative powers "seen in many things" which, therefore, "can very easily be conceived to be in air." Hobbes was not content with this response:

> It is for a philosopher to find the true or at least very probable causes of such things. How could compressed wool or steel plates or atoms of air give your experimental philosophers the cause of restitution? Or do you offer a likely cause why in a crossbow [*ballista*] the steel plate regains its usual straightness so swiftly?

The experimentalist came back with the affectation of causal nescience that Boyle enjoined: "I cannot give a very certain cause for this thing." Hobbes's insistence upon a causal enterprise troubled the experimentalist. Where do causal questions end? "Whatever true cause I told you, you would not then acquiesce to its truth, but would ask me further what was then the cause of this cause, whence it would go on to infinity." This Hobbes flatly denied; the identification of proper causes ends the inquiry: ". . . when you will

[80] Among Boyle's many statements of this type, see "Usefulness of Experimental Natural Philosophy," pp. 45-46; cf. Bechler, "Newton's Optical Controversies," pp. 132-134.

have come to some external cause, there I will leave off asking you."[81] If the profession of causal nescience was acceptable to, even celebrated by, experimentalists, to Hobbes it constituted a damning admission that the experimental programme was not philosophical. Causal inquiry *could* be concluded; it did not breed dissent but could provide the surest remedy for dissent.

Hobbes relentlessly pursued the question of the spring of the air as a causal physical explanation. Either the air's elasticity was offered as a causal account or it was not. If not, then one learned nothing of causes from the experimental programme, and the whole enterprise was truly vacuous. Moreover, even if the air's spring was advanced as a cause, then, as Hobbes endeavoured to show, what resulted was an absurdity. Hobbes argued that this concept, as Boyle used it, was fundamentally anti-mechanical, proceeding from an impossible notion of body. Philosophers "make a legitimate hypothesis from two things: of which the first is, that it be conceivable, that is, not absurd; the other, that by conceding it, the necessity of the phenomenon may be inferred." Yet the hypothesis of the spring of the air *was* absurd, "unless perhaps we concede what is not to be conceded, that something can be moved by itself. For you suppose that the air particle, which certainly stays still when pressed, is moved to its own restitution, assigning no cause for such a motion, except that particle itself."[82]

No argument against Boyle's position could have been, if accepted, more devastating. Boyle advertised his mechanical philosophy as the best way to undermine the "vulgar" and dangerous conception of self-moving matter.[83] Now, Hobbes argued that a true and self-consistent mechanism was obliged to specify the material and mechanical cause of the air's elasticity:

> For if spring were allowed by them [the Greshamites] to be something in the threads of the air, and they were to search for something by which, when somewhat curved yet at rest, the threads would be moved again to straightness: if they wish to be taken for physicists, they would have to assign some possible cause for it.[84]

[81] Hobbes, "Dialogus physicus," pp. 247-249.

[82] Ibid., pp. 254-255.

[83] On this point, see J. Jacob, *Boyle*, chap. 3; idem, "Boyle's Atomism," where the location of vulgar hylozoism among the radical sectaries is discussed; idem, *Stubbe*, chap. 3. Boyle's response to this charge is examined in our next chapter.

[84] Hobbes, "Dialogus physicus," p. 271.

And, if the experimentalists declined to assign such a cause, then how were they different from those Peripatetics and others who invoked "metaphorical terms such as *fuga vacui, horror naturae,* etc.?"[85] As it was, Hobbes was content that the experimentalists should condemn themselves out of their own mouths by admitting their ignorance of causes. He asked: "But what cause . . . do they assign?" Answer: "As yet, none, but they seek for it with this experiment itself."[86] And later: "So you admit there to be nothing yet from your colleagues for the advancement of the science of natural causes, except that one of them has found a machine" whose operations render the causal hypotheses of Hobbes even "more probable." Experimentalist: "Nor is it shameful to admit it; for it is something to advance so far, if nothing further is allowed." Hobbes attempted to force home the appropriate conclusion:

> Why *so far*? Why such apparatus and the expense of machines of difficult manufacture, just so as you could get as far as Hobbes had already progressed? Why did you not rather begin from where he left off? Why did you not use the principles he established? Since Aristotle had rightly said that *to be ignorant of motion is to be ignorant of nature*, how did you dare to take such a burden upon yourselves, and to arouse in very learned men, not only of our country but also abroad, the expectation of advancing physics, when you have not yet established the doctrine of universal and abstract motion (which was easy and mathematical)?[87]

This is why, Hobbes said, the experimental "philosophy," not being a science grounded in causal knowledge, was no better than the lore acquired by people who were mere mechanics:

> If indeed philosophy were (as it is) the science of causes, in what way did they have more philosophy, who discovered machines useful for experiments, not knowing the causes of the experiments, than this man who, not knowing the causes, designed machines? For there is no difference, except that the one who does not know acknowledges that he does not know, and the others do not so acknowledge.[88]

[85] Ibid., p. 276.
[86] Ibid., p. 261; cf. pp. 277, 287.
[87] Ibid., p. 273; also p. 236.
[88] Ibid., p. 278. Compare Wilkins, *Mathematical Magick* (1648), p. 8, which, twelve years before Boyle's *New Experiments*, identified "philosophy" with "that discipline which discovers the generall causes, effects and properties of things."

Much of the *Dialogus physicus* was, in fact, devoted to displaying this equivalence between low artisans and experimental philosophers who were ignorant of causes. Hobbes attacked the celebrated experiment of weighing the air in a bladder suspended on a balance in an exhausted receiver. Boyle could be satisfied that the bladder was indeed depressed on exsuction, but not that it therefore weighed more, since he possessed no knowledge of the efficient cause of gravity. Similarly, Hobbes pointed out that Boyle offered no cause of spring, and compared him to those who ask how many times a bell has rung, "though they have not heard the first stroke."[89] At most, what the experimentalists achieved was the enrichment of "natural history": they "make the phenomena visible."[90] These purposes were not to be despised, but they were not the objectives of the philosopher. Philosophy obliges men to give assent; natural history carries with it no such obligation.

Hobbes's Literary Technology

The scope and character of the assent at which Hobbes's philosophy was aimed is evident in his views on philosophical method, briefly sketched in the preceding chapter. More concretely, it is visible in Hobbes's literary practices. The contrast between the literary forms employed by Boyle and Hobbes is instructive. For example, both used the dialogue form in natural philosophy. However, there is a telling difference between Boyle's usage in the dialogues that constitute *Sceptical Chymist* and Hobbes's literary practices in the natural philosophical dialogues of the *Dialogus physicus, Problemata physica* and the *Decameron physiologicum*. We noted in chapter 2 that the dialogues of *Sceptical Chymist* had the character of a conversation among four participants, so structured that consensus was displayed as emerging through the conversation. Each participant was given something to contribute to the dénouement, which was seen to depend upon the free exchange of factual information. Hobbes's natural philosophical dialogues were in the traditional Socratic mould. There were only two participants: one unambiguously represented Hobbes and the other personated an antagonist (a vacuist, an experimentalist, an inductivist). Truth did not emerge through the exchanges between Hobbes and his interlocutor, for it was

[89] Hobbes, "Dialogus physicus," pp. 261, 271.
[90] Hobbes, "Mathematicae hodiernae," p. 228.

already fully contained in Hobbes's philosophy. Knowledge was portrayed as flowing from Hobbes to interlocutor, who mainly played the role of recipient.

Nevertheless, interlocutor's role was far from negligible. His participation was necessary for the literary exemplification of the conception of knowledge and its social transmission that Hobbes recommended for philosophy.[91] Interlocutor may pose a simple question or express perplexity, to which Hobbes offers satisfying solutions. He may make a statement about physical processes or align himself with certain positions which Hobbes reveals to be fallacious. Interlocutor's statement may be countered with a question from Hobbes, requesting definitions. Interlocutor may then admit that he has no adequate definition for his usage, and Hobbes may then supply the reasons why interlocutor's statement is founded upon absurd speech. Or, interlocutor may give a reply to Hobbes's query, in which Hobbes discerns a flawed logical process: "It is no good argument." Interlocutor can point out possible incompleteness in Hobbes's claims: "These assertions need demonstration," which demonstration Hobbes supplies. Interlocutor then manifests his contentment with Hobbes's proof and gives his assent: "It is very probable"; "It is like enough to be so"; "No doubt"; "It is true." As the dialogue proceeds, interlocutor ceases to represent the adversary and becomes a possible convert: "I am a narrator of other philosophies to you, not a defender." Towards the end of the dialogue, conversion is total: ". . . I agree with and approve of everything you have said." But one step remains to be taken on this philosophical road to Damascus: having given his assent, interlocutor can now act as "philosopher" himself; he can put Hobbes right and thus display the power of right method to command assent even from a master. Thus, the last line of the *Dialogus physicus*

[91] It would be valuable to have a detailed study of the uses and career of the dialogue form in natural philosophy during the sixteenth and seventeenth centuries. For some interesting remarks, see Multhauf, "Some Nonexistent Chemists"; Beaujot and Mortureux, "Genèse et fonctionnement du discours"; Hannaway, *The Chemists and the Word*; Christie and Golinski, "The Spreading of the Word," esp. pp. 238-246. Literary historians have treated the dialogue form systematically, but have had little to say about its scientific uses: Hirzel, *Der Dialog*, does not mention Boyle's dialogues and dismisses Hobbes's *Dialogus physicus* and *Problemata physica* from consideration (vol. II, p. 399n). Merrill, *The Dialogue in English Literature*, chap. 5, treats "The Philosophical Dialogue" but concentrates on Shaftesbury and Berkeley. For perceptive comments on Galileo's dialogues, see Feyerabend, *Against Method*, chap. 7.

belongs to Hobbes, and it is "I judge the same. I have erred: and you have rightly corrected my error."[92]

In these ways the Hobbesian dialogue dramatized the power of philosophical method to secure complete assent. Men may err, but the force of proper method consists in its capacity to put men right, surely and swiftly, when the nature of their error is pointed out to them. Just as philosophical knowledge is produced using the tools of logic, so it is transmitted logically and syllogistically, and this transmission is effective. Thus, in *De corpore* Hobbes described the method by which men *make* knowledge, the "method of invention," and then showed its relationship to the method by which we *demonstrate* to others:

> And seeing teaching is nothing but leading the mind of him we teach, to the knowledge of our inventions, in that track by which we attained the same with our own mind; therefore, the same method that served for our invention, will serve also for demonstration to others. . . . [This method] proceeds by a perpetual composition of propositions into syllogisms, till at last the learner understand the truth of the conclusion sought after.[93]

No man, Hobbes maintained, can continue in error in the face of proper method. The dialogues display the overpowering force of method to compel assent and to correct error. In Hobbes's dialogues, it is method, not matters of fact, that puts men right and that mobilizes consensus. When empirical evidence, whether from observation or from experiment, is given a role in the dialogues, it serves to *illustrate* the conclusions reached by method, and not to determine belief. Thus, in both Boyle's and Hobbes's writings, literary structure and process dramatize the social relations and practices deemed appropriate to the production of knowledge. Differences in theories of knowledge-production and evaluation are displayed in different literary technologies.

Similar considerations inform the character of Hobbes's philo-

[92] In his reply to the dialogues of Hobbes's *Examinatio et emendatio mathematicae hodiernae*, Wallis noted the structural roles of A and B (by which letters Hobbes designated his two participants). They were, Wallis said, "Thomas" and "Hobs," and "when Hobs hath occasion to assume what he cannot prove, Thomas, by a *Manifestum est* saves him the trouble of attempting a demonstration." Wallis, *Hobbius heautontimorumenos*, pp. 15, 103; cf. Laird, *Hobbes*, p. 38. See also the hilariously acute account of the scientific dialogue in Ellis, *So This is Science!*, pp. 45-46.

[93] Hobbes, "Concerning Body," pp. 80-81; cf. p. 87.

sophical iconography. The figures supplied to illustrate Hobbes's natural philosophical works were almost entirely geometrical in character, depicting abstract geometrical treatments of physical processes (see, for example, the diagram representing gravitation in our translation of the *Dialogus physicus*). Only very rarely did Hobbes include a pictorial representation of a physical phenomenon or process, and even more rarely did he offer a representation of an experimental system. One example appears in the *Problemata physica*: it is an engraving of the basic Torricellian experiment, but it depicts the apparatus in minimal detail, with no obvious effort at showing the particularities of any specific experimental apparatus.[94] Unlike Boyle, Hobbes never used the engraver's art so as to offer the viewer a virtual sensory experience of an experimental scene. We have already noted that Hobbes endeavoured to describe the air-pump in the *Dialogus physicus* without the aid of a picture. The mind, evidently, was not thought to require the assistance of the eye, much less of the hand. In such ways, Hobbes's philosophical iconography expressed the relative evaluations he placed upon logical and geometrical methods, on the one side, and manipulations of experimental systems, on the other. His iconographic preferences and usages were noticed by his enemies. In 1662 the geometer John Wallis wrote that

> . . . I cannot but observe, in the general, a great Resemblance between this his *Physical Hypothesis* and his *Geometrical Conclusions*. For as in these he draws a Multitude of Lines whereof there is no Use made, as to the Construction or Demonstration of his Problem, . . . so much of his Hypotheses is to no purpose as to the Effects of Nature.[95]

Cause, Convention, and Certainty in Hobbes's Philosophy

Hobbes rejected Boyle's experimental programme because he considered that it was not *philosophy*. And, because it was not philosophy, it could not generate the kind of certainty appropriate to

[94] Many copies of the Molesworth edition are missing their set of engravings, which should be bound in at the back; see also the 1668 Amsterdam edition of Hobbes, *Opera philosophica*, chap. 1, p. 19. For other engravings of experimental apparatus in Hobbes's works, see ibid., p. 27, and *De corpore*, in ibid., chap. 3, cap. xxvi, fig. 2; *White's De Mundo Examined*, p. 501, fig. 3.

[95] Wallis, *Hobbius heauton-timorumenos*, pp. 156-157.

philosophical inquiries. Where, then, did certainty reside? How was philosophical knowledge to be founded and how did method contribute to the quest for certainty? As Hobbes and Boyle proffered radically different solutions to the problem of knowledge, it is interesting to start from a position that both adopted.

Boyle reckoned that God could produce the same effect by a number of different natural causes, and on this basis he recommended methodological caution and even nescience about the ability of natural philosophers to unveil real causes. Laudan has ascribed this position, and especially Boyle's use of the clock metaphor to express it, to the influence of Descartes' methodological writings. In the *Principia* of 1644 Descartes had described two watches which told time equally well but whose internal workings were quite different. God might have ordered the world-clock in any number of different ways so as to produce the effects which we observe. Since the world of corpuscles is inaccessible to our senses, the best we can do is to produce "hypotheses" about how the world might be put together. Descartes concluded: "And I believe I shall have done enough if the causes that I have listed are such that the effects they may produce are similar to those we see in the world, without being informed whether there are other ways in which they are produced."[96] Whether or not Descartes' views "influenced" Boyle is not our concern here. However, we want to underline the conclusions that Boyle drew from this position: causal inquiry was to be tactically segregated from the main tasks of the natural philosopher; hypotheses about causes were conjectural and should be regarded as distal to fact production.

As Laudan briefly notes, Hobbes adopted an apparently identical posititon. For example, in the *Problemata physica* of 1662 Hobbes said that "The doctrine of natural causes hath not infallible and evident principles. For there is no effect which the power of God cannot produce by many several ways."[97] And earlier, in his *Six Lessons* to the "egregious professors of mathematics," Hobbes emphasized that in natural philosophy "there lies no demonstration of what the causes be we seek for, but only of what they may be."[98] Again, our concern is not the source of Hobbes's position; Laudan's suggestion about a Cartesian influence seems plausible enough, although similar views can be found in the writings of Zabarella,

[96] Quoted in Laudan, "The Clock Metaphor and Probabilism," pp. 77-78, 92-93.
[97] Hobbes, "Seven Philosophical Problems," p. 3.
[98] Hobbes, "Six Lessons," p. 184.

Galileo and others of the "Padua School," which Hobbes much admired.[99] What methodological consequences did Hobbes derive from the claim that God might produce the same effects by many different causes? In stark contrast to Boyle, Hobbes did *not* move from the admission that our knowledge of natural causes was conjectural to the tactic of bracketing off causal inquiry from the foundations of natural philosophy. For Hobbes causal statements ought to form one of the bases and starting points of any philosophical enterprise whatever.

In *De corpore* Hobbes offered two definitions of "philosophy," or, rather, he presented the philosophical enterprise in two aspects. "Philosophy," he said, "is such knowledge of effects or appearances, as we acquire by true ratiocination from the knowledge we have first of their causes or generation: And again, of such causes or generations as may be from knowing first their effects."[100] There are, then, two ways of proceeding in philosophy: the first (which Hobbes called "synthetical") goes from known causes to effects; the second (the "analytical method") goes from "sense" to the construction of causal principles.[101] Elsewhere in *De corpore*, the probabilistic character of the enterprise is emphasized: "Philosophy is the knowledge we acquire, by true ratiocination, of appearances, or apparent effects, from the knowledge we have of *some possible* production or generation of the same; and of such production, *as has been or may be*, from the knowledge we have of the effects."[102] As James rightly observes, Hobbes did not place equal value upon the synthetical and analytical dimensions of philosophy.[103] In the synthetical method men agree about the definitional foundations or principles of things and then demonstrate how effects necessarily follow. In the analytical method we note certain effects and cast about for a conceivable cause such that these effects would have occurred. The

[99] Laudan, "The Clock Metaphor and Probabilism," p. 95n; but see Watkins, *Hobbes's System*, pp. 54-59, for Hobbes and the Padua School. For similarities between Hobbesian demonstration and medieval and Renaissance demonstrative methods, see Gaukroger, *Explanatory Structures*, pp. 166-170 (for Zabarella); Hacking, *The Emergence of Probability*, chap. 3 (for Aquinas); and Schmitt, "Towards a Reassessment of Renaissance Aristotelianism."

[100] Hobbes, "Concerning Body," p. 3.

[101] Ibid., pp. 74-75. For general surveys of Hobbes's philosophical method: Watkins, *Hobbes's System*, chaps. 3-4; James, *The Life of Reason*, chap. 1; Brandt, *Hobbes' Mechanical Conception*; Madden, "Hobbes and the Rationalistic Ideal"; von Leyden, *Seventeenth-Century Metaphysics*, pp. 38-41.

[102] Hobbes, "Concerning Body," pp. 65-66; cf. pp. 387-388 (emphases added).

[103] James, *The Life of Reason*, p. 14.

cause we come up with may or may not be the real cause of these particular effects.

The distinction between these two philosophical practices highlights the differential capacity of various branches of knowledge to offer secure causal accounts. In Hobbes's view, geometry is a paradigm of both certain and *causal* knowledge. It appears strange to speak of geometry as causal, but Hobbes did so for a very important reason: men *make* both geometrical definitions and geometrical objects.[104] It is not difficult for us to see that the axiomatic and definitional foundations of geometry are human constructs, but it is less apparent how it can be said that geometrical figures are man-made. Do they not simply exist, as essences or forms, if not as physical entities, outside ourselves? Hobbes defined a line as "made by the motion of a point, superfices by the motion of a line."[105] But what moves the point is a human hand. According to Hobbes, geometry is demonstrable, "for the lines and figures from which we reason are drawn and described by ourselves."[106] But what of *space*? Is not space the object of geometry, and is it not absurd to suggest that space is a construction? Hobbes rejected the view that space, unlike matter and motion, is a real existent. Imagine, he said, that the entire world of things were annihilated. Imagine further that one man alone survived this universal destruction:

> If therefore we remember, or have a phantasm of any thing that was in the world before the supposed annihilation of the same; and consider, not that the thing was such or such, but only that it had a being without the mind, we have presently a conception of that we call *space*: an imaginary space indeed, because a mere phantasm, yet that very thing which all men call so . . . I . . . define *space* thus: *Space is the phantasm of a thing existing without the mind simply*. . . .[107]

Therefore, the idea of space itself, the very substratum of geometrical objects, is man-made.

[104] Sacksteder, "Hobbes: Man the Maker"; idem, "The Artifice Designing Science in Hobbes."

[105] Hobbes, "Concerning Body," p. 70.

[106] Hobbes, "Six Lessons," p. 184.

[107] Hobbes, "Concerning Body," pp. 93-94; see also James, *The Life of Reason*, pp. 16-17; Madden, "Hobbes and the Rationalistic Ideal," pp. 113-114 (although Madden is mainly concerned to criticize Hobbes's "queer conception of space"); Sacksteder, "Hobbes: Geometrical Objects"; idem, "Hobbes: Teaching Philosophy to Speak English," pp. 42-43 (on "phantasm").

Here we come up against an apparent paradox and an obstacle to our understanding of what was involved in the different forms of life proposed by Hobbes and Boyle. In our culture, saying that knowledge is artifactual and conventional is tantamount to saying that it is not authentic knowledge at all. This general disposition accounts for the fact that academic exercises concerned to uncover and display the conventional bases of knowledge, such as Wittgenstein's, are dealt with as if they were attempts at exposé or disparagement. In everyday life, we ourselves diminish knowledge-claims by showing their constructed nature or their conventional bases. Such practices make sense within a particular game. And that game, as we have shown in chapter 2, is one in which knowledge is, so to speak, ultimately vouched for not by human agency (individual or collective) but by reality itself. Man is not a maker but a mirror. Yet, within other language-games, the situation might be quite different. It might be that knowledge is taken as secure insofar as it is seen as constructed in a certain way and uses certain conventions. This, we argue, is the case with Hobbes: certainty was a function of convention. Here is where Hobbes as rationalist and Hobbes as conventionalist come together. This point is perhaps most evident in Hobbes's treatment of the certainty to be expected from both geometry and from *civic philosophy*. Having said that geometry is demonstrable because geometrical figures "are drawn and described by ourselves," Hobbes then claimed that "civic philosophy is demonstrable because we make the commonwealth ourselves."[108] This goes against all the intuitions of the empiricist. Hobbes was saying that one can only completely explain or understand that which one makes; the empiricist regards the man-made component of knowledge as a distortion of the mind's mirroring of reality.[109]

What, then, is the status of natural philosophy in the hierarchy of certainty? Natural philosophy, in Hobbes's view, cannot command the sort of certainty which is the prerogative of geometry and civic philosophy. This is because the causes of natural effects are not of our own construction, but must be sought from the effects themselves. As Hobbes said, "In natural causes all you are to expect, is but probability."[110] Nevertheless, the search for possible causes

[108] Hobbes, "Six Lessons," p. 184; Sacksteder, "Hobbes: The Art of the Geometricians," p. 146.

[109] On Hobbes and "the genetic principle," see von Leyden, *Seventeenth-Century Metaphysics*, pp. 39-40; James, *The Life of Reason*, pp. 25-26; on empiricism as the mirroring of reality, see Rorty, *Philosophy and the Mirror of Nature*, chap. 3.

[110] Hobbes, "Seven Philosophical Problems," p. 11; cf. p. 3: "The doctrine of

in natural philosophy is not trivial, nor is its outcome, when properly pursued, without its variety of compulsion. If the causes we adduce are not, after all, God's causes, still they should be such causes as satisfy our reason. They will be causes which *may be* fictions, but which, if granted, will show how the effects *necessarily* follow. As Hobbes said in the *Tractatus opticus*: "More is not therefore demanded of physics, than that what is supposed or feigned of motion should be imaginable, and through the conceding of these things the necessity of the phenomena should be demonstrated, and finally that nothing false can be derived from them."[111] Or, as James concisely summarizes,

> . . . it is enough for mortal ratiocination to know what might intelligibly have caused [observed effects], what may be *conceived* as a cause which would, inevitably, have *these* results. Thus Hobbes is less concerned to discover the methods of God in creation than to satisfy the rational requirements of the human mind; and whether in fact we know the real cause or another but equally intelligible cause, is immaterial to the results for human happiness. It is less reality than intelligibility that Hobbes looks for.[112]

In fact, knowledge of God's causes in creation is ruled out on the same grounds that ruled *theology* out of the philosophical enterprise: man can have no certain knowledge of God, "eternal, ingenerable, incomprehensible."[113] Thus, for Hobbes, the task of the natural philosopher was to approach as near as he could to the products of the geometer and the civic philosopher; he could not equal them in the generation of certainty but, through the application of correct method, he could do better than the Scholastic or the experimentalist. Scholastics misfired by founding their philosophy upon absurd speeches and impossible ontologies; experimentalists failed through confusing natural history with natural philosophy. A catalogue of facts, separated from causal inquiry and unstructured by correct method was *pointless*: "We cannot from experience conclude . . . any proposition *universal* whatsoever."[114]

Hobbes's attack on Boyle's programme stemmed from his con-

natural causes hath not infallible and evident principles." For discussion of Hobbes's idea of "probability": Hacking, *The Emergence of Probability*, esp. pp. 47-48.

[111] Alessio, "Thomas Hobbes: *Tractatus opticus*," p. 147.

[112] James, *The Life of Reason*, p. 13.

[113] Hobbes, "Concerning Body," p. 10.

[114] Hobbes, "Human Nature," p. 18.

tention that experimental procedures lacked the compulsory force of true philosophy. Both Hobbes's and Boyle's programmes were equally concerned with the problem of assent, but their solutions were radically different. In Boyle's view assent was to be secured through the production of experimental findings, mobilized into matters of fact through collective witness. The individual agreed with other individuals about what he had witnessed and believed. The programme was, therefore, founded upon collectivized individual sensory experience. Dissent was to be managed by bracketing off from natural philosophy those items, such as metaphysics, that were not so founded. Boyle's compulsion was only partial; there was room to differ and tolerance was essential to the maintenance of this partial and liberal compulsion. Managed dissent within the moral community of experimentalists was safe. Uncontrollable divisiveness and civil war followed from any other course.

For Hobbes civil war flowed from any programme which failed to ensure absolute compulsion. What was a judicious and liberal bracketing strategy to the Greshamites was, to Hobbes, a wedge opening the door which looked out on the war of each against each. Any working solution to the problem of knowledge was a solution to the problem of order. That solution had to be absolute. Hobbes therefore sought to bypass the individual, his unreliable sensory experience, and the category of individual belief. Where, then, did Hobbesian compulsion reside? Hobbes found his solution not in belief nor in witness but in behaviour, not in the individual but in the social. When he said that men make the commonwealth, he did not mean that *some* men did. All men make and sustain society, because all men that have natural reason can be made to see that it is in their interests that Leviathan be created and maintained. Having made civil society to protect themselves, the obligation to submit is total. The force by which submission is exacted is the delegated force of all those that enter into society and live as social beings. The intellectual enterprise which rationally demonstrates this to all men possesses an absolutely compulsory character. It is in philosophy what Leviathan is in society. Men can truly understand that which they make. And it is precisely the same in geometry. Men make the definitions, the figures and the spatial substratum of geometry, and they can be shown that they have done so. Geometry and civic philosophy are, then, on a par. But what of the force of logic?

It would be a mistake to take Hobbes to be saying that logic, or the laws of inference, compel of themselves. For the force of logic is exactly the same as the force by which Leviathan secures submission: it is the delegated force of society, working on the natural reasoning capacities of all men. In an historical treatment of the authority of kings to pronounce upon religious doctrine and to reject the interpretations of clerical exegetical experts, Hobbes made an illuminating comparison with the prerogatives of kings in geometry:

> And although priests were better instructed in nature and arts than other men, yet kings are able enough to appoint such interpreters under them; and so, though kings did not themselves interpret the word of God, yet the office of interpreting them might depend on their authority. And they who therefore refuse to yield up this authority to kings, because they cannot practice the office itself, do as much as if they should say, that the authority of teaching *geometry* must not depend upon kings, expect they themselves were geometricians.[115]

The force by which Leviathan lays down and executes the laws of the commonwealth is therefore the same force that lies behind geometrical inferences.

Finally, we can reflect upon the relations among the various bodies of intellectual culture that Boyle's and Hobbes's different strategies of assent imply. For Boyle and the Royal Society there was to be a strict boundary between natural philosophy and political discussion. This boundary manifested their evaluation of the capacity of each to secure consensus and assent. Through the matter of fact, experimental natural philosophy could mobilize effective consensus. By contrast, civic philosophy might sow the seeds of division, which would, inevitably, infect the practice of natural philosophy. However, the relationship stipulated between natural philosophy and theology was more problematic. On the one hand, theological discussions had a tendency to divide and to corrode and should not, as Sprat and others said, be meddled with. On the other hand, the practice of natural philosophy was to be subservient to the higher truths of proper Christian religion. One could and one ought to go "from Nature up to Nature's God." For Boyle, as for Newton in a later period, to discourse of God did properly belong to natural philosophy. For Hobbes, the relations were quite

[115] Hobbes, "Philosophical Rudiments," p. 247. (This is the 1651 English translation of *De cive* of 1642.) Note that Hobbes was geometry tutor to the future Charles II.

different. Civic philosophy and geometry belonged together in their assent-producing capacity. Natural philosophy, insofar as it could imitate their methods, was part of the same discourse. But theology must be set apart, for we cannot know what is unknowable and we must take for doctrine what Leviathan lays down. Leviathan's truth and the truth of the air-pump are products of different forms of social life.

· V ·

Boyle's Adversaries: Experiment Defended

Longvil:	*But to what end do you weigh this Air, Sir?*
Sir Nicholas Gimcrack:	*To what end shou'd I? To know what it weighs. O knowledge is a fine thing.*

THOMAS SHADWELL, The Virtuoso *(1676)*

WHO were Boyle's adversaries? This appears to be a straightforward question. Within three years of the publication of the *New Experiments* Boyle was confronted with critical replies on three main fronts. 1661 saw the appearance not only of Hobbes's *Dialogus physicus* but also of a hostile treatise by the Jesuit Franciscus Linus entitled *Tractatus de corporum inseparabilitate*. In the following year the Cambridge Platonist Henry More joined the attack with remarks in the third edition of his *Antidote against Atheism*, amplified in a number of tracts over the next fifteen years.

Boyle replied to each of these critics, but he did so in different ways in order to make different points and to defend different aspects of his findings and programme. In the preceding chapter we offered a detailed assessment of Hobbes's criticisms. We do not intend to follow the same plan with respect to the views of Linus and More. Our focus in this chapter is on Boyle and on the concept of the *adversary* as seen by *him*. Whom did Boyle identify as his adversaries? What was it about their various criticisms that he wished particularly to counter? What aspects of his own conceptual repertoire and enterprise did he want especially to defend? And, in his responses to his critics, what were the rules of engagement that Boyle implemented? Thus we are concerned here with Boyle's *stipulations* about his adversaries' positions and we will consider their remarks mainly insofar as he defended himself against them.[1]

Given our concerns in this book, we do not need to give equal weight or emphasis to each of Boyle's adversaries and to his dealings with them. Hobbes, Linus, and More attacked different aspects of

[1] We do not offer these accounts of the exchanges between Boyle and Linus and More as exhaustive or definitive; clearly, much more detailed attention is warranted. We especially stress the need for "charitable interpretations" of More's and Linus's points of view, as our treatment of Hobbes in chapter 4 ought to make clear.

Boyle's work. To simplify: Hobbes and More attacked crucial aspects of the experimental programme as such, while Linus did not. Each of them opposed a number of Boyle's findings and interpretations, including the explanatory status of the "spring of the air." Only Hobbes called into question the politics of experimental knowledge-production. More challenged the uses Boyle defined for experimental knowledge once it had been produced. Since this book is concerned with the contest over experimentalism, we concentrate our attention upon Boyle's reaction to those who voiced doubt about the status of experimental knowledge. We can also use this study of Boyle's reactions to his adversaries for a supplementary purpose. Historians have paid relatively little attention to Boyle's experimental researches after 1660.[2] By showing how seriously Boyle took his adversaries, and particularly their identifications of experimental troubles, we shall demonstrate that much of Boyle's *experimental* work during the decade following the *New Experiments* was undertaken to address these criticisms. In chapter 6, we go on to show how Boyle's technical adjustments to the air-pump itself were informed by a similar concern to defend the integrity of the machine he had used to guarantee experimental knowledge.

This chapter is divided into three parts. In the first we review the nature of the objections offered to *New Experiments* by Linus, and the way in which Boyle countered them; in the second part we give a more detailed account of Boyle's responses to Hobbes, and in the third we examine the exchanges between Boyle and More.

Linus's Funicular Hypothesis

Linus had been professor of mathematics at the English College of Liège. From 1658 he was in London. His *Tractatus de corporum inseparabilitate* of 1661 was the work of a committed Aristotelian.[3]

[2] Contrast Frank, *Harvey and the Oxford Physiologists*, esp. chaps. 5-6, with M. B. Hall, *Boyle and Seventeenth-Century Chemistry*. We intentionally omit historical work on Boyle's Law (e.g., Agassi, "Who Discovered Boyle's Law?") because this was not a product of the air-pump.

[3] Linus, *Tractatus de corporum inseparabilitate; in quo experimenta de vacuo, tam Torricelliana, quàm Magdeburgica, & Boyliana, examinantur . . .* (London, 1661). For an account of Linus's life, see Reilly, *Francis Line*. Linus dedicated the *Tractatus* to Henry Pierrepont, Marquis of Dorchester, a royalist scholar to whom Hobbes dedicated his *Six Lessons* in 1656. For Jesuit natural philosophers and their reaction to experimental and mechanical philosophy in the seventeenth century, see Renaldo, "Ba-

Like Hobbes, Linus stipulated that Boyle was a vacuist, and wrote against him in support of the plenum. Unlike Hobbes, the Jesuit offered a nonmechanical explanation of why a vacuum was an impossibility: "*Naturam à vacuo abhorrere*."[4] Linus's arguments in defence of plenism derived partly from Aristotelian axioms and partly from experiment. Linus said there was no vacuum in the Torricellian space. This was apparent because one could see through that space; if there were a vacuum, "no visible species could proceed either from it, or through it, unto the eye," and the Torricellian space would appear like a little black pillar.[5] Linus listed three possible valid interpretations of the Torricellian space: that of Hobbes (it was full of "common air"), of the Jesuit Noël (the "more subtle parts [of air] which he calls aether, which enters through the pores of the glass"), and of the Jesuit Zucchi (it was full of mercurial spirit). The fourth theory was that of the majority of natural philosophers, including Gassendi, Magnani, Pecquet, Ward, and Charleton, who were vacuists.[6]

Importantly, Linus's solution to the problem of the Torricellian space performed two functions: it filled the space and it constituted a nonmechanical explanation of the findings for which Boyle adduced the spring of the air. This was the *funiculus*. Linus's identification of the funiculus also proceeded from experimental phenomena. If you perform the Torricellian experiment, closing the upper orifice of the mercury-containing tube with your finger, you will feel your finger being sucked down into the tube (figure 6). Linus said that this observation contradicted Boyle's assertion that the pressure of the external air was actually pushing up the column of mercury. To Linus this phenomenon meant that there was a substance in the Torricellian space and that this substance performed the role of sustaining the mercury column in position. It was a certain internal thread (*funiculus*) whose upper extremity was attached to the finger and whose lower extremity was attached to the surface of the mercury. Linus's preferred view was that this

---

con's Empiricism, Boyle's Science and the Jesuit Response"; Middleton, "Science in Rome," pp. 139-140, 147-148; Heilbron, *Elements of Early Modern Physics*, pp. 93-106.

4 Linus, *Tractatus*, p. 47, quoted in Boyle, "Defence against Linus," p. 135. For a detailed exposition of Linus's views as commitment to the *fuga vacui*, see Sir Matthew Hale, *Difficiles nugae* (1674), p. 141.

5 Linus, *Tractatus*, p. 23; Boyle, "Defence against Linus," p. 136. (Where Boyle translated Linus's Latin we tend to use that translation; all others are by Schaffer.)

6 Linus, *Tractatus*, pp. 6-9. Linus pointed to chapter 26 of Hobbes's *De corpore*.

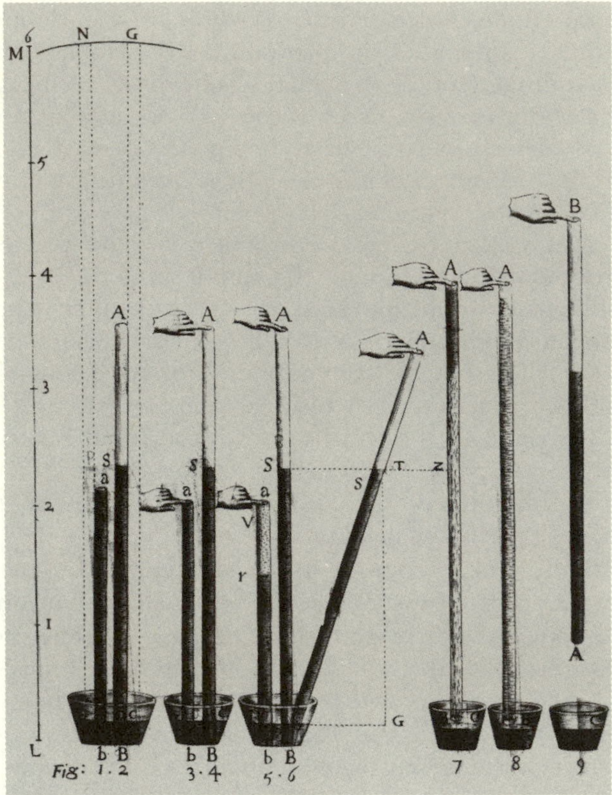

FIGURE 6

Diagram from Franciscus Linus's Tractatus de corporum inseparabilitate *(1661), showing evidence from suction for the existence of a* funiculus. *The finger is pulled down into the glass tube by the thread attached to the fluid in the tube. (Courtesy of Edinburgh University Library.)*

funiculus was composed of rarefied mercury.[7] (A funiculus could, however, be elaborated from other substances, in which cases it still performed the same functions.) The funiculus contracted in rarefied conditions and relaxed in nonrarefied conditions: hence it not only opposed the formation of a vacuum by preserving the continuity of matter in the world, but also accounted for the pull on one's finger in the Torricellian experiment.

Linus deployed the funiculus as an alternative to the air's spring

[7] Ibid., esp. p. 28. Matthew Hale suggested that the Baconian term "*motus nexus*" might make Linus's "funiculus" more acceptable to English naturalists: *Difficiles nugae*, 2d ed. (1675), "Additions," pp. 38-39.

in his systematic review and reconstruction of many of the experiments in Boyle's series. For example, Linus explained the force felt upon the retraction of the sucker in the air-pump. His account depended on two claims: (1) the retraction of the sucker rarefied the air in the receiver, and (2) the more rarefied the air, the stronger the power of its funicular contraction. As he said, the pump was indeed "well evacuated; that is, full of some very fine air." As the pump was operated "all of the air (as much as it can be) is extracted from the receiver" and "the remaining air (as much as it can be) rarefied." What remained in the receiver acquired "a greater force of contracting itself" and this was what accounted for the force on the sucker.[8] Similarly, Linus used the contractive force of the funiculus in his reinterpretation of the void-in-the-void experiment (Boyle's seventeenth; see chapter 2). When one placed the Torricellian apparatus in the receiver and exhausted the air, the mercury in fact descended almost to the level of the mercury contained in the vessel below. However, according to Linus, this was *not* owing to the air's spring.

> The quicksilver descends in the tube because of that exhaustion, since it is drawn down by the air lying on the restagnant mercury. For that incumbent air, being greatly rarefied and extended by its exhaustion, vehemently contracts itself and by this contraction endeavours to lift out the restagnant mercury from its vessel, so that (the restagnant mercury now weighing less on the bottom of its vessel) the mercury in the tube must descend.[9]

Perhaps most interestingly, Linus addressed himself to Boyle's thirty-first experiment: that concerning the cohesion of marble discs in the air-pump. He noted that the discs did not separate on the exhaustion of the receiver, as Boyle had expected and hoped they would: "When [Boyle] noticed this, he considered the reason why the lower stone did not fall, attributing it to the imperfection of the receiver, which would mean that the air could not be sufficiently exhausted, rather than abandoning the theory of the spring of the air." According to Linus, the fault was not with the machine but with the theory. The marble discs continued to cohere because of nature's abhorrence of a vacuum and because of the extreme difficulty of elaborating a funiculus from the substance of marble:

[8] Linus, *Tractatus*, pp. 101-103.

[9] Ibid., pp. 115-117.

> Since, when falling, the lower stone would have to disconnect simultaneously from the whole of the upper surface, nor could the neighbouring air insinuate itself into the whole remaining space, it is necessary that the stone would descend in no other way than by leaving after itself a fine substance, which mercury or water leave behind them when descending this way. Yet since such a substance is separated from the marble with more difficulty than from mercury, or any other fluid body, it thence follows that it adheres here so tenaciously.

Linus ventured that "if two perfectly polished marbles were so joined, that no air at all were left between them, they could not be drawn asunder by all the power of man." Linus quoted against his adversary Boyle's own accounts of the extreme difficulty of separating very smooth cohered bodies.[10]

While Linus deployed the funicular hypothesis in opposition to Boyle's mechanical spring of the air, he did not deny that the atmosphere had weight and spring. What Linus claimed was that the power of the air's spring and what it could accomplish were strictly limited. The air's spring was simply incapable of doing what Boyle said it did in his experimental system: "I do not deny there to be some weight in the air, and even elasticity, or a force of regaining its extension if it be reduced to a smaller space . . . ; however, I deny that in this way it may have enough gravity or elasticity as are imagined."[11] In challenging Boyle to show the power of the spring, Linus was challenging him to demonstrate the power of the mechanical philosophy.

We need not follow Linus's particular experimental reconstructions any further. For present purposes, two points need to be made. First, Linus did not seek to deny the role and value of experiment itself, nor to impugn the status of experimentally produced knowledge. Second, Linus never cast doubt upon the physical integrity of the air-pump. Both of these considerations are central to understanding the nature of Boyle's response. So far as experimental procedures were concerned, Linus both praised them and practised them. Linus celebrated "that very noble man, Robert Boyle, best deserving in this experimental philosophy, [who] restored the Magdeburg machine . . . to a better form; as is seen in his recently published book *New Experiments*, where indeed he dis-

[10] Ibid., pp. 124-126; partially translated in Boyle, "Defence against Linus," p. 126.

[11] Linus, *Tractatus*, pp. 11-12.

plays many very beautiful experiments, exhibited and examined in the presence of very learned men."[12] In his *Tractatus* Linus faithfully reproduced the engraving of the air-pump included in Boyle's text and illustrated his own experimental protocols.[13] He even echoed Boyle's stipulations about the relationship between experimental findings and their physical interpretations: "Since we labour in the present work to supply more suitable reasons than we have yet offered for the experiments, the prudent reader will judge of what might be preferable."[14]

Linus's zeal for experiment went so far as the replication of that most laborious of pneumatic experiments, Pascal's Puy-de-Dôme trial, in which the Torricellian apparatus was carried up a mountain. But his findings differed radically from those that Pascal reported and Boyle assimilated to his theory of the spring of the air:

> . . . I acknowledge that I do not agree with that experiment of which [vacuists] make so much, in proving that mercury is kept up to its level by the external air, suspecting some error to have happened in the operation: since, having made a similar experiment on another mountain, not indeed so high (not that there was any need for such a height, since we see that almost all the noticeable fall happens in the lower part of the mountain . . .) . . . ; I say, having performed the experiment in this manner, I certainly found no difference between the height of the mercury at the foot and at the top of the mountain.[15]

In any case, if the mercury were observed to fall as one carried the apparatus up a mountain, this might not be due to the spring but to temperature change.[16] Our second point concerns Linus's views of the air-pump. In the *Tractatus* Linus described in detail how this machine worked.[17] We have shown how his funicular hypothesis constituted an alternative explanation to the spring of the air, for example in accounting for the force felt upon the retraction of the sucker. Linus, therefore, rejected the explanatory adequacy of Boyle's spring. However, unlike Hobbes, he never identified problems with the integrity of the pump as problems for the explana-

[12] Ibid., p. 95. Quite possibly Boyle resented the suggestion that all he had done was to renovate and improve Guericke's device.

[13] Ibid., fig. 27; a detailed account is on pp. 96-98.

[14] Ibid., "Ad Lectorem," sig A5^r.

[15] Ibid., pp. 66-68.

[16] Ibid., pp. 68-69, 117-119.

[17] Ibid., pp. 96-103.

tions Boyle proffered; he did not say that the pump leaked. Indeed, in the case of cohering marbles, Linus dismissed Boyle's own explanation which pointed to residual air and leakage as a way of saving the mechanical explanation.

Boyle's reply to Linus came swiftly in an appendix to the second edition of his *New Experiments*, which was published in 1662.[18] A quite separate appendix dealt with Hobbes; the introduction to Boyle's defence against Linus made reference to his two adversaries and how he proposed to deal with them, but, in the main, the replies were distinct, manifesting Boyle's recognition that he had to defend different aspects of his programme. Interestingly, one of the very few remarks in which Boyle conjoined the criticisms of his two adversaries defended the status of the experimentally produced matter of fact. Neither Linus nor Hobbes, he said, had "seen cause to deny any thing that I deliver as experiment. . . . So that usually, without objecting any incongruity to my particular explications, they are fain, to fall upon the hypotheses themselves."[19] (However, as we have already seen in the case of Hobbes's *Dialogus*, Boyle's stipulation is at best doubtful: Hobbes did deny Boyle's matters of fact, notably in the case of stating what was contained in the "exhausted" receiver of the air-pump.)

Boyle's *Defence against Linus* took the opportunity both to exonerate his experimental work from criticisms and to reiterate and exemplify the proper way of proceeding in natural philosophy. It contained four main elements: (1) a restatement of the rules of the experimental game, including the correct manner of conducting disputes; (2) a restatement of the boundary conditions of experimental philosophy, including the demarcations between natural knowledge and theology and between matters of fact and hy-

[18] Another reply to Linus was by the Rutland natural philosopher Gilbert Clerke in his *Tractatus de restitutione corporum, in quo experimenta Torricelliana & Boyliana explicantur & rarefactio Cartesiana defenditur* . . . (1662; dated 20 November 1661). Clerke had earlier written a straightforwardly Cartesian tract against Bacon's atomism, Hobbes's version of plenism, and Seth Ward's vacuism: *De plenitudine mundi* (1660). In the later text Clerke attacked both Linus and Hobbes's *Dialogus physicus* from a perspective now heavily informed by Boyle's *New Experiments*. The air-pump trials were interpreted as exemplifying the action of a subtle aether that filled all space and penetrated glass. Linus and Clerke both maintained that the air had a spring; Linus added a contractive spring to Boyle's spring whose power, Linus said, was limited; Clerke gave an orthodox Cartesian (spongy) account of the spring. Hobbes was therefore unique in denying the spring as a matter of fact. For Clerke's Cartesianism, see Pacchi, *Cartesio in Inghilterra*, p. 88.

[19] Boyle, "Defence against Linus," p. 122.

potheses; (3) a defence of his mechanical interpretations against Linus's funicular hypothesis; and (4) a particular defence of the power of the spring of the air to account for the products of the air-pump and related pneumatic phenomena.

Boyle took pains to make clear that he generally approved of Linus's manner of constructing and delivering his criticisms. Boyle took Linus seriously as a critic "because he seems to have more diligently than some others [almost certainly Hobbes] inquired into our doctrine."[20] Boyle judged Linus to be a competent member of the experimental community. Further, Boyle applauded Linus as an active experimenter, even when, as in the case of Linus's replication of Pascal's Puy-de-Dôme trial, the results were said to contradict the accepted finding:

> But though, instead of disapproving, I am willing to commend his curiosity, to make the experiment himself, and especially since it was both new and important; and though also I like his modesty, in rather suspecting some mistake in the manner of the observation [i.e. Pascal's] than that the experimenters did not sincerely deliver it.[21]

Nevertheless, Linus's experimental observations had to be rejected, and Boyle provided Linus with a repertoire of excuses for his experimental failure. As Boyle said, "There must be an error somewhere" and "I must rather charge it upon the examiner's observation (I say, his observation, not his want of sincerity) than upon Monsieur *Paschal*'s." Pascal's observation, according to Boyle, was better witnessed than Linus's: Gassendi attested to Pascal's reliability, and the same source "relates, that the like observation was five times repeated . . . which circumstances sufficiently argue the diligence wherewith the experiment was tried in *Auvergne*." In England, Pascal's report was checked by Ball, Power, and Towneley, and Boyle himself undertook a replication by carrying the apparatus up Westminster Abbey. True, there was variation in the level to which the mercury was reported to descend in these trials, but Boyle offered a subsidiary hypothesis that might account for this without necessitating the abandonment of the spring. This variation, he told Linus, was due to the "differing consistence and other accidents of the neighbouring air, in the particular places and times of the experiments being made." So far as Linus's suggestion re-

[20] Ibid., p. 120.
[21] Ibid., p. 152.

garding the role of temperature was concerned, Boyle was less disposed to grant this, even though differences in temperature were not taken into account in any of the trials that Boyle validated.[22] Thus, while commending his adversary's experimental practice and the manner of his reporting it, Boyle saw no reason to accept Linus's findings: the balance of credibility was against Linus; the reasons Linus gave to invalidate Pascal's results were implausible; and a number of secondary assumptions could be pointed to in order to gloss away reported variations, including Linus's. (Parenthetically, a modern commentator on Pascal's Puy-de-Dôme experiment is deeply suspicious of "the high degree of accuracy" reported in successive trials and suggests that an element of what is now called "data massage" was involved.)[23]

In this way, Boyle acquitted Linus of the charges of incompetence to participate in experimental disputation and of insincerity in delivering his contradictory findings: ". . . I suspect not, that he does wilfully mistake my sense."[24] But perhaps Linus had not been quite diligent enough as an experimentalist. Perhaps he had come to erroneous conclusions through the failure to perform a sufficient number of experiments. In the case of Linus's objections to Boyle's interpretation of cohered marbles, Boyle suggested that "possibly he would have spoken less resolutely, if he had made all the trials about the adhesion of marbles, that we relate ourselves to have made." And he forcibly reminded Linus that a leaking receiver could plausibly account for the peculiarities of both the void-in-the-void experiment and the experiment on cohesion.[25] Within the

[22] Ibid., pp. 152-154. Linus was reviving many of the Jesuit attacks on Pascal in the 1640s, notably that of Etienne Noël, S.J. For these attacks, see Fanton d'Andon, *L'horreur du vide*, pp. 47-57 (on Noël's *Plein du vide* of 1648).

[23] Conant, "Boyle's Experiments in Pneumatics," pp. 8-9: "To be able to repeat the Torricellian experiment [carried up a mountain] so that there was less than a twelfth of an inch . . . difference in successive readings, as Perier [Pascal's brother-in-law] claimed, is remarkable. The accidental intrusion of a slight amount of air is very difficult to avoid. . . . It may be that Perier, persuaded of the reality of the large differences in height of the mercury column at the top and bottom of the mountain, succumbed to the temptation of making his argument convincing by recording exact reproducibility of his results on repeated trials." See also Reilly, *Francis Line*, pp. 68-69.

[24] Boyle, "Defence against Linus," p. 171.

[25] Ibid., pp. 173-174. Boyle made use of the fact that the level of mercury in the void-in-the-void experiment did not quite fall to that in the basin. This was because "there remained air enough [in the exhausted receiver] to keep up in the tube a cylinder of about one inch long of quicksilver." And in the case of the continued cohesion of the marbles Boyle reminded Linus that the reason might plausibly be

limitations presented by his understanding and his experimental diligence, Boyle reckoned that Linus had mobilized his brief as well as he could have done. Quoting St. Augustine, Boyle judged that " 'In a bad cause they can do no other; but who compelled them to undertake a bad cause.' " Linus had reasoned well enough, but he had reasoned upon bad foundations; he had "failed rather in the choice than in the management of the controversy."[26] Appropriate manners could be maintained.

In his *Defence* Boyle would therefore demonstrate not merely that Linus was wrong, but also how experimental controversies ought to be conducted. He set out by reminding his readers "how indisposed I naturally am to contentiousness" and how unhappy he was to be "publickly engaged in two controversies at once."[27] He was inclined, he said, not to make any reply. But he felt obliged to put aside his naturally irenic and retiring posture. Boyle's declared unwillingness to publish in controversy was a matter of frequent comment in the 1660s. But now Boyle felt that his adversaries had attacked not just "one or two of my conjectures" but the greatest number of them, including the central hypothesis of the air's spring. Second, a reply would give him the opportunity to restate and further illustrate, through experiment, the power of the spring of the air. Third, Boyle worried that, if he did not reply, his silence would be taken to mean that the criticisms were valid. Fourth, Boyle was not in the position of defending his own honour and reputation, which he declared himself reluctant to do. Instead, he found himself defending the collective position of the experimentalists of the Royal Society, for their collective benefit.[28] In so doing, Boyle would avoid casting aspersions on the character of individual ad-

---

"some small leak in the receiver." Linus ignored these considerations, already identified by Boyle in *New Experiments*, because the porosity and leakage of the air-pump was no part of his critical strategy. Boyle's reference to further experiments on cohesion was to early experiments performed in atmospheric conditions and reported in *The History of Fluidity and Firmness*, published in 1661 as one of the *Certain Physiological Essays*. These and other issues involving cohesion are discussed later in this chapter.

[26] Boyle, "Defence against Linus," pp. 177-178.

[27] Ibid., p. 119.

[28] Ibid., pp. 119-120, 124. For Boyle's notorious unwillingness to publish, see Southwell to Oldenburg, 20 February/1 March 1660, in Oldenburg, *Correspondence*, vol. I, p. 355 ("He is a person of . . . soe much caution withall"), and Katherine Boyle to Robert Boyle, 29 July/8 August 1665, in Boyle, *Works*, vol. VI, p. 525 ("cut them off from being published"). For Boyle as protagonist of the Royal Society, see Boyle, "Examen of Hobbes," pp. 187-191 (to be discussed below).

versaries; he would show his "custom of exercising civility, even where I most dissent in point of judgement." This controversy was not about persons, but about the interpretation of facts: "I hope there is not in my answers any thing of asperity to be met with; for I have no quarrel to the person of the author, or his just reputation." Moreover, Boyle insisted that he had no greater attachment to his doctrine or findings than was warranted by the evidence; he would give them up if proved to be wrong, and so should every experimentalist.[29] Thus, the management of the controversy with Linus was to constitute a concrete exemplar of proper discourse in the moral community of experimental philosophers.

A second aspect of Boyle's *Defence* consisted in his underlining of crucial cultural boundaries within and surrounding proper natural philosophy. As we have seen, Boyle claimed that neither Hobbes nor Linus disputed with him about matters of fact. Boyle made this point in order to preserve that which he regarded as foundational—experimentally generated matters of fact—and to bracket off that which he wished to treat as "superstructed" upon matters of fact—hypotheses. Thus Boyle judged that Linus "takes no exceptions at the experiments themselves, as we have recorded them (which, from an adversary, who in some places speaks of them as an eye-witness, is no contemptible testimony, that the matters of fact have been rightly delivered)." Indeed, Boyle noted that just because Linus attributed *some* motion of restitution to the air, he could make out "many more" of the phenomena of the pump than "most other Plenists that deny the spring of the air can deduce from their hypothesis if granted." Further, Boyle also treated those experiments to which Linus did not take exception as implicitly agreed between them.[30] The dispute could then be made visible as one over interpretations and not one over facts or over proper practices for producing knowledge. This was the ground upon which disputes could legitimately be conducted, and within which controversies could be contained and decently managed. Boyle reminded Linus that "it was not my chief design [in *New Experiments*] to establish theories and principles, but to devise experiments, and to enrich the history of nature with observations faithfully made and delivered."[31]

[29] Boyle, "Defence against Linus," pp. 124, 177. In point of fact, Boyle gave up or modified none of his opinions in the course of his controversies with either Linus or Hobbes.

[30] Ibid., pp. 163, 177.

[31] Ibid., p. 121.

Yet Linus's appeal to the "funiculus" threatened the demarcations between natural philosophy and religion that Boyle regarded as crucial, and upon which the real utility of experimental natural philosophy to religion depended. According to Boyle, Linus had produced a text in which final causes were covertly adopted in an apparently natural philosophical account. Plenism and the *horror vacui* lurked behind Linus's arguments, and behind these doctrines the erosion of the most important distinction: that between brute nature and divinity. Boyle used the image of God's action both in his response to Linus and in his response to Hobbes. He picked out his critics' illegitimate use of arguments *a priori* and then associated this illegitimacy with an equally unacceptable infringement of the proper boundaries of natural philosophy. He told Linus that "none is more willing [than myself] to acknowledge and venerate Divine Omnipotence," but that "our controversy is not what God can do, but about what can be done by natural agents, not elevated above the sphere of nature. For though God can both create and annihilate, yet nature can do neither: and in the judgment of true philosophers, I suppose our hypothesis would need no other advantage, to make it be preferred before our adversaries, than that in ours things are explicated by the ordinary course of nature, whereas in the other recourse must be had to miracles."[32] Similarly, he told Hobbes that "when Mr. *Hobbes* has recourse to what God can do (whose omnipotence we have both great reason to acknowledge) it imports not to the controversy about fluidity to determine what the almighty Creator can do, but what he actually has done."[33] Boyle's mechanism was a way of stipulating what belonged to nature and what belonged to God: Linus's account, however, endangered that distinction at all levels, and therefore threatened both proper natural philosophy and proper theology.

There was one subject in particular that involved both sorts of boundaries that Boyle wished to maintain: those between "physiology" and "metaphysics," and those between the power of God

[32] Ibid., p. 149. For Boyle's position on the "ordinary course of nature," see McGuire, "Boyle's Conception of Nature."

[33] Boyle, "Examen of Hobbes," p. 236. Both Hobbes and Linus raised the question of what powers could in principle be attributed to natural action in the cohesion trials. Linus (*Tractatus*, pp. 123-126), as we have seen, stated that the funiculus connecting the marbles could not be "pulled apart by all the power of man," and Hobbes ("Dialogus physicus," p. 245) asked "why I should think it more difficult for almighty God to create a fluid body less than any given atom whose parts might actually flow, than to create the ocean."

and the power of nature. This was the subject of the void. Boyle reiterated that he had not and would not "declare myself for or against a vacuum."[34] He said he could not understand why Linus, like Hobbes, had attacked him as a vacuist when he had explicitly declared his nescience on the matter and had identified the question as metaphysical in character. Nevertheless, Linus's plenism had to be combatted because the explanatory resources he used to reject a vacuum were anti-mechanical. Linus's funicular hypothesis, according to Boyle, masked the actual strategy the Jesuit employed to deny a vacuum. Linus made the funiculus "the immediate cause of the phenomena," yet "if you pursue the inquiry a little higher, he resolves them into nature's abhorrency of a vacuum."[35] Even the alleged contractive behaviour of the funiculus manifested Linus's fundamental anti-mechanism and his vitalistic conception of nature. The funiculus was said to restore itself to its naturally relaxed state; it was self-moving. But, as Boyle drily remarked, "I am not very forward to allow acting for ends to bodies inanimate."[36]

Finally, Boyle defended the spring of the air from the limitations Linus would place upon its power. Boyle recognized that Linus did not deny that the air possessed some elasticity: the challenge was to demonstrate that the spring was powerful, that it was a sufficient explanation of pneumatic phenomena.[37] Boyle seized this opportunity to display the explanatory power of his central mechanical concept by showing the power of its action. Further experiments were undertaken for this purpose. Linus "denies not, that the air has some weight and spring, but affirms, that it is very insufficient to perform such great matters as the counterpoising of a mercurial cylinder of 29 inches. . . . [W]e shall now endeavour to manifest by experiments purposely made, that the spring of the air is capable of doing far more than it is necessary for us to ascribe to it, to solve the phaenomena of the Torricellian experiment."[38] This was the precise context in which "Boyle's Law" was elaborated. The work Boyle undertook in reply to Linus was not done with the air-pump, but with a specially constructed J-shaped tube in which pressures higher than atmospheric could be attained. Using this apparatus, Boyle showed that if he compressed air twice as strongly as usual he could produce twice as strong a spring. He concluded that the

[34] Boyle, "Defence against Linus," p. 135; see also p. 137.

[35] Ibid., p. 135.

[36] Ibid., p. 143.

[37] Ibid., pp. 124, 134, 156.

[38] Ibid., p. 156.

process could go on indefinitely, so that there were no limits to the power of the air's spring. The mathematical regularities displayed in the resultant tables relating spring to compression were labelled as a specific refutation of Linus's attempted restriction of the spring: ". . . here our adversary may plainly see, that the spring of the air, which he makes so light of, may not only be able to resist the weight of 29 inches, but in some cases of above a hundred inches of quicksilver, and that without the assistance of his Funiculus."[39] There was no need of the funiculus to explain all the phenomena: "[The] two main things, that induced the learned examiner to reject our hypothesis, are, that nature abhors a vacuum; and that though the air have some small weight and spring, yet, these are insufficient to make out the known phaenomena."[40] Linus's funicular hypothesis was therefore rejected on two grounds; it was "unintelligible" (as were all Scholastic explanatory resources) and it was unnecessary (since the spring and weight of the air were all-powerful).

Hobbes as Adversary

Boyle's conception of Linus as adversary was shaped by a concern to defend the physical and explanatory power of the air's spring. There was only one thing that Boyle guarded more vigilantly than the doctrine of the spring of the air, and that was the form of life which generated such an item of knowledge: the experimental programme. Hobbes threatened the experimental form of life for natural philosophers. Boyle's response to Hobbes was fundamentally a defence of the integrity and value of experimental practices. Thus it was different in tone and in substance to Boyle's reply to the Jesuit experimentalist. With Linus, Boyle shared a commitment

[39] The discrepancies in the tables between observation and hypothesis were ascribed to "some such want of exactness as in such nice experiments is scarce avoidable." Ibid., p. 159.

[40] Ibid., p. 162. Experiments to test Linus's funiculus continued for some time and engaged Sir Robert Moray and Christiaan Huygens as well as Boyle: see Huygens to Moray, 4/14 July 1662, in Huygens, *Oeuvres*, vol. IV, pp. 171-173; Boyle to Moray, July 1662, ibid., pp. 217-220; and, for a report of an experiment on the cause of the adhesion of one's finger on the top of an open tube of mercury, ibid., pp. 295-299 and Birch, *History*, vol. I, p. 166. For Henry More's graphic use of the pull felt on one's finger to evince some "spirit of nature," see More, *Remarks upon Two Late Ingenious Discourses* (1676), p. 93, commenting on Hale, *Difficiles nugae*, p. 118.

to experiment while radically differing on the question of mechanism. With Hobbes, Boyle shared a broadly mechanistic conception of nature, while radically differing on the means by which knowledge was to be produced.

Boyle's *Examen of Hobbes's Dialogus* was published in 1662, together with his *Defence against Linus*, as an appendix to the second edition of *New Experiments*. By this time, Boyle had largely finished with the Jesuit, but the *Examen* was only the opening shot in a salvo of explicit and implicit responses to Hobbes's *Dialogus* which continued for over a decade. We can categorize Boyle's reply to Hobbes under four main headings: (1) a technical response, involving modifications to the design and operation of the air-pump; (2) a reiteration of the rules of the experimental game, and a stipulation that, within these rules, Hobbes had failed as a natural philosopher; (3) an experimental programme devoted to clearing up the troubles which Hobbes had pointed to in his comments on *New Experiments*; and (4) an ideological response, identifying theological grounds for rejecting Hobbes's natural philosophy.

Hobbes had said that the air-pump leaked, that it did so massively, and that, therefore, Boyle's experimentally produced matters of fact did not have the status Boyle claimed for them. In order to defend the integrity of his matters of fact Boyle was obliged to defend the physical integrity of the air-pump. Boyle recognized that the air-pump did leak, although in *New Experiments* he denied that this leakage seriously undermined the evidential status of his findings. Within the experimental programme in pneumatics there were good reasons to seek to improve the efficiency of the machine. Confronted with Hobbes's attack, Boyle now had two reasons to do so; leakage was a recognized source of trouble *within* the experimental programme; through Hobbes, leakage became a publicly visible trouble that might destroy the experimental programme itself.

At the same time that he was composing his reply to Hobbes's *Dialogus* late in 1661, Boyle was energetically redesigning his pneumatic engine. (We present a detailed account of this redesign work in the 1660s in the next chapter. The abbreviated account here glosses over some technical details and discrepancies in Boyle's story, and we also defer to chapter 6 the consideration of relations between English and Continental air-pump technologies during the period.) Figure 7 shows the modified air-pump that Boyle certainly constructed by 1667, and quite probably as early as January 1662.[41]

[41] Described in Boyle, "Continuation of New Experiments," pp. 176-182; see also

The major changes Boyle made to the original design of 1659 were threefold: (1) the pumping apparatus was immersed in a tank of water; (2) the receiver was now indirectly connected to the pump proper, being placed on a flat wooden and iron board; (3) various improvements were made in the cement and seals used to secure the device against the intrusion of external air. All of these modifications were intended to enhance the physical integrity of the machine—its ability to remove and keep out atmospheric air.

Referring to the design as it appeared in Boyle's *Continuation of New Experiments*, here are some details of these changes in the air-pump; (1) the brass barrel of the pump itself (figure 7, NO) was placed with its mouth upwards in a wooden trough which was filled with water. Unlike the original design, this new barrel had no valve in it. Instead there was an aperture (PQ) in the sucker that could be opened and closed by means of a long stick (R). Boyle said that the advantage of this arrangment was that "the sucker, lying and plying always under water, is kept still turgid and plump, and the water being ready at hand to fill up any little interval or chink that may happen to be between the sucker and the inside of the barrel, together with the newly mentioned plumpness of the sucker, very much conduce to the exact keeping out of the air."[42] Thus this aspect of the new design was intended to seal off the major avenue for the entry of the air that Hobbes had identified. (2) The indirect connection between pump and receiver appears to have been made for three reasons. First, this is by far the most convenient arrangement, given the immersion of the pump upside-down and, especially, given the consequent protrusion of the rack and pinion device. Second, it allowed for an easier and more effective seal around the receiver. There was no need for a stopcock attached to the receiver; stopcocks had in the past proved leaky and the new combination of self-sealing by pressure and cement constituted an improvement to the integrity of this part of the machine. Third, this arrangement allowed for the easy interchange of receivers; one could simply use whatever size of glass receiver one wished and was appropriate for the experiment at hand. The experimental apparatus could be placed on the board and the receiver arranged over it. Further, there was no need for an aperture in the top of the receiver—another potential avenue for leakage. (3) Finally, Boyle was constantly searching for better cements to seal his ma-

Wilson, "Early History of the Air-Pump," pp. 336-338. For the dating of this pump, see Robert Moray to Christiaan Huygens, 24 January/3 February 1662, in Huygens, *Oeuvres*, vol. IV, pp. 27-28.

[42] Boyle, "Continuation of New Experiments," p. 181.

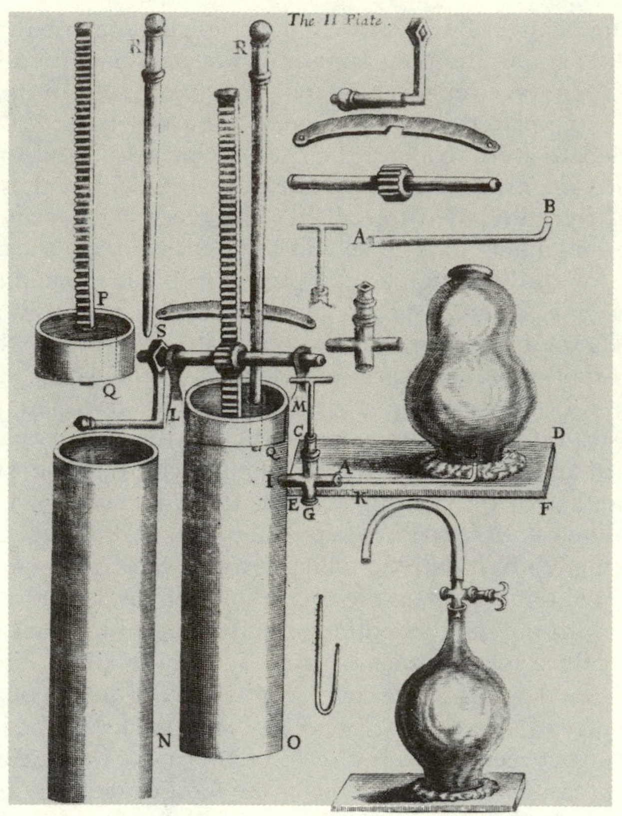

FIGURE 7
Boyle's revision of the air-pump, as it appeared in Continuation of New Experiments *(1669). (Courtesy of Edinburgh University Library.)*

chine against the intrusion of air. The new design, to an extent, alleviated the problem of poor cements by reducing the number of apertures that might leak. Yet, as Boyle said, finding a "good cement to fasten the receivers to the . . . plate, is a thing of no small moment," and he employed different mixtures for different experimental purposes. A new cement was extensively used in the experimental series published in *Continuation*: "A well wrought mixture of yellow bees-wax and turpentine, which composition, as it serves better than most others to keep out the air, so it has the conveniency, which is no small one, of seldom needing to be heated, and seldomer to be much so; especially if we imploy a little more turpentine in winter than in summer."[43]

[43] Ibid., p. 182.

Although details of Boyle's air-pump were not published until 1669, as early as 1662, in his *Examen of Hobbes*, Boyle was using its new features as a resource to counter his adversary's criticism that the machine massively leaked: "Though the pump be all the while kept under water, yet the exhaustion of the cylinder and receiver will be made as well as in the open air. I demand then of Mr. *Hobbes*, how the pure air gets in by the sides of the sucker that is immersed in water?" Boyle imaginatively rehearsed his adversary's possible riposte: "I presume . . . he will here say . . . that the air passes through the body of water to fill up that deserted space, that must otherwise be void."[44] Boyle rightly sensed that the issue of the integrity of his engine was quite directly connected to the integrity of the experimental form of life as a way of generating assent. He rejected Hobbes's potential reply on grounds of inherent implausibility: ". . . I appeal to any rational man, whether I am obliged to believe so unlikely a thing upon Mr. *Hobbes*'s bare affirmation. If I be, I must almost despair to prove things by experiments."[45] And in a later text, first published in 1674, Boyle offered his interpretation of Hobbes's animus: Boyle said his adversary desired "to be revenged on an engine that has destroyed several of his opinions."[46]

The "Way of Discoursing"

As in his *Defence against Linus*, Boyle justified himself for entering into public dispute. However, in the case of the response to Hobbes, these justifications involved far more than the defence of the spring of the air. They also concerned the "way of discoursing about natural things," and the protection of experiment itself.[47] Aspects of Boyle's reply are already familiar from our account of the response to Linus. Boyle attempted to get his adversary onto the ground of the experimental philosophy, its practices, and crucial boundaries: "Mr. *Hobbes* does not, that I remember, deny the truth of any of

[44] Boyle, "Examen of Hobbes," p. 208; cf. p. 193. Of course, the redesigned pump, with its sucker immersed in water, did not exist when Hobbes wrote his *Dialogus physicus*. In later tracts against Boyle—the *Problemata physica* and *Decameron physiologicum*—Hobbes did not refer to any device but the original 1659 design.

[45] Boyle, "Examen of Hobbes," p. 208. Compare idem, "Some Considerations about Reason and Religion," pp. 166-167 (1675): "As for the new Somatici, such as Mr. *Hobbes* . . . , hitherto I see not, that he hath made any new discovery either of new truths, or old errors."

[46] Boyle, "Animadversions on Hobbes," p. 114 (originally published in 1674 as part of *Hidden Qualities of the Air*).

[47] Ibid., p. 106.

the matters of fact I have delivered." Boyle asked readers to consider "what new experiment or matter of fact Mr. *Hobbes* has . . . added to enrich the history of nature, what new truths he has discovered, or what errors . . . he has well confuted." Boyle identified Hobbes's criticisms as denials of interpretations, not of facts. Hobbes had rejected the weight and spring of the air—"our two grand hypotheses themselves."[48] By assimilating their disagreements to conflicts over interpretation, Boyle here suggested that he and Hobbes were playing the same game and that Hobbes played it badly.

But the main reason, in Boyle's view, that Hobbes had not produced any new "matters of fact" was that this adversary systematically denigrated the performance of experiments. Hobbes appeared as the great scourge of all experimentalists: ". . . my adversary, not content to fall upon the explications of my experiments, has (by an attempt, for aught I know, unexampled) endeavoured to disparage unobvious experiments themselves, and to discourage others from making them." This was highly dangerous, for, if Hobbes's anti-experimentalism were credited, "I dare be bold to say, he would far more prejudice philosophy by this one tract, than he . . . can promote it by all his other writings."[49] The "thing" Hobbes "seemed to aim at," was, Boyle said, "prejudicial to true and useful philosophy."[50] Hobbes's writings were no less dangerous for the fact that he had not troubled to perform experiments himself, for this kind of attack might demonstrate how easy it was to erode the foundations of proper philosophy; laboratory labour might be undone by armchair criticisms. Boyle expressed his hope that in a properly governed philosophical community the authority of a critic would be gauged by his known facility in making experiments. To criticize experiments one should be an experimentalist.[51]

Boyle recognized that Hobbes had specially criticized the worth of "unobvious experiments," those that were, as Hobbes had said, "forced by fire." "Unobvious" or "elaborate" experiments produced artificial phenomena through the mediation of purposely designed

[48] Boyle, "Examen of Hobbes," pp. 197, 233. Compare Boyle, "Some Considerations about Reason and Religion," p. 167: ". . . I am not much tempted to forsake any thing, that I looked upon as a truth before, even in natural philosophy itself, upon the score of what he (though never so confidently) delivers."

[49] Boyle, "Examen of Hobbes," p. 186.

[50] Boyle, "Animadversions on Hobbes," p. 105.

[51] Boyle, "Examen of Hobbes," p. 186.

apparatus and with the application of collective human labour. These were contrasted with "obvious experiments," the simple observation of naturally produced or common phenomena.[52] Although Boyle had no interest in denigrating such "obvious experiments," the labour and the discipline of "elaborate experiments" were among their chief recommendations. Hobbes, however, had exclusively relied on "obvious experiments," and Boyle was now concerned to display their severe limitations. Boyle suggested that if the philosopher were genuinely in pursuit of causal knowledge (as Hobbes purported to be), then "obvious experiments" were ill-suited to that enterprise:

> . . . obvious experiments are by no means to be despised, yet it is not safe in all cases to content one's self with such; especially when there is reason to suspect, that the phaenomenon they exhibit may proceed from more causes than one, and to expect, that a more artificial trial may determine, which of them is true.[53]

In this way Boyle commended the experimental programme as a valuable resource in causal inquiry, while elsewhere he radically segregated the experimental production of matters of fact from the causal enterprise.

The infant experimental programme was potentially at risk from Hobbes's attack for reasons that, in Boyle's view, had little to do with the substantive merit of those criticisms. This danger stemmed from Hobbes's reputation and following, from his standing as a leading exponent of the *mechanical* philosophy (varieties of which might not be readily distinguished by neophytes), and from the manner of proceeding in philosophical discourse that Hobbes recommended and exemplified. Despite his professed "natural indisposedness to contention," Boyle claimed that he was obligated to respond, principally because Hobbes's "fame and confident way of

[52] Ibid., p. 241: ". . . Mr. *Hobbes* appears offended at me and others for troubling ourselves to make unobvious experiments." Boyle here quoted Hobbes's *Mathematicae hodiernae*, but omitted the latter's distinction between natural history (which is "enriched" by such experiments) and natural philosophy (which is not). Cf. Boyle, "Animadversions on Hobbes," p. 105, and "Some Considerations about Reason and Religion," p. 190: "Those men, that have an instrument of knowledge, which other men either have not, or, (which is as bad) refuse to employ, have a very great advantage."

[53] Boyle, "Examen of Hobbes," p. 193. In exemplification, Boyle reckoned that he had provided superior alternative accounts of the "unobvious experiments" that Hobbes had adduced; cf. pp. 191-192.

writing might prejudice experimental philosophy in the minds of those who are yet strangers to it." Compared to Hobbes, Boyle referred to himself as but "a young writer" (he was almost forty years younger) with nothing like his elder's weight in the natural philosophical community.[54] Unless Hobbes were publicly refuted, readers might "mistake confidence for evidence."[55] The necessity of a reply to Hobbes was greater than that of a reply to Boyle's Jesuit adversary, since the former had conducted philosophical controversy in a way that would disrupt the concord and harmony the experimentalists aimed to secure.

The appearance of *Dialogus physicus* therefore gave Boyle a valuable opportunity "to give an example of disputing in print against a provoking though unprovoked adversary, without bitterness and incivility." Incivility would be met with decorum; passion would be countered by sense; personal abuse would be turned aside by substantive argument; proud dogmatism would be laid low by modest experiment. Thus Boyle would not only defeat Hobbes's particular criticisms of his experiments but at the same time give a concrete demonstration of how disputes ought to be managed and brought to a conclusion. In Boyle's view, Hobbes wrote as a Restoration *wit*. He had, according to Boyle, attacked persons as a way of attacking their ideas. In the *Examen* Boyle would, however, eschew "quick and smart expressions, which are wont to be employed in disputes, to expose or depreciate an adversary's person or cause." If Hobbes were inclined to respond to Boyle's defence, the experimentalist desired that "his reply be as inoffensive as I have endeavoured to make my examen." Remember, Boyle lectured, the Law of Vespasian.[56]

In protecting the proper procedures of experimental philosophy against the beast of deductivism, Boyle was not entirely above wit and innuendo. He acidly censured Hobbes for his handling of Descartes' philosophy, which, he suggested, "is thought to be in some particulars not so unlike his own." Indeed, in 1675 Boyle published the view that the "grand position of Mr. *Hobbes*" on the motion of bodies had been "cautiously proposed, . . . by *Des Cartes*" and, furthermore, was now "crudely proposed by the favourers of Mr. *Hobbes*." In 1662 Boyle commented wrily that such treatment of Cartesian doctrines would cast a poor reflection "upon the Eng-

[54] Ibid., p. 186, 190.

[55] Boyle, "Examen of Hobbes's Doctrine about Cold," p. 687 (originally published in 1665 as part of *New Experiments and Observations touching Cold*).

[56] Boyle, "Examen of Hobbes," pp. 188-189.

lish civility in the opinion of strangers." Boyle argued that Hobbes had fixed upon the wrong target. He had attacked the Royal Society instead of the views of Boyle as an individual. Despite Boyle's efforts to build up an experimental collective, despite his stress upon the public and consensual nature of experimental knowledge, Boyle instructed Hobbes to remember the difference between the individual author of a text and the community. Dispute should remain within that community, which itself must never be challenged as a whole. Boyle told Hobbes that, so far from adopting his experiments as unquestioned truth, the Society had correctly proposed that they be repeated and that it was for this reason he gave the air-pump to the Society. Why, then, attack the "Greshamites"? Moreover, why attack a society "whereof [Hobbes's] own great patron, and my highly honoured and learned friend, the Earl of *Devonshire* himself, is an illustrious member"?[57]

If Boyle had expected that Hobbes would be silenced or converted by the *Examen*, he was mistaken. In 1668 the Amsterdam edition of Hobbes's collected philosophical works appeared, containing the unaltered text of the 1662 version of *Problemata physica* and a slightly revised version of the *Dialogus physicus*. This elicited a response from Boyle, the 1674 *Animadversions upon Mr. Hobbes's Problemata de Vacuo*.[58] Thus, given Boyle's attacks on Hobbes in texts such as *Considerations about the Reconcileableness of Reason and Religion* (1675) and Hobbes's own publication of *Decameron physiologicum* in 1678, the exchanges between the two continued almost to Hobbes's death. Again, Boyle proffered justifications for engaging in public controversy. First, he judged that the criticisms contained in *Problemata physica* were "but some variations of, or an appendix to" the *Dialogus physicus*. Since he had replied to that text some twelve years earlier, and since Hobbes had apparently not been satisfied with that reply, a further effort was indicated. Second, Boyle noted Hobbes's gall in dedicating this further attack to the King: all such attempts to sway the royal patron against the

[57] Boyle, "Some Considerations about Reason and Religion," p. 168; "Examen of Hobbes," pp. 197, 190-191. Both the Earl of Devonshire and his son were at this time Fellows of the Royal Society. Devonshire was often absent from meetings, but was evidently a regular subscriber; see Hunter, *The Royal Society and Its Fellows*, pp. 164-165.

[58] This account of the occasion for Boyle's *Animadversions* is an informed guess, based on circumstantial evidence and Boyle's own statements. It is possible that there was a long delay in Boyle's reading of the original 1662 text, or between the composition and publication of Boyle's essay. See Boyle, "Animadversions on Hobbes," p. 104.

experimental programme needed to be resisted. Finally, Hobbes continued to raise the issue of the overall validity of experiment, and Boyle's response, fourteen years after the founding of the Royal Society, showed that experimentalism still could not be taken for granted. Boyle said that Hobbes

> has been pleased to speak very slightingly of experimentarian philosophers (as he stiles them) in general, and, which is worse, to disparage the making of elaborate experiments; [therefore] I judged the thing, he seemed to aim at, so prejudicial to true and useful philosophy, that I thought it might do some service to the less knowing, and less wary sort of readers, if I tried to . . . shew, that it is much more easy to under-value a frequent recourse to experiments, than truly to explicate the phaenomena of nature without them.

Interestingly, Boyle took particular note of Hobbes's use of the dialogue form to couch his criticisms, and seized the opportunity of ironic imitation: "My adversary having proposed his problems by way of dialogue between *A* and *B*; it will not I presume be wondered at, that I have given the same form to my animadversions."[59] As in Hobbes's dialogues, the interlocutor played the role of offering absolute assent to the protagonist's arguments and evidence.

The Constitution of the Air

That the proper "way of discoursing" in natural philosophy was a moral matter is easily seen. Moral considerations also informed Boyle's conceptual and technical defence of the integrity of his pump. The moral threat posed by Hobbes to the community of natural philosophers consisted, on the one hand, of his violations of the rules and conventions of experimental discourse: his "dogmatic" way threatened to sow seeds of divisiveness in the garden of experimental harmony. But the particular nature of his criticisms also sinned against the boundary conditions laid down by experimental philosophers. By violating the boundaries between "physiology" and "metaphysics," between "matters of fact" and causal items, Hobbes threatened to destroy the classifications that permitted amicable discourse and that generated uncoerced assent.

[59] Ibid., p. 105.

Nowhere are these interconnections between the conceptual, the technical, and the moral more visible than in Boyle's treatment of the constitution of the air, as this related to specific experimental performances and Hobbes's criticisms thereof. Hobbes had to be put right on two heads: first, he had attacked Boyle as a vacuist when Boyle had declared no position on this "metaphysical" matter;[60] second, he had deployed notions of the constitution of the air which were themselves either "metaphysical" or manifestly erroneous.[61] In either case, Hobbes's claims about the make-up of the air were specifically addressed to that most important issue: the physical integrity of the air-pump and the legitimacy of its findings. Boyle's treatment of these criticisms took both a programmatic and an experimental form.

Hobbes had argued that Boyle's machine produced no vacuum. We have already given an account of the technical response that Boyle made: he had designed a modified version of the air-pump that was arguably less likely to leak with respect to atmospheric air.[62] However, Boyle recognized that a decisive defence against Hobbes would involve him in a more explicit discussion of the constitution of the air than he had previously provided. There are two points to be made in this connection: first, that Boyle's treatment of the air's constitution in those texts was always addressed to the specific question of the physical integrity of the air-pump and the manner in which Hobbes had impugned that integrity; second, that Boyle offered a variety of versions of Hobbes's claims about the constitution of the air. The refutation of each of these versions allowed Boyle to defend the pump and the legitimacy of its experimental products in different and in equally important ways.

Boyle said that Hobbes had been "unclear" in his statements about the constitution of the air. But Hobbes's fundamental contention in the *Dialogus physicus* had been this: the atmosphere consisted of a mixture of terrestrial particles, which possessed a "congenital" simple circular motion, and of a pure air (*aer purus*), sometimes compared with an "aethereal body." The former fraction mechanically performed the functions which Boyle attributed to spring; the latter fraction filled the space in the receiver at all times. The subtler and purer fractions of air were still *air*, and so the

[60] Boyle, "Examen of Hobbes," p. 191; see also p. 207: ". . . I have not here taken upon me the person of a Vacuist."

[61] Boyle, "Animadversions on Hobbes," p. 112.

[62] See, for example, Boyle, "Continuation of New Experiments," p. 250.

receiver could never legitimately be said to be empty. This portion was the one which, Hobbes claimed, intruded on the integrity of the air-pump.[63] Nevertheless, when Boyle defended his pump's integrity from Hobbes's aspersions, he was concerned to refute three positions, all of which he attributed to Hobbes: (1) that the pump leaked with respect to Hobbesian "pure air"; (2) that it leaked with respect to "common air"; (3) that it was the alleged aether which intruded into the machine. All of Boyle's stipulations and accompanying defences permitted him to say that the pump worked as it was claimed to do and that Hobbes's technical criticisms of its operation were without foundation.

We have noted above that the modified pump produced by Boyle in 1662, with its pumping mechanism immersed in water, was regarded as an adequate riposte to one of Hobbes's tactics for impugning its integrity. Boyle asked Hobbes how "pure air" could get in "by the sides of the sucker that is immersed in water?"[64] Boyle was fully aware of the Hobbesian category of "pure air" and its role in the argument against a vacuum in the air-pump. However, both in the *Examen of Hobbes* and in the later *Animadversions*, Boyle ascribed to his adversary the position that the pump was massively violated by the intrusion of "common air." In Boyle's view this was an entirely insupportable claim, and could be easily dealt with by pointing to a range of experimental findings. In the *Examen* Boyle said that if Hobbes were indeed claiming that the air-pump was always full of "common air," then this was manifestly untrue: the "engine is in great part devoid of it."[65] And again, in the *Animadversions*, Boyle argued that his opponent was striving to "prove our receiver to be always perfectly full, and therefore as full at any one time as at any other of common or atmospherical air." It was, he said, nonsensical for Hobbes to maintain that "our receiver, when we say it is almost exhausted, is as full as ever (for he will have it perfectly full) of common air."[66] Although Boyle never insisted that his receiver was ever absolutely free of atmospheric air, the overwhelming majority of his experimental performances were not seriously discommoded by the presence of a small amount of residual air.[67] So far as Boyle was concerned, he had obtained

[63] Hobbes, "Dialogus physicus," pp. 244, 246, 253; and chapter 4, above.

[64] Boyle, "Examen of Hobbes," p. 208.

[65] Ibid., p. 207.

[66] Boyle, "Animadversions on Hobbes," pp. 112-113, 119; cf. p. 127.

[67] The notable exception was the experiment on the cohesion of marble discs, which is discussed in detail in the following section of this chapter.

an *operational* vacuum: the almost total absence of atmospheric air. Whatever air remained in the exhausted receiver had been taken account of and did not embarrass the proper interpretation of his pneumatic experiments.

Boyle's most consequential stipulation about Hobbes's views on leakage and the constitution of the air concerned the role of an aether. Boyle agreed with his adversary that the air might be regarded as a heterogeneous mixture. He allowed that one of its fractions might be an aether, and even that this aether might intrude itself (or be always present) in the receiver. But he took exception to Hobbes's alleged use of the aether to impugn the findings of the pump: "Mr. *Hobbes* seems to think he has sufficiently confuted me" if he shows

> that there is a subtile substance, which he calls aether (but which I wish he had better explained) in some places, which I take not to be filled with air; and that the aether has or has not some accidents, which I deny not, but that the atmosphere or fluid body, that surrounds the terraqueous globe, may, besides the grosser and more solid corpuscles wherewith it abounds, consist of a thinner matter, which for distinction-sake I also now and then call etherial.

This aether must either be demonstrated by experiment to exist or it was to be regarded as a *metaphysical* entity. Plainly, Hobbes's introduction of an aether in the context of the air-pump trials was, for Boyle, based on no such experimental evidence. Boyle preferred to speak common-sensically of the *air*—a substance whose presence or absence in the air-pump could be addressed experimentally. According to Boyle, Hobbes

> seems to have misapprehended my notion of the air. For when I say, that the air has gravity and an elastical power, or that the air is, in great part, pumped out of the receiver, it is plain enough, that I take the air in the obvious acceptation of the word, for part of the atmosphere, which we breathe, and wherein we move.[68]

[68] Boyle, "Examen of Hobbes," p. 196. For Boyle's most explicit discussion of the constituents of the air, see "The General History of the Air," esp. pp. 612-615. Here, he speculated that "our atmospherical air may consist of three differing kinds of corpuscles": earthy exhalations, more subtle portions (including emissions from the Sun and other stars), and corpuscles responsible for the air's elasticity. All these were distinguished from the aether.

The concession that there might be an aether, and that the exhausted receiver might contain aether, served two purposes for Boyle. First, it provided him with yet another reason not to plump for either a Cartesian-plenist or an atomical-vacuist ontology; second, it might form the basis for further experimental researches that could test whether or not the existence of aether had any consequences that were meaningful within the experimental programme. If the receiver was permeable to aether, Boyle could potentially identify the alleged leakage of the pump with the more tractable and less vexed question of the aether porosity of all materials and all barriers.[69]

One of the major uses of the redesigned air-pump in the 1660s was in testing the putative aether for its physical and experimental effects. In his *Continuation of New Experiments Physico-Mechanical*, Boyle described a series of experiments involving a receiver fitted with a type of bellows (see figure 8, bottom left). When the receiver was evacuated in the usual way, these bellows could be worked so that whatever then remained might be formed into a jet or stream. This jet could then be directed onto a detector, such as a lightly constructed windmill or feather.[70] Any movement of these detectors could be taken as an indication that the exhausted receiver still contained either a significant amount of atmospheric air or of an aether that had physical consequences. The first possibility was eliminated by a replication of the void-in-the-void experiment (number 17 of the original *New Experiments*). With the new machine Boyle got the mercury to fall *exactly* to the level of the lower basin, allowing him to conclude that the leakage problem was, for all practical purposes, solved. No detectable atmospheric air remained.[71] The aim of the series of experiments employing the bellows was to make the putative aether the subject of "sensible experiments (for I hear what has been attempted by speculative

[69] Boyle, "Examen of Hobbes," pp. 208-209. For Boyle's continuing interest in the porosity of glass to various subtle effluvia, see "Essays of the Strange Subtilty . . . of Effluvia," esp. p. 726 (1673); for a demonstration that glass is not pervious to air, see "An Essay of the Intestine Motions," esp. pp. 454-457 (1661); cf. "Experiments and Considerations about the Porosity of Bodies," esp. pp. 787-792 (1684).

[70] Boyle, "Continuation of New Experiments," esp. pp. 250-258. For a good modern account of these experiments, see Conant, "Boyle's Experiments in Pneumatics," pp. 38-49. Although Boyle claimed that these trials dated from around 1666-1667, internal evidence indicates that some, at least, were performed as early as 1662—that is, from roughly the date of Boyle's *Examen of Hobbes*.

[71] Boyle, "Continuation of New Experiments," p. 216.

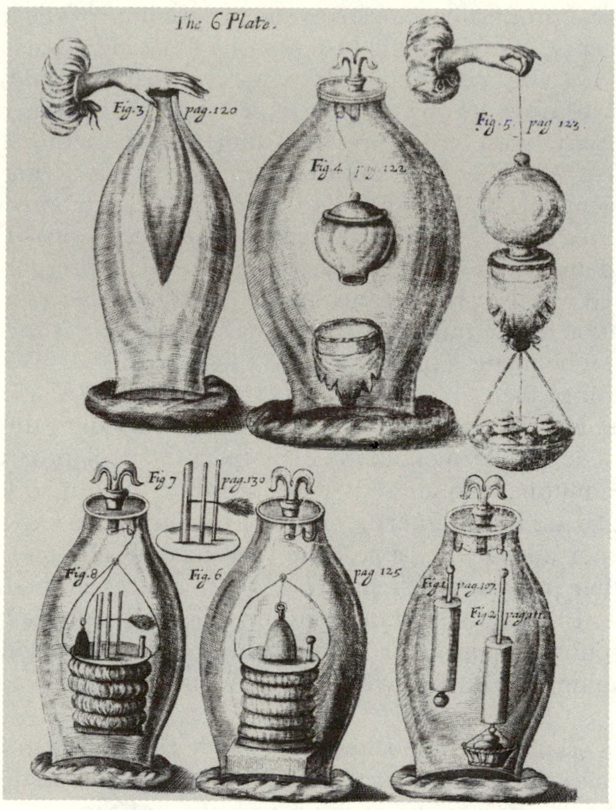

FIGURE 8

Diagram of Boyle's experiments to manifest the effects of an "aether-wind" forced out of a bellows in the exhausted receiver (from Continuation*). Note especially the bottom left figure, where a feather was used as a detector. Boyle claimed to detect no motion that would indicate the existence of an "aether-wind." (Courtesy of Edinburgh University Library.)*

arguments)" and to "discover anything about the existence [and] the qualifications of this so vast aether" posited by the Cartesians and other plenists.[72] In the event, Boyle found that the "jet" supposedly emitted from the bellows did not move any of his detectors. What could be concluded?

If there was an aether, if it was "really so subtle and yielding a matter" that could penetrate wood, leather and glass, and if it was present in the exhausted receiver, then it was not "sensible." It had

[72] Ibid., p. 250.

no physical properties relevant to the programme of the air-pump experiments. A vacuist could then argue "that either the cavity of the bellows was absolutely empty, or else that it would be very difficult to prove by any sensible experiment that it was full." For Boyle's part, these trials provided him with a further basis for nescience: "We, that have not declared for any party, may by our experiment be taught to have no confident expectations of easily making [the aether] sensible by mechanical experiments." The plenist could maintain belief in the aether, even its presence in the evacuated receiver, but he must not deploy the aether as a physical explanation of the pump's phenomena: this aether "is such a body as will not be made sensibly to move a light feather."[73] What Boyle had accomplished through the experiments in *Continuation* was to shift plenist criticism of his air-pump researches onto the ground of "sensible" experimentation. Boyle permitted plenists to retain their commitment to an all-pervading subtle matter, so long as it did not figure in the interpretation of the air-pump experiments. The aether was not "sensible," and therefore it was a metaphysical entity that had no legitimate role within the experimental programme. In this way, although not mentioning Hobbes by name, the air-pump experiments reported in *Continuation* belong to the programme of research which his criticisms prompted and made urgent.

In this section we have shown that Boyle responded to Hobbes's attack on the integrity of his machine by making several stipulations about the latter's notions of the constitution of the air and by offering systematic technical, experimental, and conceptual rebuttals. Each line of reply had the effect of protecting the value of experiment and of reinforcing crucial boundaries between practices that belonged to experimental philosophy and those that did not. Boyle's technical and experimental response to the assertion that his pump was violated by atmospheric air was important because, if this were so, the matter of fact upon which his interpretation rested would be no fact at all. Boyle's technical, experimental, and conceptual responses to the claim that the pump was always full of an aether were important because his handling of this assertion allowed him to police the demarcations dividing the legitimate discourse of "physiology" from the illegitimate discourse of "metaphysics." Experimental philosophy dealt with what the pump actually contained and with what could then be rendered "sensible."

[73] Ibid., pp. 250-251, 256.

Only on that basis could the experimental programme generate assent and order in the natural philosophical community. Only if the air-pump were seen to be practically inviolable could the experimental programme perform the epistemological and moral functions expected of it. For these reasons, Boyle's replies to Hobbes on the constitution of the air were defences of the moral order which the apologists of experimental practices aimed to establish and secure.

Boyle and Hobbes: Men of Marbles

Could an experiment fail? In an obvious sense the answer is yes. Any outcome of an experiment other than the one expected could count as a failure. For example, if one performed the void-in-the-void trial and the level of mercury in the cylinder did not fall almost to the level of the "restagnant" mercury, the experiment could be deemed a failure: the expected result was not achieved. The problem of "unsuccessful" experiments was a matter of intense interest to Boyle in the early stages of his published experimental researches. The *Two Essays, Concerning the Unsuccessfulness of Experiments* dates from about the same time as the air-pump trials. In these essays Boyle made two points vital to the performance, interpretation and reporting of experiments generally. First, he offered a repertoire of excuses for the failure of particular experiments. He provided a list of plausible reasons why the outcome expected (specifically of "unobvious" or "elaborate" experiments) might not be obtained. The materials might be adulterated or impure; there might be natural diversity in their composition; tradesmen might be inept in their handiwork; experiments that succeeded in small scale might not work in large scale; and so forth.[74] Thus Boyle proffered a set of factors that could be brought into play so that a given expectation or theory need not be invalidated by experimental failure. There might be nothing at all wrong with the theory; the fault might lie in the materials, the apparatus, the inherent intractability of the experimental design, or in a range of factors

[74] Boyle, "Unsuccessfulness of Experiments," esp. pp. 334-335. For analyses of Boyle on unsuccessful experiments: Stieb, "Boyle's Medicina Hydrostatica and the Detection of Adulteration"; Wood, "Methodology and Apologetics," pp. 6-7, citing Sprat, *History*, pp. 243-244.

that could not be easily specified.[75] Second, both here and in related essays published early in his career, Boyle stressed the importance of candid and total reporting of actual experimental outcomes, whether these successfully fulfilled expectations or not.[76] As we have argued in chapter 2, such complete and "circumstantial" reporting was basic to the credibility of experimental practices as the grounds of a consensual natural philosophy. Only if the circumstances of failures were reported would like retailing of successes be fully credited. Moreover, it was a key element of the experimentalist's "modesty" to admit failure and to enlist the assistance of the collective in remedying its causes.

So, for Boyle the notion of "unsuccessfulness" was a *positive* resource in validating the experimental programme. There was no reason why experimental failure should dictate the jettisoning of any particular physical hypothesis, much less a highly valued or fundamental one. While in practice Boyle reported numerous instances of failed pneumatic experiments, he never once took any of these failures as reasons to abandon, or even to cast significant doubt upon, the "doctrine" of the spring of the air. (In the next chapter, we shall examine Boyle's response to "success" and "failure" in the problem of one particularly uncomfortable finding, "anomalous suspension.") If we shift our attention from Boyle's programmatic writings to his actual experimental and discursive practices, we might even advance the apparently paradoxical contention that his experiments could *not* fail.

In Boyle's practice, there were no impersonal and abstract criteria that identified success or failure. A successful experiment was one that was *counted* as successful. The judgment that an experiment had been successfully performed was a judgment that it had met expectations for all practical purposes. Any physical hypothesis could be saved from an admitted experimental failure either by pointing to a range of subsidiary hypotheses or by modifying the key hypothesis so that this could be seen as unaltered for all practical purposes. Any experiment judged to be "successful" could be judged by someone else to be "unsuccessful." A critic so minded could make trouble for any experiment whatever. This is to say that there was no such thing as a "crucial experiment." All judgments about the performance and the meaning of experiments

[75] In chapter 4 we noted that Hobbes possessed a sensibility similar to that of the "modern" Duhem-Quine thesis about the defeasibility of all experimental conclusions. Hobbes employed this sensibility to erode the value and credibility of experiments; Boyle used a comparable perspective to *defend* experiments.

[76] See, for example, Boyle, "Proëmial Essay."

were contingent judgments.[77] Nevertheless, as a contingent matter, Boyle made greater investments in the successful outcome of some experiments than he did in others: experiments such as the void-in-the-void trial were *labelled* as crucial confirmations of the spring of the air and the judgment that they had in fact been successfully performed was treated as vital evidence in favour of that doctrine.

In this section, we examine an experiment Boyle did not label as entirely successful—one that Hobbes seized upon as crucial disconfirmation of Boyle's explanatory resources. We display the nature of Boyle's responses to Hobbes's criticisms and the means by which critical failure could be transmuted into a stipulated crucial success. We will see what investments were involved in the success or failure of this experiment and the way in which the integrity of the experimental programme hinged upon the allaying of doubts about the experiment in question.

The experiment is one we have treated in earlier chapters: the cohesion of smooth marble discs when placed in an air-pump that was then evacuated. Boyle had been experimenting on the problem of cohesion even before the first air-pump trials. In *The History of Fluidity and Firmness* (1661) Boyle reported on cohesion experiments performed in about 1659.[78] His purpose was to develop the cohesion phenomenon as a centre-piece of experimental natural philosophy and of his doctrine of the spring and weight of the air in particular. As Boyle freely conceded, the whole series of experimental trials on cohesion was troubled: the mere production of the phenomenon was far from easy, and it commonly failed. He supplied potential replicators with a set of reasons why they might fail, making use of the principles outlined in the essays on unsuccessful experiments published in the same volume. One excuse involved problems in securing marbles (or glasses) of sufficient smoothness: irregularities in their surfaces would permit the entry between them of a small amount of air, with its attendant spring, and they would either not cohere or would quickly fall apart:

> . . . experience has informed us, that it is extremely difficult, if at all possible, to procure from our ordinary tradesmen either glasses or marbles, so much as approaching such and

[77] For recent literature on the problem of crucial experiments, see Worrall, "The Pressure of Light"; Pinch, "Theory Testing in Science."

[78] Boyle, "History of Fluidity and Firmness" (originally published, like "Proëmial Essay" and "Unsuccessfulness of Experiment," in *Certain Physiological Essays* [1661]). The preface inserted in this collection stated that "these are my recentest composures of this nature (having been written but the last year save one)," i.e., the "History" was written in 1659; Boyle, *Works*, vol. 1, p. 357.

such an exquisiteness: for we could very hardly get either experienced stone-cutters or persons skilled at grinding of glasses, to make us a pair of round marbles, though of an inch or two only in diameter, that would for so much as two or three minutes hold up one another in the air by contact.[79]

Success in these experiments was regarded as a weapon against Scholastic invocation of the *horror vacui*. Replicators were warned not to let initial failure prompt any compromise with Scholasticism. Similarly, success was also regarded as crucial to the doctrine of the spring and weight of the air. Replicators were cautioned not to let such difficulties suggest doubt about this doctrine:

> . . . we have never yet found any sort of experiments, wherein such slight variations of circumstances could so much defeat our endeavour; which we therefore mention, that in case such experiments be tried again, it may be thought the less strange, if others be not able to do as much at the first or second, or perhaps the tenth or twentieth trial, as we did after much practice had made us expert in this nice experiment, and suggested to us divers facilitating circumstances, which could not here in a few words be particularly set down.[80]

Persistence through these troubles was worth the labour, for the experiments promised much to the experimental natural philosopher. First, if an account of the phenomenon of cohesion could be given that was consistent with Boyle's ontology, then, since the phenomenon was a prize specimen of Scholastic physics, this would weigh heavily in the choice of systems that philosophers made:

> I know [Boyle said], that the Peripatetics, and the generality of the school-philosophers, will confidently ascribe the sticking of the marbles, not to the cause we have assigned, but to nature's abhorrency and fear of a vacuum. [However], without having recourse to any such disputable principle, a fair account may be given of the proposed phaenomenon, by the pressure or weight of the air.

Second, the separation of the marbles, once cohering, could count as a demonstration that any such horror of a vacuum either did not exist or, at least, that its power was strictly limited:

[79] Boyle, "History of Fluidity and Firmness," pp. 406-407.
[80] Ibid., pp. 407-408.

> . . . if nature did so violently oppose a vacuum as is pretended, it is not likely, that any force whatsoever that we could employ, would be capable to produce one; whereas in our case we find, that a little more weight added to the lower of the marbles is able to surmount their reluctancy to separation, notwithstanding the supposed danger of thereby introducing a vacuum.[81]

Because of the exemplary character of the cohesion phenomenon, its explanation was contested by natural philosophers. There was very little agreement among them on its interpretation. The physical explanation of the phenomenon was therefore a highly prized accomplishment.[82]

In the *Examen* of 1662, Boyle referred to his earliest trials on cohesion that were published in *The History of Fluidity and Firmness*, specifically the experiments that investigated the cohesion of smooth marbles and their separation in air. In the *later* text, however, Boyle spelt out the technical difficulties of producing the phenomenon, and, in particular, developed the technique of using purified spirits of wine (a recipe was provided elsewhere in the book) to help prevent air coming in between the smooth planes. When he found that alcohol was "too fugitive and subtile," he used almond oil, assuring his readers that these liquids could not be considered as a glue, since the cohering bodies could still slide over each other.[83] Two important lessons might be learnt from these trials. First, they showed that "this pressing or sustaining force of the air, as unheeded as it is wont to be, is very great, but it may also assist us to conjecture how great it is." If one could obtain a set of marbles of the requisite smoothness and standard diameter, the weights needed to cause that separation might be treated as a measure of the air's pressure. Boyle experimented with marbles of various diameters, variously lubricated. He attached weights to the

[81] Ibid., p. 409.

[82] For a survey, see Millington, "Theories of Cohesion."

[83] For reference to the "History of Fluidity and Firmness," see Boyle, "Examen of Hobbes," p. 224, citing Boyle, "New Experiments," p. 69. Boyle had to take care lest his lubricants act as glue, since if they did he would be dealing with the phenomenon of adhesion, not cohesion, and could be criticized on those grounds. The recipe for the purified alcohol is in Boyle, "Unsuccessfulness of Experiments," pp. 332-333. See also idem, "Experiments and Considerations about the Porosity of Bodies," esp. 779, where Boyle conceded that "no art could polish the sides of a component body," such as marble, "so that they should be perfectly smoothed," and that "marble itself, as it is marble, abounds with internal pores." Such claims could be used to contain the damage of failure in the cohesion trials.

lower marbles of the cohering pairs to see at what point they could be made to fall apart. In this series he said he was able to suspend up to 1344 ounces from a set of cohering marbles of about three-inch diameter. But again Boyle was struck by the variability of this experimental system and with the difficulty of replication.[84]

Second, these experiments were clearly regarded as vitally important to Boyle's experimental natural philosophy, since they fitted into his manifestation of the power of the air's pressure in the context of the doctrine of firmness. In this essay, Boyle acknowledged that the causes of firmness in bodies were "the grossness, the quiet contact and the implication of their component parts." Nevertheless, Boyle emphasized the *superior* explanatory importance, in the actual case of cohesion, of "the pressure of the atmosphere, proceeding partly from the weight of the ambient air . . . and partly from a kind of spring by virtue of which the air continually presses upon the bodies contiguous to it."[85] In fact, Boyle was more concerned to display the explanatory role of the air's pressure than he was to specify a single mode by which the air might produce cohesion. So Boyle offered two alternative accounts of the role of the air in cohesion. In both cases it was necessary to show how an isotropic pressure could be directed at the cohering bodies suspended in air, since cohesion could be produced whether the bodies were suspended horizontally or vertically. It was also necessary to show how any pressure whatsoever could be produced at the bottom surface of the cohering pair of glasses or marbles.

First, Boyle appealed to the air's spring, "a kind of recoil (though not properly so called) from the terrestrial globe upwards," which "may strongly press any body, upon which it can bear, against any other, which has no such elastical power to repel from it a body so pressed against it." Second, Boyle appealed to "the pressure of the air considered as a weight," as a completely alternative account. This account was complex. Because air was "not without some

[84] Boyle, "History of Fluidity and Firmness," pp. 407, 409. Compare this enterprise of quantification with that in "New Experiments," pp. 33-39, where Boyle hoped to use the experiment of the void-in-the-void by relating the height of mercury in the barometer to the degree of evacuation in order to give "a nearer guess at the proportion of force betwixt the pressure of the air . . . and the gravity of quicksilver." Compare also the enterprise elicited by Linus's criticisms, which resulted in "Boyle's Law." All these attempts aimed to establish a measure of the power of the air's pressure and to demonstrate its strength against those who doubted it.

[85] Boyle, "History of Fluidity and Firmness," pp. 401-403.

gravity," it must fall to Earth. There it would be prevented going further, and would thus "press as well upwards as in any other way." Then, simultaneous separation of marbles would be very difficult, since to achieve this the force of separation must be "capable to surmount the power of the weight of the above-mentioned cylinder of the atmosphere." Finally, Boyle claimed that there really was a "weight or pressure of the lateral air" that prevented the separation of marbles whose plane of contact was vertical.[86]

Given the importance of the air's pressure in both these accounts, the ultimate prize for Boyle would be the successful performance of the experiment of marbles in the air-pump. This could be represented as a decisive instance of the pressure of the air as a physical power. We recall from chapter 2 that Boyle expected his cohering marble discs to fall apart spontaneously when the air was removed from the receiver. Boyle performed the trials reported in *The History of Fluidity and Firmness* by 1659; before the end of that year he was able to repeat these experiments in the receiver of his new air-pump. He reported the results in *New Experiments* in 1660. However, even with small weights attached to the lower marble, the thirty-first trial of the *New Experiments* did not "succeed," that is, the marbles did *not* separate. Boyle indicated two possible reasons for this continued cohesion. First, he conjectured that the alcohol might act as a glue, rather than merely as a means of excluding intervening air. But he had declared his doubt that alcohol did act "after the manner of a glutinous body." Second, he recuperated the apparent *failure* as an actual *success* by arguing that "the not falling down of the lowermost marble might, without improbability, be ascribed to the pressure of the air remaining in the receiver." He appealed to the nineteenth experiment, where water contained in a barometer within the receiver could not be made to fall below a height of one foot on evacuation. This was used to show that air capable of maintaining a height of one foot of water remained in an apparently evacuated receiver, and thus, Boyle argued, this air could exercise enough pressure to keep two smooth marbles cohering. So Boyle presented experiment 31 as "a strange proof of the strength of the spring of the air."[87]

In the *Dialogus physicus* Hobbes fastened upon this case of Boyle's experimental failure. He had the interlocutor wager that a suc-

[86] Ibid., pp. 404-406.

[87] Boyle, "New Experiments," pp. 69-70; and see our accounts in chapters 2 and 4.

cessful outcome (in Boyle's terms) would mean "it would not be possible to doubt that [Boyle's] assigned cause was true." In Hobbes's view, the experiment had in fact *succeeded*: its outcome (the nonseparation of the marbles) was expected within the Hobbesian physical system, which was plenist, and which, as we have seen, ruled out the spring of the air in this instance. Hobbes contested Boyle's accounts given in *The History of Fluidity and Firmness* and in *New Experiments*. He noted that, in the former text, Boyle offered two mutually exclusive accounts. He also pointed out Boyle's claims that air was reflected from the Earth's surface because of its own weight, that this pressure was exercised through a cylinder, and that such pressure could be lateral as well as vertical. Hobbes denied each of these three claims. Hobbes also pointed to the "failure" of experiment 31 and celebrated this report. The experiment was defined as crucial by both parties, and the issues were sharply distinguished.[88]

Boyle had, so to speak, laid a bet on an experimental result; that result had not been produced. Hobbes had, to continue the metaphor, "seen his bet and raised it." The challenge to Boyle was to make the experiment work. His great adversary had intimated that he would be persuaded of Boyle's assigned cause if the experimental outcome were as Boyle had expected it to be. For once, the grounds of the contest were the ones of Boyle's declared preference: physical explanations would hinge on the findings of "unobvious experiments." Now it was no longer good enough for Boyle to save the experiment by pointing to subsidiary assumptions about the minimal leakage of the pump. Boyle had regarded that leakage as an unfortunate, but tolerable, trouble, and occasionally as a useful one. This would now have to be repaired (along with other features of experimental design) if the experiment were to "succeed" and assent to be secured from his critic. In experiments 17, 19, and 31 Boyle used leakage as a means by which findings could be recuperated as successes. Elsewhere he treated it as a minor problem. Now Hobbes had made the pump's leakage into a perceived source of grave trouble. Over the next fifteen years Boyle bent his efforts to making the experiment of the marbles into a success.

When Boyle came to reply to Hobbes in the *Examen* of 1662, he was not in a position to report such a successful performance. First, he presented the trial as evidence that the marbles cohered strongly, and that the air's pressure was powerful. The receiver was merely

[88] Hobbes, "Dialogus physicus," pp. 267-269.

"not sufficiently exhausted" after the exsuction of air, so the marbles could not fall apart. Boyle pointed Hobbes's attention to the further experiments on cohesion which did not use the pump. These trials had been done in reply to Linus. Since the spring of the air was strong enough to overcome an attached weight of 400-500 ounces in free air, Boyle asked why it was difficult to accept that the spring of a small quantity of residual air in the pump was sufficient to overcome a weight of 4-5 ounces.[89] Second, Boyle deliberately linked Hobbes's attack on the experiment of marbles in the air-pump with the original philosophical context in which the phenomenon had been discussed: the doctrine of fluidity and firmness. Boyle claimed that the thirty-first experiment was in fact marginal to *New Experiments*. Since the text of that experiment included a covert reference to *The History of Fluidity and Firmness*, Boyle could argue that "I handle this matter in this place but incidentally, and may make use of what I have delivered, where I treat of it more expressly; as I have since done in print in the *History of Fluidity and Firmness*, which Mr. *Hobbes* appears to have seen by those censures of some passages of it, which I shall hereafter examine." So Boyle minimized the role of the marbles trial in the text under attack, censured Hobbes for his illegitimate use of citations of a different text, and, finally, added a lengthy appendix to the *Examen* in which Hobbes's attack on the whole doctrine was disputed.[90] In this appendix, Boyle worked to link the attack on the marbles experiment with a much more general debate on the definition of terms, specifically that of "firmness." For Boyle "fluidity" and "firmness" were wholly empirical categories.[91]

Boyle's final tactic in his defence of the cohesion trials involved putting the burden of proof back onto his adversary. Boyle was content to show that his own account was not inferior to that of Hobbes. Here four linked points were made. First, Boyle acknowledged that he had offered accounts both in terms of *spring* and in terms of *weight* of the air. The term "pressure" was used to cover both of these accounts. Neither cause was unassailable, but both were legitimate. However, Hobbes had not offered any legitimate cause. Boyle invoked his well-worked criteria of acceptable causal accounts in natural philosophy. Second, Boyle appealed to the success of trials in free air: "How chance a sufficient weight hung to the lower marble can immediately draw them asunder?" These were

[89] Boyle, "Examen of Hobbes," p. 225; cf. idem, "Defence against Linus," pp. 139-143, 173.

[90] Boyle, "Examen of Hobbes," p. 224.

[91] Ibid., pp. 235-236.

not necessarily relevant to trials with the pump, but they must be dealt with alongside these latter experiments. Third, Boyle observed that Hobbes's claim that the weight lines due to air pressure would form a pyramid, rather than a cylinder, was pettifogging: effectively the lines were parallel. Finally, and most importantly, Boyle pointed to Hobbes's account as a prime example of his adversary's lack of clarity: "I confess [Boyle wrote] I do not well see what he drives at . . . ; nor am I the only person, that complain of his writing often enough obscurely." Again and again he protested against Hobbes's illegitimate ascription to Boyle of views he did not hold. This applied to Boyle's doctrine of fluidity, and it applied, also, to Boyle's doctrine of the reflection of the air at the Earth's surface. Boyle claimed that Hobbes had attacked the notion that air particles could be reflected at the Earth and impinge on the lower surface of the marble. But Boyle also claimed his own real view was that the air, "diffusing itself upon the surface of the terrestrial globe, because its descent is there resisted, does, like water and other liquors, press almost equally every way." Thus the responses Boyle presented in the *Examen* centred on the acceptable language of natural philosophy, its *manners*, and correct experimental *practice*.[92]

Boyle next returned to this troubled experiment in experiment 50 of the *Continuation*. The exact date at which this experiment was performed remains uncertain, but there are good reasons to believe that it was done in 1662, just after the reply to Hobbes.[93] Boyle finally reported success.[94] Perhaps no pneumatic experiment he ever performed was recounted in such elaborate technical detail, and to such dramatic effect. The drama resided in the moment-by-moment interplay between Boyle's expectations and the results witnessed. He had made several modifications to the original experimental system in order to secure success. The new apparatus and trials were significantly different in almost all respects from those described in 1660-1661. The differences included the use of

[92] Ibid., pp. 225-227.

[93] Boyle, "Continuation of New Experiments," pp. 274-276. The preceding experiment of this series (number 49), drew on notes (*Adversaria*) and was dated April-May 1662. As we show in chapter 6, Boyle was working on his new pump and the composition of the reply to Hobbes from autumn 1661. There is a citation of the 1662 edition of Henry More's *Antidote against Atheism* in the text of experiment 50 as "lately published."

[94] Boyle considered this success sufficiently important to report it again in the second edition of *The History of Fluidity and Firmness*: see *Certain Physiological Essays*, 2d ed. (London, 1669), p. 227; also *Works*, vol. 1, pp. 410n-411n.

the redesigned air-pump, with its smaller receiver, capable, as Boyle said, of achieving a more complete working vacuum. Another change was the use of very different lubricating substances, whose function, he claimed, was to ensure perfect smoothness of the marbles by filling up any irregularities in their facing surfaces. Alcohol had little adhesive effect, but was too volatile to use. All the other lubricants undeniably had an adhesive effect which had to be allowed for. Finally, Boyle weighted the lower marble much more heavily than he had in 1659-1660: the weight could be as much as a pound, instead of the 4-5 ounces of the original trial, and it was justified as a way of surmounting "the cohesion which the tenacity of the oil and the imperfect exhaustion of the receiver might give" the marbles. (A critic with a mind to make trouble could have objected that the attached weights served to ensure the marbles' separation, and not merely to correct for the forces working against a "natural" outcome.)

Boyle presented the trial in detail, acknowledging that even "some recent favourers of our hypothesis have declared themselves to be troubled with" the experiment. Here is an extract from Boyle's account of his success:

> [W]hen the engine was filled and ready to work, we shook it so strongly, that those that were wont to manage it, concluded that it would not be near so much shaken by the operation. [The normal process of exhausting the pump apparently produced a certain amount of vibration.] Then beginning to pump out the air, we observed the marble to continue joined, until it was so far drawn out, that we began to be diffident whether they would separate; but at the 16th suck . . . the shaking of the engine being almost, if not quite over, the marbles spontaneously fell asunder, wanting that pressure of the air that formerly had held them together.

The reader was assured that Boyle "foretold that it would cost a great deal of pains so far to withdraw the air, as to make them separate" and that this was "not the only time that this experiment succeeded with us." Sometimes separation was said to occur at the eighth suck, or even sooner, and Boyle also found that "a greater exhaustion" was needed if half a pound was hung from the marble rather than one pound.[95]

95 Boyle, "Continuation of New Experiments," p. 275. Note how, in the indented quotation, Boyle included his physical *explanation* as part of the *description* of the

Despite all these modifications, and his report of successful separation, Boyle foresaw "objections or scruples" that "some prepossessions might suggest." These would concentrate on the function of the weight suspended from the marbles, and the relation between exhaustion and separation, that is, the use of separation as a means of *calibration* of the receiver's contents. In the experiments of the void-in-the-void, Boyle had transformed the fall of mercury or water as evidence of the action of air pressure into the use of the barometer as a measure of the contents of the receiver. He followed the same strategy here: he crafted an experiment that could display the presence or absence of cohesion as dependent on the relative presence or absence of air. The experimental apparatus is shown in figure 9. It represents a complete transformation of the marbles experiment. (Arguably, it is not "the same" experiment at all.) The apparatus consists of a turning-key with a string fixed to the upper marble, and a device to keep the lower

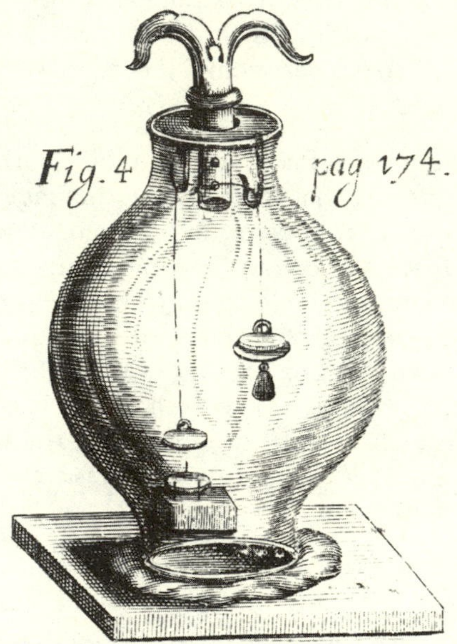

FIGURE 9

Diagram of Boyle's experiment to examine the behaviour of cohering marbles in the receiver (from Continuation*). (Courtesy of Edinburgh University Library.)*

---

experimental outcome. While Hobbes was nowhere named here, the *Animadversions* later made it clear that this trial was to refute him. Henry More, by contrast, was the chief "favourer of our hypothesis" who had focused on these experiments.

marble in place when it fell away (see left of picture). The turning-key and the string allowed the experimenter accurately to raise and lower the marbles. Boyle reported that the marbles could be made to separate on evacuation, as before. This new experiment permitted him to replace the separated discs on each other, and then to let in the external air, whereupon the marbles cohered as strongly as they had at the beginning of the experiment. He could vary the amount of air in the receiver, and observe varying degrees of cohesion: "The little and highly expanded air that remained in the receiver having not a spring near strong enough to press them together" until enough air had been readmitted. Thus the fact of cohesion had been moved from an accomplishment (to evince the competence of Boyle's doctrine of firmness) to a measure (to calibrate the receiver).[96]

With this report Boyle regarded experimental success as achieved beyond question—a success that "will not, I presume, be unwelcome, since it supplies us with no less than matters of fact."[97] In principle, this experiment supplied much more than a matter of fact: Hobbes's bet on its outcome would be lost, and with it, the credibility of his physical system. However, Hobbes did not recant. In the *Decameron physiologicum* of 1678, Hobbes still did not make any mention of Boyle's "successful" experients on cohesion *in vacuo*. He continued unconvinced and unrepentant. He still cited the cohesion of marbles as paradigmatic support for the plenist account. Boyle did try again, returning to this experimental problem in the *Animadversions on Hobbes* of 1674. He recalled his early failure: the interlocutor in this dialogue commented that Boyle "would have put fair for convincing Mr. *Hobbes* himself, . . . if you had succeeded in the attempt you made, . . . to disjoin two coherent marbles, by suspending them horizontally in your pneumatical receiver." But then Boyle mobilized his subsequent successes, reported in *Continuation*, and claimed that they had now shown that when "the spring of the little, but not a little expanded air, that remained," had "grown too weak to sustain the lower marble," the marbles really did separate.[98] In 1674, unlike 1660, it was necessary that Boyle appeal to the *weakened* spring of expanded air: fourteen years earlier he had appealed to the remarkably *powerful* spring of expanded air to explain failure.

[96] Ibid., pp. 275-277.
[97] Ibid., p. 274.
[98] Boyle, "Animadversions on Hobbes," p. 111.

In addition, Boyle recapitulated his attack on the Hobbesian doctrine of firmness. Hobbes had maintained that the separation of the marbles was only possible by an actual bending of the stones, so that air could successively fill the spaces made by flexion: the analogy of the separation of two pieces of wax was used here.[99] If Hobbes were correct, then marbles of greater diameter might be easier to separate than those of lesser diameter. Boyle undertook experiments to test this, found no such connection, and announced further support for the theory of the air's pressure. There was, he said, no experimental evidence to sustain Hobbes's claim that the marbles eventually separated bit by bit from their line of contact: they were "severed in all the points at the same instant." Importantly, Boyle stipulated the options which he claimed were left open to Hobbes after this set of trials. There were only two choices: either Hobbes would have to join the vacuists, or else would at least have to abandon his doctrine that "common air" leaked into the receiver. This last point still irked Boyle. He accepted that Hobbes *could* argue legitimately that "some aetherial or other matter more subtil than air, and capable of passing through glass" was present in the receiver. But Boyle would *not* allow Hobbes to remain committed to the complete leakage of the pump. Boyle read Hobbes as arguing that the receiver was porous to "common air"; Hobbes always argued that the receiver was porous to a fraction of that air, which was, nevertheless, *air*. Ultimately, this debate came to focus on the structure of the pump and its competence as a producer of matters of fact about the air and its pressure.[100]

At the beginning of this section we asked whether, and in what way, an experiment by Boyle could fail. The experiment of the marbles was initially judged by Boyle to be unsuccessful. Its subsequent performances were deemed to be successful. But success in persuading his adversary was not forthcoming. In this very important sense the experiment was a failure. In practice, this "crucial" experiment was not crucially decisive. Nevertheless, our examination of this particular experiment and its career through the 1660s and 1670s illustrates a number of points of general interest to the understanding of the experimental programme and the

[99] For Hobbes's views on hardness, see "Seven Philosophical Problems," p. 32; idem, "Concerning Body," p. 474; Millington, "Theories of Cohesion," pp. 259-260.

[100] Boyle, "Animadversions on Hobbes," pp. 110, 113. For Boyle's speculations on hardness and the internal cohesion of bodies, see his "An Essay of the Intestine Motions," p. 444; "Experimental Notes of the Mechanical Origin of Fixedness," pp. 306-313; and "History of Fluidity and Firmness," p. 411.

problem of assent. First, it demonstrates the role of criticism in generating experimental refinement. We have argued that it was the fact of Hobbes's objections that made the successful performance of this experiment vital and urgent to Boyle. There is no reason in principle why Boyle should not have left the original "failure" of 1659-1660 alone. He had, after all, assimilated its failure to the doctrines whose credibility he wished to establish. The failure could have been treated as venial, rather than mortal, and dismissed from further consideration. Yet the objections of Hobbes and others made this course of action practically untenable. The labour and the technical elaboration expended in making the experiment into a success were elicited by critics' judgments that the original outcome was a source of grave trouble to Boyle's experimental philosophy.

Second, we have the range of investments that different participants made in the outcome of this experiment. Historically, the phenomenon of cohesion was a paradigmatic instance for a variety of philosophies of nature. It was used to prove the existence of a vacuum; it was used to establish the role of anti-mechanical and of mechanical principles; and it was contested by different kinds of mechanical philosophers. The sense that it was a crucial experiment therefore derived from a shared agreement that its explanation was fundamental to the establishment of any system of natural philosophy. Boyle meant to appropriate cohesion as a vivid demonstration of the spring and weight of the air, and his account was disputed by those who controverted those doctrines. Moreover, cohesion was to be appropriated as a demonstration of the power and integrity of experiment. Any perceived trouble with the experiment of cohesion in the air-pump was a source of trouble for the pump and for the experimental form of life. This experiment in particular had to be seen to succeed, not just in Boyle's judgment but in that of his rival philosophers: it had to win their assent. Thus, the "closure" of this experimental problem was to be a decisive tactic in establishing the integrity of experimental practices.

The experiments with cohering marbles were invaluable resources, and Boyle could not dictate their proper use. The "successful" separation of marbles reported in 1669 was a deliberate response to Hobbes. In a different context the earlier "failure" could become much more useful for Boyle's readers. For example, both Newton and Huygens found the trial reported in *New Experiments* an indispensable argument for a subtle aether which penetrated glass. The failure of marbles to separate in an evacuated receiver stayed in the literature *after* Boyle announced he had achieved separation. In early 1664 Newton made notes on *New Experiments*: Boyle's report showed that "ye pressure of ye air is not verry strong" and the cohesion of "ye 2 pollished sides of 2 marbles" must be due to "ye pressure of all ye matter" between the Earth and the Sun. Boyle's failure was thus a successful account of the pressure of an interplanetary aether. In the 1670s, Boyle sent Newton most of his publications on mechanical philosophy, and met Newton on a number of occasions. Newton also acquired a copy of *Animadversions on Hobbes* (1674) in which the separation of marbles *in vacuo* was reported. Yet in February 1679 Newton sent Boyle a lengthy comment on "ye Physicall qualities we spake of." In this celebrated letter, Newton posited "an aethereall substance" which was certainly present in an evacuated air-pump. Newton suggested that the pressure of this aether was why "two well polished metalls cohere in a Receiver exhausted of air."[101]

Similarly, in July 1672, Huygens sent a paper on such an aether to the Académie Royale in Paris and to the Royal Society in London. Huygens insisted that, apart from air pressure, "I conceive that there is yet another pressure, stronger than this, due to a matter more subtle than air, which penetrates glass . . . without difficulty." A critical phenomenon that evinced this pressure was, again, the failure of marbles to separate in an evacuated receiver. Boyle's report in *Continuation* (1669) was ignored. In February 1673 Leibniz pointed out the apparent contradiction between the recent accounts of Boyle and Huygens. Huygens' aether was, however, unaffected. Thus the phenomenon of cohesion was not "closed" within the problematics of late seventeenth-century natural philosophy. In the next chapter, we examine the very similar career of the phenomenon of capillarity in the air-pump. Both sets of phenomena proved difficult to solve within a completely mechanical framework, and it was in this setting that Newton suggested the nonmechanical theory of attractions in his final Query to the *Opticks* of 1706: "The

[101] McGuire and Tamny, *Certain Philosophical Questions*, pp. 292-295, 348-349; J. Harrison, *The Library of Newton*, pp. 107-109; Oldenburg to Newton, 14/24 September 1673, in Newton, *Correspondence*, vol. I, pp. 305-306; Newton to Oldenburg, 14/24 December 1675, ibid., p. 393; Newton to Oldenburg, 18/28 November 1676, ibid., vol. II, p. 183; Newton to Oldenburg, 19 February/1 March 1677, ibid., pp. 193-194; Newton to Boyle, 28 February/10 March 1679, ibid., pp. 288-290. For references to the cohesion of marbles in texts available to Newton in 1679, see Boyle, "Animadversions on Hobbes," pp. 111-112, and "Experimental Notes of the Mechanical Origin of Fixedness," p. 307. For the context of the 1679 letter, see Westfall, *Force in Newton's Physics*, pp. 369-373.

same thing I infer also from the cohering of two polish'd Marbles *in vacuo*." In such arguments, Boyle's "failure" of 1660 was much more valuable than his "success" reported in 1669.[102]

HOBBES, IDEOLOGY, AND THE "VULGAR" CONCEPTION OF NATURE

Boyle identified Hobbes as an enemy of the experimental programme in all its most important aspects, and his treatment of his adversary consisted of a systematic defence of that programme. Hobbes's criticisms were to be rejected because they threatened to undermine the technical basis of experimentation, the conceptual resources of the experimental philosophy, and the social organization of men thought requisite to produce matters of fact and consensus. There was, however, another dimension to the threat that Hobbes was seen to pose. Boyle invited natural philosophers to reject Hobbes's views because they were dangerous to good religion and to the conception of nature that was required by proper Christianity. In this section we show that Boyle argued this case on what *might* be called purely theological grounds: Hobbes, he said, recommended a conception of the Deity and His role that was manifestly insufficient for assuring men of His existence and Providence, and which invited atheism. For that reason alone, the Christian natural philosopher ought to be gravely suspicious of Hobbesian principles in the study of nature. In Boyle's view the role of the natural philosopher included the terms of reference of the Christian apologist: one could not evaluate a philosophy of nature without considering its perceived implications for religion and the moral order.[103] However, Boyle's identification of the religious dangers of the Hobbesian programme was *not* purely theological. Certain of the threats Hobbes posed to right religion were simulta-

[102] Huygens to Gallois, published 15/25 July 1672, in Huygens, *Oeuvres*, vol. VII, pp. 205-206; "An Extract of a Letter . . . Attempting to Render the Cause of that Odd Phaenomenon of the Quicksilvers Remaining Suspended far above the Usual Height . . . ," *Philosophical Transactions* 7 (1672), 5027-5030; Oldenburg to Huygens, 5/15 September 1672, in Huygens, *Oeuvres*, vol. VII, p. 220; Leibniz to Oldenburg, 26 February/8 March 1673, in Oldenburg, *Correspondence*, vol. IX, p. 490. For Newton, see *Opticks*, pp. 390-391; Westfall, *Force in Newton's Physics*, p. 383; Millington, "Theories of Cohesion," p. 268; idem, "Studies in Capillarity and Cohesion," pp. 361-363.

[103] See, for example, Boyle, "The Christian Virtuoso"; idem, "Usefulness of Experimental Natural Philosophy," esp. pp. 15-59.

neously threats to the centre-piece of Boyle's natural philosophy: the doctrine of the spring of the air and its use as an explanatory item in that philosophy.

Boyle's views on the conception of nature most agreeable to proper Christianity were developed in 1666 in his *Free Inquiry into the Vulgarly Received Notion of Nature*, a text not published until 1686.[104] What was "vulgar" and dangerous about this notion of nature was that it ascribed to nature and to matter attributes that were only properly attributes of God. Nature was not to be treated as an agent; nature could do nothing by itself; it was devoid of purpose, volition and sentience. Material nature by itself was inanimate, "brute and stupid"; the only ultimate source of agency in the world was God Himself. Central to Boyle's demonstration was the contention that "motion does not belong essentially to matter." Neither parcels of matter, nor nature as a whole, were capable of self-movement. To maintain otherwise was at once wrong and dangerous. It was dangerous because the notion of self-moving matter and self-sufficient nature dispensed with the active superintendence of the Deity. And an idea of God without active superintendence of nature was tantamount to no God at all:

> [T]he excessive veneration men have for nature, as it has made some philosophers . . . deny God, so it is to be feared, that it makes many forget him. . . . [T]he erroneous idea of nature would, too often, be found to have a strong tendency to shake, if not to subvert, the very foundations of all religion; misleading those, that are inclined to be its enemies, from over-looking the necessity of a God, to the questioning, if not to the denial of his existence.[105]

[104] The preface is dated 29 September 1682, but it was not published until 1686. James Jacob has argued that Boyle wrote this initially against the pantheism and pagan naturalism of Henry Stubbe and the radical sectaries; Nicholas Steneck has identified the target with some form of Cambridge Platonism; and Keith Hutchison has suggested a wide range of naturalisms; see J. Jacob, *Boyle*, pp. 167-169; idem, *Stubbe*, pp. 143-153; Steneck, "Greatrakes the Stroker"; Hutchison, "Supernaturalism and the Mechanical Philosophy," pp. 300-301, 328n-329n. Whatever Boyle's main target, *Free Inquiry* undoubtedly collected together a number of positions that could be said to contribute to a "vulgar" conception of nature, including those associated with Aristotelians, Cartesians, and with Hobbes himself.

[105] Boyle, "Free Inquiry," pp. 210, 192; also "Usefulness of Experimental Natural Philosophy," p. 47; "The Christian Virtuoso," p. 520. On the denial of God's active providence as perceived practical atheism, see Hunter, *Science and Society*, chap. 7; Shapin, "Of Gods and Kings." Compare the now classic statement by Newton in the late 1660s: "Hence it is not surprising that Atheists arise ascribing that to

There were a number of common phenomena of nature that seemed to support the notion of self-moving or sentient matter. Among the most important and widely cited of these were suction and restitution. Throughout his life, Boyle was concerned to argue against an explanation of suction that attributed to matter the abhorrence of a vacuum. One of the virtues of the mechanical account and of the spring of the air was precisely that they showed the explanatory inadequacy of "vulgar" conceptions that ascribed qualities of sentience and volition to mere matter.[106] In this way, the doctrine of the air's spring was an important resource in arguing against a notion of nature and matter which, Boyle said, was pernicious to right religion. A proper understanding of restitution was equally vital. It was tempting to attribute the restitution of "springy bodies" to their inherent possession of sense and self-movement. If one compressed a spring or a pile of wool fleece, how else did these bodies "know" the original shape they were to reacquire and how did they manage to achieve this restitution? Boyle did not offer a single definitive account of the cause of restitution, but argued strongly against the adequacy of the "vulgar" notion of this phenomenon.[107]

Restitution occupied a central place within Boyle's mechanical philosophy, for the spring of the air was the most important explanatory resource in his pneumatics. As we have seen in chapters 2 and 4, Boyle said he wished to ban the search for causal notions from mechanical and experimental philosophy. Specifically, he wished to use the idea of the air's spring without identifying its cause. It was a nescience that was intolerable to Hobbes: an enterprise that did not identify the cause of the spring could not be *philosophy*. Moreover, Hobbes publicized his suspicions that Boyle's nescience masked an anti-mechanical cancer in the body of his adversary's experimental philosophy. As he wrote in the *Dialogus physicus*, the doctrine of the air's spring is an absurdity, "unless

---

corporeal substances which solely belongs to the divine. Indeed, however we cast about we find almost no other reason for atheism than this notion of bodies having, as it were, a complete, absolute and independent reality in themselves." (Newton, *Unpublished Scientific Papers*, p. 144.)

[106] For examples, see Boyle, "Cause of Attraction by Suction"; idem, "Free Inquiry," pp. 193-194; idem, "Usefulness of Experimental Natural Philosophy," pp. 37-39; idem, "History of Fluidity and Firmness," pp. 409-410; idem, "New Experiments," p. 75.

[107] Boyle, "Free Inquiry," pp. 205-210.

perhaps we concede what is not to be conceded, that something can be moved by itself."[108]

The contest between Hobbes and Boyle was, among other things, a contest for the rights to mechanism. The charge that Boyle's philosophy was anti-mechanical at its core was potentially fatal. Moreover, this charge, unless refuted, identified Boyle's philosophy of nature as "vulgar." According to Hobbes, the spring of the air was just such a notion of matter that Boyle regarded as dangerous to good religion: it conceived of matter as self-moving. Boyle replied to this charge in the *Examen of Hobbes*. His tactics were to hoist Hobbes on his own petard. Whose natural philosophy was it, Boyle inquired, that was truly mechanical? And whose philosophy actually contained a "vulgar" conception of nature? In Boyle's opinion Hobbes's philosophy was far more vulnerable to this indictment than his own. Hobbes, he said, seemed to believe that "motion is natural to some, if not all parts of matter"; he "ascribe[s] a motion of their own to multitudes of terrestrial corpuscles."[109] For example, Boyle pointed to the Hobbesian account of gravity in terms of the "simple circular motion" of earthy particles. What is the cause of this particulate motion? Hobbes had not produced a satisfactory mechanical causal account, resulting in a contradiction to "his fundamental doctrine, that *Nihil movetur nisi à corpore contiguo & motu*." In 1675 Boyle spelt out Hobbes's dilemma. If "every body needs an outward movent, it may well be demanded, how there comes to be any thing locally moved in the world?" Hobbes would need to appeal to some external prime mover, such as God. But if God were immaterial, then Hobbes would be compelled to admit that motion was generated by the interaction of matter and something immaterial. On the other hand, if God were material (and "Mr. *Hobbes*, in some writings of his, is believed to think the very notion of an immaterial substance to be absurd"), then Hobbes would be compelled to attribute inherent motion to this form of matter. Thus, either Hobbes would concede that motion was a product of spirit or that motion was innate in matter. In 1662 Boyle made sure that the theological implications of such "natural motion" were not missed:

> That which some will, I doubt not, peculiarly wonder at in Mr. *Hobbes*'s hypothesis, is, that he makes this regular motion of each atom *naturae suae congenitus*: for philosophers, that are

[108] Hobbes, "Dialogus physicus," pp. 254-255.
[109] Boyle, "Examen of Hobbes," pp. 194-195.

> known to wish very well to religion, and to have done it good service, have been very shy of having recourse, as he has, to creation, for the explaining of particular phenomena.[110]

What Boyle's response to Hobbes did not contain was a specific counter to Hobbes's innuendo about the spring of the air. Boyle had been put in an extremely difficult position. If he could not supply a mechanical cause for restitution and the spring, how could he exculpate himself from the charge that the motions of restitution were inherent to matter? If he did venture a causal account, he violated one of the most vigilantly policed demarcations of the experimental philosophy. In the event, Boyle opted to maintain his causal nescience, although to the end of his life he continued to offer a variety of ways in which the spring might possibly act.[111]

These exchanges between Boyle and Hobbes illustrate the difficulty of identifying, so to speak, the "real position" of each author. Evidently, nothing could be clearer than that both Boyle and Hobbes were *mechanical* philosophers, that each abominated anti-mechanical notions like the *horror vacui*, and that each avoided anything like the attribution of self-movement to matter. Yet each was in a position plausibly to charge the other with serious violations of mechanism. Hobbes was able to make out Boyle's spring of the air as anti-mechanical; Boyle was likewise able to make out Hobbes's simple circular motion as invoking a conception of self-moving matter. In this way, the category of *mechanical philosophy* was an interpretative accomplishment; it was not something which resided as an essence in the texts or in the intentions of their authors. If Boyle could accomplish and make credible an interpretation of his adversary as a "vulgar" violator of mechanism, he could achieve two important ends: first, he could show that Hobbes was inconsistent—inconsistency counting as a vicious defect in any authored system of philosophy; second, he could associate Hobbes with a conception of nature that was recognized to be dangerous to religion. Boyle did not invite two separate assessments of Hobbes's philosophy of nature: one theological and one technical. Rather, he urged one assessment that included both sorts of criteria. What

[110] Ibid., p. 205; idem, "Some Considerations about Reason and Religion," pp. 167-168. Boyle used "creation" here in the sense of *created nature*.

[111] Boyle, "The General History of the Air," pp. 613-615; cf. Hooke, *Lectures Explaining the Power of Springing Bodies* (1678), pp. 39-40; B. Shapiro, *Probability and Certainty*, pp. 50-51.

was conducive to right religion was also the basis for correct natural philosophy.

Boyle's attack on Hobbes as a subverter of Christian religion was not restricted to his adversary's "vulgar" notion of self-moving matter. Boyle directly addressed Hobbes's conception of the Deity, His nature, place, and role in the world. It was in this context that Boyle publicized the connection between Hobbes's plenism and his assault on "incorporeal substances" that we discussed in chapter 3. In the *Animadversions on Hobbes* Boyle noted both Hobbes's professed belief in God and his objections even to the idea of "interspersed vacuities." But Hobbes's God was corporeal, and where, Boyle asked, was there any space left in Hobbes's plenum for such a Deity?

> For since he asserts, that there is a God, and owns Him to be the Creator of the World; and since, on the other side, the penetration of dimensions is confessed to be impossible, and he denies, that there is any vacuum in the universe; it seems difficult to conceive, how in a world, that is already perfectly full of body, a corporeal Deity such as he maintains in his *Append. ad Leviathan, cap. 3* can have that access, even to the minute parts of the mundane matter, that seems requisite to the attributes and operations, that belong to the Deity, in reference to the world. But I leave divines to consider, what influence the conjunction of Mr. *Hobbes*'s two opinions, the corporeity of the Deity, and the perfect plenitude of the world, may have on theology. . . . Mr. *Hobbes*'s gross conception of a corporeal God is . . . unwarranted by sound philosophy.[112]

If Boyle could show that Hobbes's physics was unsound, he could hope to undermine public support for his religion and civic philosophy. Thus Boyle justified his initial response to Hobbes through the benefits that victory might bring to sound religion:

> [T]he dangerous opinions about some important, if not fundamental, articles of religion, I had met with in his *Leviathan* . . . , having made but too great impressions upon divers persons . . . , these errors being chiefly recommended by the opinion they had of Mr. *Hobbes*'s demonstrative way of philosophy; it might possibly prove some service to higher truths than those in controversy between him and me, to shew, that in the Physics themselves, his opinions, and even his ratiocinations, have no

[112] Boyle, "Animadversions on Hobbes," pp. 104-105. The reference to *Leviathan* is to the appendix in the 1668 Amsterdam edition of Hobbes's *Opera philosophica*.

such great advantge over those, of some orthodox Christian Naturalists.[113]

Boyle invited natural philosophers to reject Hobbes's physics because it conduced to irreligion, and he suggested that his adversary's civic philosophy and theology could be invalidated if it were shown that his physics was unsound. Thus Boyle's defence manifested the typical seventeenth-century network of calculation whereby criteria used to evaluate good religion were built into the evaluation of good natural philosophy and *vice versa*. This in itself is an unoriginal finding. However, the precise manner in which this network of calculation functioned, and the way in which it bore upon the technical findings of experiment, have been less well understood. For example, by attacking Hobbes's "vulgar" conception of nature, Boyle was also protecting the integrity of the experimental programme in general and of the air-pump in particular. Hobbes's plenism and his assumption of simple circular motion in the plenum had been used to demonstrate that the receiver of the air-pump was always full. The same conceptual resources that, according to Boyle, threatened a proper idea of the Deity and His attributes simultaneously threatened the perceived integrity of the machine whose workings were the key to proper natural knowledge. Boyle's ideological assault on Hobbes was, therefore, an integral part of his defence of experiment and of the engine that was its powerful and emblematic device.

Henry More: "Talking with the Naturalists in their own Dialect"

The third of Boyle's adversaries to be considered is Henry More, divine and philosopher at Christ's College, Cambridge. In the third edition of his *Antidote against Atheism* (1662) and in his *Enchiridion metaphysicum* (1671), More helped himself to Boyle's reports of the air-pump trials to show the incompetence of matter and the power of a spirit in the world. Boyle responded to these texts in 1672. Two important aspects of the work which established experimental philosophy emerged in this dispute. First, both Boyle and More

[113] Boyle, "Examen of Hobbes," p. 187. For similar strategies, see John Wallis to John Owen, 10/20 October 1655, in Owen, *Correspondence*, pp. 86-88 (dedication of Wallis, *Elenchus geometriae Hobbianae* [1655]), and Wallis, *Hobbius heauton-timorumenos*, p. 6.

represented their publications as exemplary of good manners in natural philosophical argument, since they were held to show how writers could contest the use of matters of fact without generating enmity. Second, the contest with More tested Boyle's own model of proper experimental work and, even more importantly, its proper use. Boyle charged More with improper appropriation of experimentally generated matters of fact, and improper mobilization of those matters of fact for his own purposes. We recall that Boyle had displayed his reply to Linus as an ideal case of the way in which experimentalists could argue fairly, even with Jesuits, provided the rules and boundaries of natural philosophy and theology were observed. Similarly, Boyle charged Hobbes with subversion, since Hobbes had failed to offer experimental work to contest that of Boyle, and had denied the foundational status of matters of fact. Here, too, in disputes with More, Boyle defended a model of the natural philosophical community and its behaviour, which he carefully displayed for More, and, simultaneously, denied that this amounted to an attack on More's position in metaphysics.[114]

More published his *Collection of Several Philosophical Writings* in 1662. The texts assembled here revealed the range of political and theological targets which More had assailed during the Interregnum. They also summarized the philosophical enterprise that More claimed to have mounted during those years. The *Collection* defined More's enemies: the radical sectaries, enthusiasts, and Hobbist mechanists. It laid out More's ambition for a Restoration settlement based on toleration and passive obedience, to be reinforced by a clergy trained at Cambridge. It described a "spirit of nature," given its first exposition in More's *Immortality of the Soul* (1659), republished here. Such a directive "spirit" formed part of an ontology which would effectively counter those of mechanist atheists and enthusiasts. Finally, More explained the need for a priesthood proficient in experimental philosophy. In the reissue of the *Antidote against Atheism* (originally published in 1653), he argued that such experimental work must display the real action of spirit in nature. The form of persuasion used in experimentation was a powerful

[114] For More's career at this point, see P. Anderson, *Science in Defense of Liberal Religion*, pp. 152-164; Lichtenstein, *Henry More*, pp. 128-135; Pacchi, *Cartesio in Inghilterra*, chap. 4; Cristofolini, *Cartesiana e sociniani*. For recent surveys of More and natural philosophy, see Staudenbauer, "Platonism, Theosophy and Immaterialism"; Gabbey, "Philosophia Cartesiana Triumphata"; also Boylan, "More's Space and the Spirit of Nature."

weapon against the atheists; products of such experimentation would reinforce a proper theology: "Every *Priest* should endeavour, according to his opportunity and capacity, to be also as much as he can, a *Rational* Man or *Philosopher*." More's Cambridge ally Simon Patrick agreed. In 1662 he asked how "to free Religion from scorn and contempt, if her Priests be not as well skilled in nature as the People?" In such a "rational and philosophical age," More insisted, his interests and those of his colleagues *demanded* that they "talk with the Naturalists in their own Dialect."[115]

Boyle was the most eminent and the most accessible of the naturalists whom More and Patrick used. He was also very close to More and his allies in the early 1660s. Boyle cited More's own attacks on Hobbes in his *Examen of Hobbes*; Patrick cited Boyle's *New Experiments* in the same year. Ralph Cudworth sent Boyle a copy of his celebrated treatise on the Last Supper, and in June 1665 More acknowledged Boyle's *History of Cold* with the comment that Boyle's matters of fact were a "true copy" of "the constancy of nature"; "future appeals will be made to them amongst the learned, as to the judicature of nature herself."[116] Boyle was also a very close collaborator in at least two areas of immediate concern to the Cambridge theologians: they consulted Boyle about the political campaigns mounted within the University, and they worked with Boyle in the collection of reliable spirit testimonies to use against Hobbists.[117] Boyle's support for the publication of the account of the "Demon of Mascon" matched that of More for the reports of the "Drummer of Tedworth." In the areas of spirit phenomena and priestly competence, Boyle's work and the rules of experimental natural philosophy were seen as openly available to Henry More and his colleagues.[118]

[115] More, *Immortality of the Soul*, book 3, chaps. 12-13, esp. pp. 463-465; idem, *Collection*, "Preface General," pp. iv-v, xv-xix; idem, *Antidote against Atheism*, 3d ed. (1662), in *Collection*, pp. 40, 142. See Patrick, *Brief Account of the Latitude-Men*, p. 24.

[116] Boyle, "Examen of Hobbes," p. 187; Patrick, *Brief Account of the Latitude-Men*, p. 21; Cudworth to Boyle, 27 May/6 June 1664, in Boyle, *Works*, vol. VI, p. 510; More to Boyle, 5/15 June 1665, ibid., pp. 512-513.

[117] On the Cambridge context, see Nicolson, "Christ's College and the Latitude Men"; M. Jacob, *The Newtonians*, pp. 39-47; Biography of Simon Patrick, Cambridge University Library, MSS Add 20 ff 25-26; More to Boyle, 27 November/7 December 1665, in Boyle, *Works*, vol. VI, p. 513; More to Anne Conway, 31 December 1663/10 January 1664, in *Conway Letters*, p. 220.

[118] Spirit testimonies and Boyle's role in them are discussed in chapter 7. See Boyle to Du Moulin, in Boyle, *Works*, vol. I, pp. ccxxi-ccxxii; Hartlib to Boyle, 14/24 September 1658, ibid., vol. VI, pp. 114-115; Boyle to Glanvill, 18/28 September

The 1662 edition of *Antidote against Atheism* included More's first public use of Boyle's experiments. More cited two important instances. In experiment 2, Boyle described the difficulty experienced when raising the cap ("stopple") of the evacuated receiver. "[H]e that a little lifts up the stopple, must with his hand support a pressure equal to the disproportion betwixt the force of the internal expanded air, and that of the atmosphere incumbent upon the upper part of the . . . stopple." Boyle interpreted this trial as strong evidence of the air's pressure and the evacuation of air from the receiver.[119] In experiment 32 a variant of this phenomenon was developed. The receiver was evacuated, removed from its place above the cylinder, and replaced by a valve inserted in the tap above the cylinder and sealed there with diachylon. Boyle reported that it was very difficult to remove the valve and that the air pressing against the valve would close it. As in the marbles experiments, he referred to *The History of Fluidity and Firmness*, where he had ascribed "a great force, even to such pillars of air, as may be supposed to begin at the top of the atmosphere, and recoiling from the ground, to terminate on the bodies on which they press."[120] Boyle used this measure of the pressure of the air against the doctrine of the *horror vacui*. Boyle often attempted the quantification of this pressure to attack the illegitimate Scholastic account and to show there was no place for purposive action in passive matter. "Our experiments seem to teach that the supposed aversation of nature to a vacuum is but accidental," he wrote, and offered a range of true causes. Boyle spelt out this point in the commentary on experiments 32 and 33:

> [T]hat in those motions which are made *ob fugam vacui* (as the common phrase is) bodies act without such generosity and consideration, as is wont to be ascribed to them, is apparent enough in our 32d experiment, where the torrent of air, that seemed to strive to get in to the emptied receiver, did plainly prevent its own design, by so impelling the valve, as to make it shut the only orifice the air was to get out at.[121]

1677 and 10/20 February 1678, ibid., pp. 57-60; Du Moulin, *The Divell of Mascon* (1658); More to Anne Conway, 31 March/10 April 1663, in *Conway Letters*, p. 216. For the uses of such testimony, see More, *Antidote against Atheism* (1662), pp. 86-142; Labrousse, "Le démon de Mâçon."

[119] Boyle, "New Experiments," pp. 15-16.

[120] Ibid., pp. 70-71; idem, "History of Fluidity and Firmness," pp. 403-406.

[121] Boyle, "New Experiments," p. 75.

Boyle used this kind of evidence against Linus's ascription of purpose to inanimate nature. Here, in his annotations to the thirty-second experiment, Boyle distinguished between two levels of teleology in natural philosophy. One range of qualities could *not* be ascribed to matter, because of the boundary between natural philosophy and theology: "Hatred or aversation, which is a passion of the soul" could not be "supposed to be in water, or such like inanimate body." Boyle satirized those who "ascribed to dead and stupid bodies" the "care of the public good of the universe." But it *was* necessary for natural philosophers to show that "the universe and parts of it are so contrived, that it is as hard to make a vacuum in it, as if they studiously conspired to prevent it."[122] So Boyle provided two resources for Henry More in this commentary. First, he outlined the proper place for teleology in natural philosophy, and explained how brute matter was not itself capable of rational design. Second, Boyle claimed that these trials with the air-pump gave powerful evidence of such a distinction. More used both these resources.

More's text of 1662 argued (1) that matter itself was passive, inert and stupid; (2) that its motion was guided by "some Immaterial Being that exercises its directive Activity on the Matter of the World"; (3) that mechanism alone was an inadequate way of accounting for Boyle's phenomena. More's first point was evinced in the closure of the valve by the motion of the air. Experiments 2 and 32 of Boyle's book "are evident arguments of an earnest endeavour in Nature to fill the Receiver again with Aire, as it was naturally before." But consider the result of this endeavour: the valve was closed by the endeavour, and, just as Boyle had argued, this motion was self-defeating. More glossed Boyle's comment:

> [N]either the *Aire* it self, nor any more *subtile* and *Divine* Matter (which is more throngly congregated together in the Receiver upon the pumping out of the Aire) has any *freedome of will*, or any *knowledge* or *perception* to doe any thing, they being so puzzel'd and acting so fondly and preposterously in their endeavours to replenish the Receiver again with Aire.

The same point was obvious from the difficulty in removing the cap of the evacuated receiver: the air's pressure, a consequence of

[122] Ibid. Compare Boyle, "Disquisitions on Final Causes," p. 413 (written before 1677, published 1688); Lennox, "Boyle's Defense of Teleological Inference."

its endeavour to reenter the receiver, prevented its reentry.[123] For empirical evidence of the competence of his "Immaterial Being," More appealed to Boyle's experiments 32 and 33, in which large weights were lifted by the sucker reascending into the cylinder. More claimed that these trials showed the limited applicability of any mechanical law of gravity, and that "there is a *Principle* transcending the nature and power of *Matter* that does *umpire* and *rule* all." Here, a trial that Boyle read as evidence of the power of spring was read by More as evidence of the limited explanatory power of mechanism.[124] Finally, More prefaced his treatment with arguments against naturalists who would exclude the competence of spirit, and limit all to pure mechanical law:

> [T]o conclude that to be by *Sympathy* that we can demonstrate not to be by mere *Mechanical Powers*, is not to shelter a mans self in the *common Refuge of Ignorance*, but to tell the *proximate* and *immediate cause* of a *Phaenomenon*, which is to philosophize to the height.[125]

More rewrote Boyle's experimental reports for his own purposes. He accepted their value as matters of fact. He pointed out those places where Boyle acknowledged the apparent purposiveness of natural action. He read Boyle's exclusion of the attributes of soul from natural philosophy as a limitation on the attributes of matter in nature and in proper theology. More refused to acknowledge or to use the power of spring, which he persistently made out as a mark of the ultimate limits on the mechanical philosophy. Finally, he insisted that natural philosophy's products be used as weapons in theology: this was, in fact, their best and only proper function. These contrasts became explicit conflicts after 1662. They centred on different conceptions of the function of Boyle's programme, different patterns of exploitation of matters of fact, and therefore different forms of life in experimental philosophy and in religion.

"That Monstrous Spring of the Ayre": Boyle's Response to More

The events of the 1660s turned complementary models of the utility and the pattern of work in experimental natural philosophy into

[123] More, *Antidote against Atheism* (1662), p. 44.

[124] Ibid., pp. 43-46; cf. Boyle, "New Experiments," pp. 70-72.

[125] More, *Collection*, "Preface General," p. xv.

exclusive and conflicting models. Boyle and More made a conflict out of the reports of the air-pump trials. Both writers claimed to use this conflict as exemplary of proper manners, persistently claiming that their lack of enmity was a mark of the right ordering of dispute about the use of matters of fact. Yet the issues were crucial for the survival of experimental practice, and for the different powers vested in the priesthood and in the philosophical community. More appealed to Boyle's reports as to the "judicature of nature"; Boyle refused to allow the legitimacy of an appeal that, he claimed, destroyed the authority of that judgment.

In considering the responses to Hobbes and Linus, we have seen that Boyle's reports on pneumatics of the later 1660s were designed to reinforce the status of the spring and weight of the air as experimentally produced matters of fact. These reports also established the propriety of an autonomous and authoritative experimental practice. Boyle's adversaries were told that experimental natural philosophers formed an independent community. His publications included the *Hydrostatical Paradoxes,* composed as a report to the Royal Society on Pascal's hydrostatics in May 1664 and then revised and published after the Plague in 1666, and the *Continuation* of the air-pump reports, published in 1669. Henry More read these texts as examples of the dangers inherent in a project that refused to acknowledge its subservience to greater struggles in theology and metaphysics. He observed that while Boyle's *Hydrostatical Paradoxes* was "a very pleasant discourse," nevertheless "There will be a Spiritt of Nature for . . . anything that ever will be alledg'd to the contrary, or excogitated to evade the unrelish of that principle." In his *Divine Dialogues* (1668), More again used the air-pump trials to show that "there is no purely-Mechanicall Phaenomenon in the whole Universe." For More, the spring of the air was a name for powers that surpassed the competence of pure matter in motion; it was not just another matter of fact evinced with the pump. More referred glancingly to Boyle as one devoted "to the pretence of pure mechanism in the solving of the Phaenomena of the Universe, who yet otherwise [has] not been of less Pretensions to Piety and Vertue."[126]

At the end of the 1660s, More and his correspondents saw a close connection between these developments in experimental phi-

[126] More to Anne Conway, 17/27 March 1666, in *Conway Letters*, p. 269; More, *Divine Dialogues* (1668), vol. 1, sig A6^{v}, pp. 34, 41. See Greene, "More and Boyle on the Spirit of Nature"; and Applebaum, "Boyle and Hobbes."

losophy in England and the growth of an atheistical Cartesianism in Holland and elsewhere. John Worthington pressed More with his duty to defend proper philosophy: More had, in the past, commended these positions, but now, when many "derive from thence notions of ill consequence to religion," More should "remember this evil" by "putting into their hands another body of Natural Philosophy, which is like to be the most effectual antidote." More's replies included the *Divine Dialogues*, in which he coined the term "nullibists" for those who found nowhere in nature for divinity; his *Enchiridion metaphysicum*, completed by 1670 and published in 1671; and a set of *Remarks* (1676) on the writings of Matthew Hale, who published his own criticisms of Boyle's pneumatics between 1673 and 1675. In these books More linked the dangers of an autonomous experimental philosophy with the dangers of a debased metaphysics and theology, using Boyle's matters of fact as part of a rival and corrective natural philosophy. Above all, as More wrote in 1676, "Elastick Philosophers" were suspect because they "make experiments for experiments sake, or to pass away the time, or to be thought great *natural* or rather *mechanical* Philosophers, and that in hope to shew, that all the Phaenomena of Nature may be performed without the present assistance or guidance of any *immaterial Principle*."[127]

Two chapters in *Enchiridion metaphysicum* presented More's reinterpretation of the air-pump trials in detail. Boyle replied to his adversary in a *Hydrostatical Discourse*, which was inserted in a typically chaotic collection of *Tracts* (1672). More sent Boyle a copy of *Enchiridion* and also visited him in person. Through the mediation of Ezekiel Foxcroft, a Fellow of King's College and colleague of More, he was informed of Boyle's reply. More then wrote to Boyle on 4/14 December 1671, outlining three aspects of his own use of the air-pump trials. First, More wrote that it was necessary to dissociate proper philosophy from mechanic atheism, particularly that exemplified by the Dutch Cartesians and in the work of Spinoza: "I have . . . always expressed my opinion, that this mechanical way would not hold in all phaenomena, as I always verily thought: but this would not save us from being accounted amongst the wits, one of their gang." Second, this task was blessed since it sought to prove true religion, "which is a design, than which nothing can be more

[127] John Worthington to More, 29 November/9 December 1667, in Worthington, *Diary and Correspondence*, vol. II, p. 254; More, *Divine Dialogues*, vol. I, "Publisher to the Reader"; idem, *Remarks on Two Late Ingenious Discourses*, pp. 188-189; for the reports from Holland, see Gabbey, "Philosophia Cartesiana Triumphata," pp. 243-247.

seasonable in this age; wherein the notion of spirit is so hooted at by so many for nonsense." More invited Boyle to collaborate in this worthy design: "Truly I expected, that all, whose hearts are seriously set upon God and religion, would give me hearty thanks for my pains: but, however, my reward is with him, that set me on work." Finally, because of the character of this task, and the character of the enemy, More insisted that the rules of dispute should not generate enmity with Boyle as a potential ally. Matters of fact were available to any interpreter whose purposes were worthy, "having acted according to the royal law of equity." More told Boyle that, when creatively exploited, the products of the air-pump trials "were of that excellent importance for the design of my book, and the demonstration of the grand point in hand, that I could not by any means omit them, having mentioned them also in several places of my writings, as a main pledge of the tenets I so much contend for." So "though there be an infinite disparity betwixt your experience and experimenting and mine," More wrote to Boyle, there was no danger here, "so little hurt is there in philosophical oppositions amongst the free and ingenuous!" More claimed that the "standing records" of experimental work were legitimately useful outside natural philosophy, and indeed only fulfilled their purpose inside true religion; Boyle contested the availability of his own products outside the boundaries of the experimental community.[128]

Boyle defended the autonomy and the status of this community throughout his reply to More. He did this in the structure of the text, in his insistence on proper manners in dispute, in his comments on the proper place of spirit in experimental philosophy, and in his detailed treatment of specific experiments, notably that of the cohesion of marbles. The structure of the text was crafted to refute More's claims about the interpretation of matters of fact. Boyle presented three further series of *New Experiments* on pneumatics and hydrostatics to act as resources for a separate *Hydrostatical Discourse* that responded to More's own interpretation. To this discourse he appended a letter to the Scottish mathematician George Sinclair, with whom Boyle was conducting a vituperative priority dispute. By separating the new experiments from his reply to More, Boyle exemplified his claim that matters of fact were not open to indefinite reinterpretation by those outside the community of experimental philosophers.

Boyle also defended his decision to publish in terms of the boundaries of the experimental project. He had done the same in his

[128] More to Boyle, 4/14 December [1671], in Boyle, *Works*, vol. VI, pp. 513-515.

publications against Linus and Hobbes, which he recalled here. More's texts were not "unanswerable"; the dispute was not about such issues as the *fuga vacui*; Boyle wrote that "it is not necessary, that a great scholar should be a good hydrostatician. And a few hallucinations about a subject, to which the greatest clerks have been generally such strangers, may warrant us to dissent from his opinion, without obliging us to be enemies to his reputation." Boyle made much use of these areas of competence. Experiment was a practice that broke the bounds of privileged learning: "I take that which the doctor contends for, to be evincible in the rightest way of proceeding by a person of far less learning than he, without introducing any precarious principle." At the same time, even though More possessed such privileged learning, this privilege did not entitle him to dispute in experimental philosophy: "A man may be very happy in other parts of learning, and of greater moment, that has had the misfortune to mistake in hydrostaticks." So although Boyle did not dispute with "Cartesians" or "spiritists," he insisted that such metaphysicians should not extend their power over naturalists, since "what is with great solemnity delivered for a demonstration in a book of metaphysicks, can be other than a metaphysical demonstration," and so accessible to natural philosophical attack.[129] Even though Boyle conceded that he "never was a gown-man," he could not be denied the right to reclaim authority over the illegitimate use of experimental matters of fact.[130]

Boyle carefully defined the proper site at which talk of spirit could take place. His most graphic illustration was in the form of a parable:

> [If] I had been with those Jesuits, that are said to have presented the first watch to the king of *China*, who took it to be a living creature, I should have thought I had fairly accounted for it, if, by the shape, size, motion &c. of the spring-wheels, balance, and other parts of the watch I had shewn, that an engine of such a structure would necessarily mark the hours, though I could not have brought an argument to convince the Chinese monarch, that it was not endowed with life.[131]

[129] Boyle, "Hydrostatical Discourse," pp. 596-598, 614, 625, 628 (1672). On Sinclair, see *Philosophical Transactions* 8 (1673), 5197, and H. W. Turnbull, ed., *James Gregory Memorial Volume*, pp. 510-513.

[130] Boyle, "Hydrostatical Discourse," p. 596.

[131] Ibid., p. 627. Boyle used the same image in his "Disquisitions on Final Causes," p. 443.

Boyle played the Jesuit to More's Chinaman. Boyle denied that he had set out to show that "no angel or other immaterial creature could interpose in these cases." He wrote of "the doctor's grand and laudable design, wherein I heartily wish him much success of proving the existence of an incorporeal substance." However, Boyle did contest the claim that More's "hylarchic principle" was evinced experimentally, and could be spoken of as part of experimental philosophy. It could not. The spring and weight of the air were Boyle's chief products: they were entities that could perform all the functions More derived from his "knowing" principle. Since "the force" exerted by this principle was "not invincible," Boyle said that "I see no need we have to fly to it, since such mechanical affections of matter, as the spring and weight of the air, . . . may suffice to produce and account for the phaenomena, without recourse to an incorporeal creature." Boyle did not seek to show that "there can be no such thing, as the learned doctor's *principium hylarchicum*, but only to intimate, that, whether there be or not, our hydrostaticks do not need it." This principle was "a mere Hypothesis advanced without any clear positive proof"; it was a "principle that, to say here no more, is not physical."[132]

Boyle argued that because More's spirit was not a physical principle it could not be part of the language of organized experimenters. To make this claim he had to define the proper criteria for good experimental language. Once again, therefore, Boyle pointed to organized nescience and collective witnessing in securing matters of fact. These two features marked the boundaries of the experimental community. Boyle explained to More which items could become the objects of experimental discourse and which could not. The power of the air's pressure was "not a thing deduced" from "doubtful suppositions or bare hypotheses, but from real and sensible experiments." So this power was a matter of fact whose cause remained in doubt: "We ought rather to acknowledge our ignorance in a doubtful problem, than deny what experience manifests to be a truth." Such was the case with trials of magnetism (where the cause was equally uncertain) or of the effects of air pressure on living bodies (where testimony was not yet reliable).[133]

Witnessing was a central resource in Boyle's definition of acceptable experimental items. Boyle charged that More had not

[132] Boyle, "Hydrostatical Discourse," pp. 608-609, 624, 627-628; compare Boyle's precisely parallel strategy for excluding the aether from experimental discourse.

[133] Boyle, "New Experiments about Differing Pressure," p. 643 (1672).

accepted this rule. Boyle consulted "some learned members of the Royal Society, whereof two are mathematicians, and one [More's] particular friend [probably Sir Robert Moray]." They told Boyle that More "did indeed deny the matter of fact to be true. Which I cannot easily think, the experiment having been tried both before our whole Society, and very critically, by its royal founder, his majesty himself." Here Boyle used the status of witnesses against More's brutal denial of a matter of fact. The same argument was used when Boyle argued against More's claim that divers underwater experienced *no* crushing pressure. Boyle wrote that "I am not entirely satisfied about the matter of fact." Laboratory experiments were always more authoritative than testimony which was uncorroborated by reputable witnesses:

> [T]he pressure of the water in our recited experiment having manifest effects upon inanimate bodies, which are not capable of prepossessions, or giving us partial informations, will have much more weight with unprejudiced persons, than the suspicious, and sometimes disagreeing accounts of ignorant divers, whom prejudicate opinions may much sway, and whose very sensations, as those of other vulgar men, may be influenced by predispositions, and so many other circumstances, that they may easily give occasion to mistakes.[134]

Boyle mobilized the full weight of the authority of the experimental community against More. Well-witnessed trials in hydrostatics could not be denied; unsupported testimony by nonmembers could not count against such experiments. Boyle referred to his comments in *New Experiments* (1660) and in the sets of experiments appended to his *Hydrostatical Discourse*. More claimed that the water's pressure was undetectable—but Boyle had made this manifest for other experimenters. More made much use of the closure of the valve of the air-pump by incoming air—but Boyle had explicitly stated in 1660 that the properties of a soul could not be attributed to inanimate bodies. In Boyle's community, inanimate bodies and privileged witnesses had the most authority in experiments. Boyle built his own model of the "general concourse" of active nature as a rival to the vitalism of Scholastic physics, and pointed out that the admission of "nature's abhorrence of a vac-

[134] Boyle, "Hydrostatical Discourse," pp. 614-615, 624, 626; idem, "New Experiments about Differing Pressure," p. 647. For an excellent example of Boyle's own use of a "credible" account "not written by a philosopher . . . but by a merchant," see "A Letter concerning Ambergris" (1673).

uum," or "substantial forms," or other "incorporeal creatures," would frustrate the workings of experimenters. He had written as much against previous adversaries like Linus and Hobbes. He now implied that admitting "hylarchic spirit" within the experimental vocabulary would also have this effect. Oldenburg told Boyle in 1666 that his main achievement was the expulsion of "yt Divell of Substantiall Forms," which "has stopt ye progres of true Philosophy, and made the best of Schollars not more knowing as to ye nature of particular bodies than ye meanest ploughman."[135] Boyle could not and did not deny the work of spirit in nature: he was forced to ban the use of *this* spirit from the language of experiment.

Finally, the chief prize of the air-pump trials was the establishment of the weight and spring of the air. More's interpretation of the experiments challenged this achievement at the deepest level. He considered two of Boyle's most important experiments, that on the rapid and powerful ascent of the sucker towards the evacuated receiver when released, and the experiment on the cohesion of marbles in the receiver. More argued that the cause of the rapid retraction of the sucker must be in the sucker, or in the exhausted cylinder, or in the air itself. Boyle's account was the last of these. But Boyle had given an account in terms of gravity without *explaining* gravity itself. More asked that "if this solution were truly mechanical, then what truly mechanical cause could be given for the gravitation of single particles or of the whole atmosphere?" Boyle insisted that in an experimental report it was unnecessary to offer such general causes: "Having sufficiently proved that the air we live in is not devoid of weight and is endowed with an elastical power or springiness, I endeavoured by those two principles to explain the phaenomena exhibited in our engine . . . without recourse to a *fuga vacui* or the *anima mundi* or any such unphysical principle." Yet again the limited and humble character of experiment was invoked to reject More's insistence that Boyle produce a complete and causal philosophy.[136]

More also appealed to the sucker's ascent to argue that

> if the elastic spring had such a force that it could propel upwards more than one hundred pounds of lead, then indeed

[135] Boyle, "Hydrostatical Discourse," pp. 608, 627 (on More, *Enchiridion metaphysicum*, p. 161 [1671]); Oldenburg to Boyle, 24 March/3 April 1666, in Boyle, *Works*, vol. VI, p. 223; also Oldenburg, *Correspondence*, vol. III, p. 67.

[136] More, *Enchiridion metaphysicum*, p. 138; Boyle, "Hydrostatical Discourse," p. 601.

> all terrestrial bodies which are connected together would be compressed with such a violence that none of them could resist such a compression, so that they would either break apart or they would crumble in this way by the collision of parts, and in a short time they would perish.

More linked this reading of the experiment with a set of claims about hydrostatics. He asserted that bodies have no weight when they fail to descend in a fluid; that such pressure could not be isotropic; that different volumes of air should exert different pressures ("which would be a manifest indication that the whole ingenious hypothesis would be in this respect a fiction"). Above all, if taken as exemplifying the power of spring, the air-pump trials were vulnerable, since elsewhere the spring was "unobvious" and obscure. Other adversaries had made this claim. More turned the spring of the air into an hypothesis. Boyle insisted his prize was a fact.[137]

Boyle's publication responded to this sustained assault on his own reading of the experiment on the ascent of the sucker. He used the experiments collected in his sets of *New Experiments* of 1672, published with the *Hydrostatical Discourse*. These new trials showed that "in the balance of nature the statical laws are nicely observed." In one set of these experiments, Boyle wrapped a bladder round the mouth of a vial of mercury, and then put the vial in a basin of water. The bladder changed shape: Boyle interpreted this as the result of water pressure on the mercury. This weight was displayed again by using glasses under water. When evacuated, these glasses were smashed. An analogy was then drawn with the destruction of glasses in the receiver of the air-pump.[138] Boyle used these analogies to show that the air-pump was not a special case, that it was congruent with other natural phenomena, that the spring and weight of the air were universally competent. Importantly, Boyle responded to More's claim that the spring was an "unobvious" phenomenon, since its effects were only visible in the constrained environment of the pump, and invisible elsewhere. In the attached *New Experiments about the Differing Pressure of Heavy Solids and Fluids*, Boyle pointed to the need for sensitive analogy between water and air, and between the air-pump trials and inaccessible, obscure phenomena. These analogies did allow discussion of the alleged ability

[137] More, *Enchiridion metaphysicum*, pp. 139-140.

[138] Boyle, "Hydrostatical Discourse," p. 612; idem, "New Experiments of Positive or Relative Levity," pp. 638-639 (1672); idem, "New Experiments about the Pressure of the Air's Spring," pp. 640, 642 (1672).

of divers to sustain great pressures, and of the lack of phenomenal evidence for the air's weight: "If there were any place about the moon . . . that has no atmosphere, or equivalent fluid about it, and . . . those men [i.e., divers] should be supposed to be transported thence, and set down upon our earth, there might be made an experiment fitted for our controversy." There was not, and Boyle condescended to rely on something "analogous in our pneumatical engines." The verdict of future experimenters must take precedence over metaphysical quibbles and reports.[139]

More also considered the celebrated trial with cohering marbles. In the 1660s More and Hobbes both pointed out Boyle's failure to separate these marbles in an evacuated receiver. More used two tactics in his discussion of this experiment. Boyle used the experiment as evidence of the power of air pressure, but More used it as evidence of the ignorance of the true principles of weight. The failure of the marbles trial showed that there was no current mechanical account of weight, and so it was necessary to appeal to "the impressed force of the Principle of Hyle or Spirit of Nature." Once again, as in the cases of Newton and Huygens in the 1670s, Boyle's early failure was much more useful than his later success. More used the failure to show that Boyle's experimental reports failed to provide a complete natural philosophy.

More suggested a modification of the cohesion trial that would test the role of air pressure: one of the marbles should be substituted by a piece of wood to examine the true role of smoothness in cohesion.[140] Boyle's reply referred to his *Continuation*, where the successful separation of marbles had been reported. We may recall that Boyle made this trial with some modifications. But More's modifications were unacceptable. In the *Continuation* the separation was presented as a "confirmation" of the doctrine of the spring of the air. In his reply to More, Boyle argued that any critic must first repeat the successful experiment now made public: "Since the experiment, as I proposed it, did upon trial succeed very well, it had not been amiss, if [More] had considered it as it was really and successfully made, and shewed why the pressure of the ambient air was not able to hinder the separation of the marbles." Because this trial had been a "success," the "substitution of a wooden plate" was "needless." Boyle amplified his discussion of the experiment in his *New Experiments of the Positive or Relative Levity of Bodies under Water*. He spent some time on the effect of the spring of air-bubbles

[139] Boyle, "New Experiments about Differing Pressure," pp. 644, 647.

[140] More, *Enchiridion metaphysicum*, pp. 146, 178.

which would lie between nonsmooth bodies, and would thus prevent their cohesion. Then he cited these trials in the main text, and reminded More of the need to stick to matters of fact: "I have discoursed upon supposition, that the doctor experimentally knows, what he delivers concerning the non-adhesion of an exactly smooth wooden plate to a marble one." If he did not, his comments could not be admitted in debate on experiment.[141]

In these contests with his adversaries Boyle constantly pointed to the boundary between the community of experimenters and external critics. Insofar as More or Hobbes failed to produce fresh matters of fact or reproduce those of Boyle, they broke the rules of experimental dispute. Insofar as Linus accepted the experimental form of life, he was a worthy opponent. In all cases, Boyle claimed that debate within the experimental community could be closed and settled. His replies to Hobbes did not win that adversary's assent; neither did his replies to More. On the contrary, in May 1672 More wrote that "Mr. Boyle does not take my dissenting from him in publick so candidly as I hoped, which I am very sorry for." During the next eighteen months, he met and talked with Boyle on this issue. At the same time, the Lord Chief Justice, Matthew Hale, produced a set of texts that attacked both protagonists. In 1673 Hale drew on the authoritative texts of Stevin and Mersenne to argue that "gravitation is either Motion itself, or the *conatus* or *nisus ad motum*." He rejected the explanatory capacity of the hylarchic principle and of the spring of the air. The following year, Hale extended this argument to a detailed critique of Boyle's experiments. Hale declared his allegiance to the views of Linus and other Jesuits; he performed many of the free-air trials reported by Boyle and Linus; he accepted plenist accounts of the Torricellian space and the contents of the receiver; and he gave to the air the function of "common cement and connecter of the different parts of this inferior world." Thus Hale accepted the rules of the experimental game, but he denied the superior explanatory power of Boyle's celebrated spring of the air.[142]

[141] Boyle, "Continuation of New Experiments," p. 274; idem, "Hydrostatical Discourse," pp. 604-608; idem, "New Experiments of Positive or Relative Levity," pp. 636-637.

[142] More to Anne Conway, 11/21 May 1672, in *Conway Letters*, p. 358; Hale, *Essay touching the Gravitation or Non-Gravitation of Fluid Bodies*, pp. 10, 42-44, 87; idem, *Difficiles nugae*, pp. 140-141, 249 (on Linus), 97-116 (on suction trials), 136-137 (on plenism), 240 (on air as cement); idem, *Difficiles nugae*, 2d ed., "Additions," p. 43 (on the void-in-the-void trial).

Responses to More and to Hale manifested the importance of such experimental practices. Boyle decided not to continue the debate by replying directly to Hale. John Wallis and John Flamsteed both did, via the secretary of the Royal Society. Wallis told Oldenburg in June 1674 that Hale actually "grants in effect what is contended for: yt ye Air hath a Gravity & a Spring; & yt by these ye Phaenomena may be solved." Flamsteed suggested variations of the air-pump trials that might more surely evince the "gravitation of ye aire." The air's spring was now the mark of membership of the experimental community. While Wallis claimed Hale did admit the spring of the air, Henry More claimed that Hale *denied* the spring. More answered Hale in *Remarks* that Boyle considered "had better never have been printed." More tried to recruit Hale as his ally. He corrected Hale's Scholastic language to make it less theologically suspect. Hale wrote of a *conatus* in matter as the cause of weight: More argued that such an endeavour must not be innate, but an effect of the superintendent hylarchic principle. Hale echoed Linus's appeal to the *fuga vacui* in analyzing the marbles experiment: More wrote that this principle was "but the final cause," and that the true efficient cause of cohesion was "the spirit of Nature and its Hylostatick Laws, whereby it governs the matter." So, if properly glossed, Hale's books could be used by More as further attacks on an experimental philosophy that excluded spirit and refused to offer its phenomena for proper religion. In February 1676, Robert Hooke read a lecture to the Royal Society that explicitly contested More's doctrine of the hylarchic spirit, and argued forcefully for the autonomy of experimental work. Thus, while Hooke and Boyle made the air's spring an indispensable matter of fact in experimental philosophy, More continued to see it as the chief obstacle to the proper use of the air-pump trials. He wrote that

> If I be mistaken in my Experiments, I suppose Mr. Boyle will show me my mistakes, and all the world besides, which makes me conclude it is better that I have printed them that I and such as I may be undeceived by him or some he may employ against my Lord Hailes and me. For that pinches Mr. Boyle most in those Remarks which my Lord Hailes and myself agree in which is the exploding that monstrous spring of the ayre.[143]

[143] Oldenburg to Huygens, 9/19 July 1674, in Oldenburg, *Correspondence*, vol. XI, p. 49: "Perhaps someone else, who has more leisure and more taste for disputation,

Throughout this chapter, we have seen how Boyle made acceptance of the spring of the air as a matter of fact into a test of membership of the experimental community. For More and Hobbes this community was made up of "Elaterists" or "Elastick philosophers." We have pointed out Boyle's claim that, within the community, debate could be safely pursued and surely resolved. We have seen how he dealt with adversaries who seemed to infringe the boundaries and break the rules of this practice. In the next chapter, we move on to examine how experimenters behaved when troubles emerged *inside* their own community.

---

will reply; Mr. Boyle pursuing his experimental way and modest reflections thereupon, which do not permit him to make any diversion for replying to this sort of author." See Wallis to Oldenburg, 22 June/2 July 1674, ibid., p. 37; Wallis to Oldenburg, 15/25 October 1674, ibid., p. 109; Flamsteed to Oldenburg, 25 January/4 February 1675, ibid., p. 168; Anne Conway to Henry More, 4/14 February 1676, and More to Anne Conway, 9/19 February 1676, in *Conway Letters*, pp. 420, 423; More, *Remarks upon Two Late Ingenious Discourses*, pp. 5-6, 47, 119, 150-151, 171-177; Hooke, *Diary*, pp. 214-216; Gunther, *Early Science in Oxford*, vol. VIII, pp. 187-194 (from *Lampas* [1677]; the lecture is not recorded in Birch, *History*).

· VI ·

Replication and Its Troubles: Air-Pumps in the 1660s

For the AIR-PUMP weakens and dispirits, but cannot wholly Exhaust.

CHRISTOPHER SMART, Jubilate Agno (1759)

IN previous chapters we established that the matter of fact was the fundamental category with which experimental philosophers proposed to solve the problems of order and assent. Hobbes denied that experiment could produce matters of fact that were indefeasible, and that such facts could, or ought to, form the foundations of certain knowledge. One of the tactics Hobbes used to make his case was the display of the *work* required to make a fact. When this work was publicly identified, it could be used to explode any such fact. In principle Hobbes's argument is correct. Establishing matters of fact did require immense amounts of labour. Here we endeavour to recover this labour for our historiographic purposes: to show the inadequacy of the method which regards experimentally produced matters of fact as self-evident and self-explanatory. Any institutionalized method for producing knowledge has its foundations in social conventions: conventions concerning how the knowledge is to be produced, about what may be questioned and what may not, about what is normally expected and what counts as an anomaly, about what is to be regarded as evidence and proof. In the case of Boyle's experimental philosophy, some of the most important conventions concerned the means by which the matter of fact was to be generated. A fact is a constitutively social category: it is an item of public knowledge. We displayed the processes by which a private sensory experience is transformed into a publicly witnessed and agreed fact of nature. In this way, the notion of *replication* is basic to fact-production in experimental science. Replication is the set of technologies which transforms what counts as belief into what counts as knowledge.

In chapter 2, we began to develop a view of replication as a complex set of technologies: not just the physical reiteration of the practice, but, alternatively, the *virtual witnessing* offered by literary technology. In chapter 5, we showed the relationship between the boundaries of the experimental community and accepting the status of the matter of fact. In this chapter we bring these discussions to a focus. We examine the relation between replication and the establishment of a specific matter of fact. We add weight to our view of matters of fact as social conventions. We show the intensely problematic nature of replication, and, therefore, of what was to count as a matter of fact. In replication, there is no unambiguous set of rules that allows the experimenter to copy the practice in question. The multiplication of witnesses demands the dissemination of specific sets of techniques and instruments. In our case, these techniques centred on the air-pump itself. The claims Boyle made about his phenomena could be turned into matters of fact by replication of the pump. But then other experimenters had to be able to judge when such replication had been accomplished. The only way to do this was to use Boyle's phenomena as *calibrations* of their own machines. To be able to produce such phenomena would mean that a new machine could be counted as a good one. Thus, before any experimenter could judge whether his machine was working well, he would have to accept Boyle's phenomena as matters of fact. And before he could accept those phenomena as matters of fact, he would have to know that his machine would work well. This is what H. M. Collins has called the "experimenters' regress." In this chapter, we describe the negotiations between experimenters that were supposed to lead them out of this regress. These negotiations took place at all levels of the replication process. We show the social conventions at play in the following judgments: of the moment when skill in making pumps had been transmitted, when a replica of a pump could be said to have been produced, when that replica had produced the same phenomenon as that reported by Boyle, and when a phenomenon could count as a challenge to Boyle's own claims. A range of commitments and investments bore on judgments whether replication had or had not been achieved, of whether or not a claimed phenomenon authentically existed as a fact of nature.[1]

We therefore pose a series of problems to document in the career

[1] Collins, *Changing Order*, chap. 4. For case studies of replication and its troubles in modern experimental science: Collins, "The Seven Sexes"; Harvey, "Plausibility and the Evaluation of Knowledge"; Pickering, "The Hunting of the Quark"; Pinch, "The Sun-Set"; for historical instances: Farley and Geison, "Science, Politics and Spontaneous Generation"; Ruestow, "Images and Ideas" (on Leeuwenhoek); and, for a range of examples, see Kuhn, "The Function of Measurement."

of air-pumps in the 1660s. We describe their dissemination, construction, and modification. We list the phenomena that were used to calibrate a pump's success. We point out those places where transmission of skill in making pumps could happen, and how that skill was judged. Finally, we indicate the boundaries in the experimental community: we point out what phenomena and techniques could count as challenges to Boyle, what phenomena could count as failures to support Boyle or as failures to match his technical competence, and how closure could be imposed on the open process of replication and manufacture of matters of fact. Considering the importance attached to the air-pump by historians of science, we know remarkably little about its career in the decade following its invention; and, given the importance attached to Boyle's insistence that his written reports allowed experimenters to build their own machines and replicate his findings, we know almost nothing about the processes by which this might have taken place. In this chapter we assemble information about the number, design, and location of air-pumps both in England and on the Continent (see figure 10).

No original Boyle air-pump of the designs reported in *New Experiments* or in *Continuation of New Experiments* survives.[2] During the early nineteenth century, it was widely believed that an air-pump in the possession of the Royal Society was the Boyle and Hooke original design of 1658-1659. This error was corrected by George Wilson, who pointed out that the surviving Royal Society machine had *two* barrels, whereas early Boyle pumps had only one barrel.[3] It seems that the original air-pump Boyle gave to the Royal Society was lost or destroyed between the 1670s and the end of the eighteenth century. The earliest surviving air-pumps are, in fact, of early eighteenth-century manufacture. These are the double-barrelled Hauksbee pumps, dating from circa 1703-1709. Dutch pumps from the same period also survive. Several modern replicas of Boyle's first pump have been constructed. We have very little information

[2] A piece of metal labelled as part of the original Boyle air-pump in the Oxford Musuem of the History of Science clearly does not belong to that device.

[3] For examples of these errors in the nineteenth century: Baden Powell, *History of Natural Philosophy* (1842), p. 235; Thomas Young, *Course of Lectures on Natural Philosophy* (1845), vol. 1, p. 278 ("Hooke's air-pump had two barrels."); Weld, *History of the Royal Society*, vol. 1, pp. 96n-97n ("the original Air pump . . . constructed by Boyle, was presented to the Society by him in 1662, and still remains in their possession. It consists of two barrels."). Grew, *Museum Societatis Regalis* (1681), p. 357, mentions "an Aire pump; or an Engine to exhaust the Air out of any vessel" then in the Society's possession. For Wilson's correction, see his "Early History of the Air-Pump" (1849) and *Religio chemici*, p. 212.

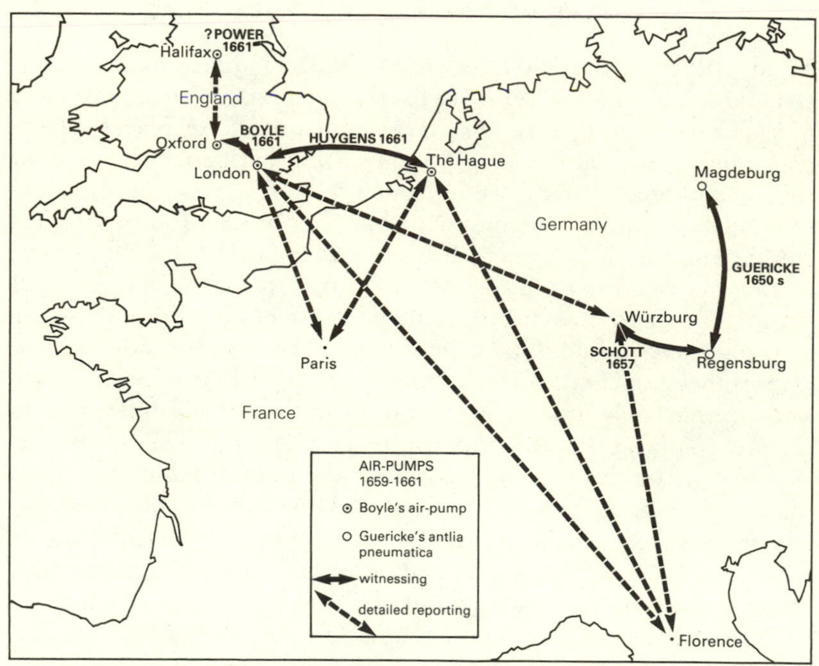

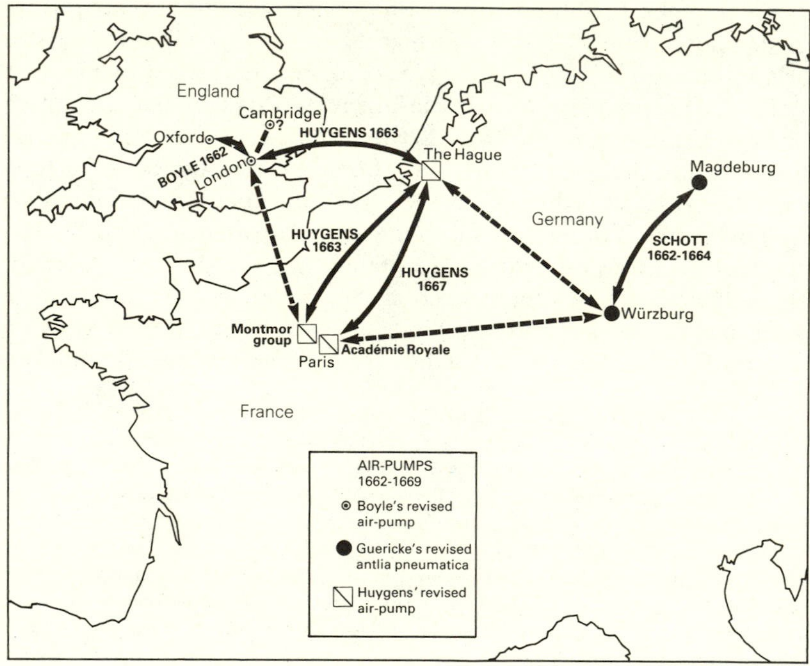

FIGURE 10

Distribution of air-pumps in Europe in the 1660s, indicating various forms of machine and contacts between centres of pneumatic experimentation. Note the contrast between contacts based on direct witnessing of air-pumps in operation and contacts based on written accounts.

about how these replicas were built, even though several important issues relating to replication are raised by their design and construction. These machines have, to our knowledge, never been operational, and none of Boyle's air-pump experiments of the late 1650s and 1660s has been repeated in modern times.[4]

We therefore rely upon the exchanges between Boyle and his contemporaries for our information on the career of the air-pumps. We follow these pumps at each major location where they were built and used during the period: Oxford, London, Holland, and France. We concentrate upon the detailed transformations in the structure of the pumps, and we focus upon those phenomena which acted as gauges of the competence of experimenters and the adequacy of their machines. To clarify the findings reported here, we summarize the implications of the information we have gathered:

1. Every pump was always in trouble, either directly because of leakage or indirectly because of rival attacks. There was no moment at which any pump was considered to be genuinely secure, nor is it possible to write down a fixed description of each pump and its working.
2. There were very few pumps working at any moment. We know of a pump or pumps in Boyle's care at Oxford; a pump or pumps at Gresham College in London; one pump in a state of reconstruction in Huygens' care in Holland (from autumn 1661); a pump belonging to the Montmor group in Paris (from November 1663); and, after its establishment in 1667-1668, a machine at the Académie Royale des Sciences. There are some references to an air-pump in Halifax in 1661 and one at Cambridge in the mid-1660s. The number of those who were able to work on air-pump trials was therefore very limited.
3. Since there were few pumps and since they were always being redesigned, we can see how difficult it was for operators to build their own *machina Boyleana* or to perform Boyle's experiments without visual experience of those trials in England. No one built a version of Boyle's machine without such experience. No one relied on Boyle's textual description

[4] There are examples of Hauksbee pumps in the Royal Scottish Museum, Edinburgh; the Deutsches Museum, Munich; Longleat House, Wiltshire; the Museum of the History of Science, Oxford; and the Science Museum, London (this last being on loan from the Royal Society). See R. Anderson, *The Playfair Collection*, pp. 67-70, and Daumas, *Les instruments scientifiques*, pp. 83-84, 115-117.

alone. Transmission of craft skill was sustained by Huygens, who was present at the air-pump trials in London in spring 1661, and who then built his own pump in autumn 1661. Huygens' presence was essential for the construction of the Montmor pump in 1663. Neither Otto von Guericke in Germany, nor the members of the Accademia del Cimento in Florence, built such a machine, even though they both possessed full textual accounts of Boyle's pumps.

4. The English pumps and Huygens' pump were very substantially redesigned over time. These changes were just those necessary to respond to criticisms which Hobbes made in his *Dialogus physicus*: air would leak between the sucker and the cylinder, the contact between leather and metal was not secure, and the cylinder would deform. At the same time, Hobbes was rigorously excluded from access to the experimental debate. His criticisms were fundamental but they did not allow him membership of the community of experimental philosophers.

5. The one sustained attempt to replicate Boyle's experiments was that of Huygens. Yet, just because Huygens wanted to calibrate his pump in comparison with those in England, he produced the first and most important *trouble* for Boyle's account of the contents of the receiver. This was the phenomenon of *anomalous suspension*, which we discuss in detail below. Differences between pumps and differences between the interpretative schemes that operators used were deployed to defend experimental competence against attack. Thus, when this phenomenon was used by Huygens to argue that his pump worked *better* than that of Boyle, Boyle denied its status as a matter of fact. Differences in practical skills and pump designs determined whether this phenomenon counted as an anomaly, as a standard of calibration, or as a fixed matter of fact. Negotiations about calibration were therefore intimately connected with stipulations about the competence of experimenters and the contents of the air-pump. We saw in chapter 2 that Boyle had said he would welcome any reliably produced matter of fact. Yet he never published any account of anomalous suspension. He kept silent because anomalous suspension resisted the recognized explanatory competence of the spring of the air. The phenomenon only became a matter of fact in England when Huygens visited Gresham College and performed experi-

ments on the phenomenon with Robert Hooke. Similarly, between 1661 and 1663 only Hooke was able to make the pump at Gresham College work well. We now move on to the detailed account of these sets of negotiations in England and in Europe.

Constructing the Pump: London and Oxford

The experimental context in which Boyle set out to obtain an air-pump is described in the essays that Boyle composed on saltpetre in the 1650s. These essays drew on texts by his Oxford colleagues in which the activity of the air was linked with nitre, and thus with *elater, antitupy,* or *spring*. Before 1655 Boyle also encountered reports by Mersenne, Gassendi, and others on the weight of the air and on engines such as the wind gun and the Torricellian tube. By January 1658 Boyle had heard of Guericke's work in Germany as reported in Caspar Schott's *Mechanica hydraulico-pneumatica* (1657).[5] In late 1658 Boyle contacted the London instrument maker Ralph Greatorex and also Robert Hooke. Hooke had arrived in Oxford in 1655 and had worked as Boyle's assistant there since 1657. Boyle demanded an instrument that could offer a space that would be secure—that is, air-tight—yet open to experimental manipulation. The collaboration resulted in the construction of an air-pump by early 1659. The machine was to be used for inquiring into the further uses of the air, now seen as "enriched with variety of steams from terrestrial . . . bodies."[6]

This pump is the one described in Boyle's *New Experiments*. Boyle took the completed machine from London to Oxford in March 1659. As we have seen, it consisted of a large glass receiver with an opening at the top, above a brass cylinder in which a sucker could be made to rise and fall by turning a geared handle. A stopcock was inserted at the bottom of the receiver, and a valve at the top of the cylinder (figure 1). In November 1659, Boyle told Hartlib that "I am now prosecuting some things with an engine I formerly writ to you of." Boyle hoped his work "would not, perhaps,

[5] Frank, *Harvey and the Oxford Physiologists*, chaps. 4-5; Webster, "Discovery of Boyle's Law"; Turner, "Robert Hooke and Boyle's Air-Pump." See also Boyle to Hartlib, 19/29 March 1647, in Boyle, *Works*, vol. I, p. xxxviii; Hartlib to Boyle, 7/17 January 1658, ibid., vol. VI, p. 99; Boyle, "New Experiments," pp. 2-6.

[6] Ibid., pp. 116-117; idem, "A Physico-Chymical Essay . . . touching the . . . Red-integration of Salt-Petre," p. 371; Hooke, *Posthumous Works*, pp. iii-iv.

be unacceptable to our new philosophers, wherever they are. But we have not yet brought our engine to perform what it should." In December, Boyle responded to requests from Oldenburg and Dungarvan in Paris for information about these experiments, and completed his book on 20/30 December 1659. It was published in summer 1660 and immediately translated into Latin. By summer 1660 Boyle had carried a machine from Oxford back to London, where he began public demonstrations of its effects.[7]

There were now two sites at which air-pump trials could be performed. The subsequent career of the air-pumps in Boyle's possession at Oxford is not clear. He was in London very often during the early 1660s, but spent more time at his house in Oxford. He left there permanently in 1668 to take up residence at the house of his sister Lady Ranelagh in London. Until then, he certainly possessed more than one pump at Oxford. John Mayow worked as an operator for him there, and in the autumn of 1667 a visitor was shown "Mr. Boghils [=Boyle's] Air pump, and ye Torricellian Experiment both att Mr. Boghils Lodging."[8] These engines were very different from that described in *New Experiments* in 1660. In March 1661 Boyle was instructed by the Royal Society "to hasten his intended alteration of his air pump," and on 15/25 May 1661 he gave his original pump to the Society in London. By December 1661 Boyle certainly began to design a completely new pump at Oxford. It is very likely that this reconstruction was prompted by Hobbes's attack.[9]

Hobbes's *Dialogus physicus* appeared in August 1661, and Boyle read it immediately in Oxford. In early October Boyle began writing his *Examen* of Hobbes's book, again in Oxford. In that text Boyle wrote that his pump was now placed underwater. He used this fact to reject Hobbes's supposed claim that common air would leak into the receiver. We also know from reports by Robert Moray to Christiaan Huygens that this new pump was planned from December 1661—that is, soon after Boyle began writing his reply to Hobbes. Moray told Huygens that Boyle was

[7] Walter Pope to Boyle, ? 10/20 September 1659, in Boyle, *Works*, vol. VI, p. 636; Hartlib to Boyle, 15/25 November 1659, ibid., p. 131; Boyle to Hartlib, 3/13 November 1659, in Worthington, *Diary and Correspondence*, vol. I, p. 161; Sharrock to Boyle, 9/19 April 1660, in Boyle, *Works*, vol. VI, p. 319.

[8] For Mayow as assistant, see Boyle, "Continuation of New Experiments," p. 187; for John Ward's visit in September-October 1667, see Frank, "The John Ward Diaries," p. 170.

[9] Birch, *History*, vol. I, pp. 8, 16, 19, 23.

> . . . making another machine, even more exact than his first. And not yet having been able to think of all the other details, he has charged me to tell you, that his design is to place its cylinder parallel to the horizon, in a vessel full of water, better to prevent the entry of the air. When he tells me more you shall also know it.[10]

Hobbes argued that in the original design leaks would occur between the contact of the leather washer and the inner wall of the cylinder, between the edges of the leather and the brass, and because of changes in shape of the cylinder cavity. These sources of trouble might plausibly be combatted by immersing the pump. It is also important that Huygens himself immediately rejected the virtue of these changes, noting that Guericke had done much the same and had also encountered enormous problems. Huygens suggested to Boyle, for example, that it would be better to improve the design so that "the piston would be better adjusted." We shall see that these problems made the replication of Boyle's original experiments as reported in his earlier book a matter of some difficulty.[11]

We have noted that by March 1661 the Royal Society had already begun to suggest changes to Boyle's pump. From May 1661 this pump, Boyle's original design, was at Gresham College. Most of the experiments of which we have any details were performed there, and we know that this pump, too, was in an almost permanent state of reconstruction. This reconstruction often involved the replacement or enlargement of the glass receiver, which could be easily broken or easily removed. During the reconstructions, the London pump was gradually transformed into a model like that which Boyle now had in Oxford from early 1662. In October 1661, Moray told Huygens that "we intend to reform [the pump which Boyle] gave to the Society, principally in that which touches the exclusion of the air from the pump": once again, the Royal Society planned changes to meet criticisms that the pump leaked.[12]

[10] Boyle, "Examen of Hobbes," p. 208; for the composition of this text and Boyle's new engine, see Hartlib to Worthington, 14/24 February 1662, in Worthington, *Diary and Correspondence*, vol. II, part 1, p. 109; Moray to Huygens, 9/19 August, 27 August/6 September, 9/19 October and 13/23 December 1661, in Huygens, *Oeuvres*, vol. III, pp. 312, 317, 368-370, 425-428.

[11] Hobbes, "Dialogus physicus," pp. 244-246; Huygens to Moray, 20/30 December 1661, in Huygens, *Oeuvres*, vol. III, pp. 439-440.

[12] For Boyle's presentation of the engine, see Birch, *History*, vol. I, p. 23; Moray to Huygens, 9/19 October 1661, in Huygens, *Oeuvres*, vol. III, pp. 368-370.

Three concerns dominated these discussions in London. One was the possible enlargement of the receiver; a second, securing the pump against leaks; the last, a set of attempts to repeat fresh phenomena which Huygens was reporting from Holland. The result of the first of these concerns was the attempt to construct a receiver large enough to contain a man; the result of the second was the transformation of the Royal Society's pump design; the result of the last was a prolonged dispute about the identity of any air-pump. From March 1662 the Society discussed its pump in detail. Moray told Huygens that the London machine was not yet "adjusted as well as all the others which Mr. Boyle has made; and furthermore we have resolved to make a machine of such a size that a man may enter it." By the end of March, the Society received a copy of Hobbes's *Problemata physica* in which further criticisms were made of the porosity of the pump, and they commissioned a glass-maker at Radcliffe to construct a larger receiver.[13] During April and May 1662 the Royal Society conducted trials with its pump using a receiver large enough to receive a man's arm. As late as July 1662, the "operator" was "ordered to carry Mr. Boyle's engine to Mr. Oldfield, in order to make the top of the cylinder of it and the sucker to meet together." This was, of course, just the problem Hobbes had pointed to in August 1661 and again in March 1662. It is clear that during this period, at least, the London pump could not be made to work well for any extended period, and that leakage remained a severe trouble.[14]

Fresh troubles also emerged, notably those associated with the reports from Holland. In March 1662 Moray told Huygens that "our machine being less well adjusted than yours, we have not yet done anything worthwhile." On 5/15 March, Moray read the Society the letters he had received from Huygens, and Oldenburg sent a colleague to Holland to examine Huygens' pump. Croune, Goddard, and Rooke were charged with repeating Huygens' work. Then in June 1662 Rooke died, and Moray used this to explain to Huygens why the Society had not been successful in working its air-pump. Relations with Hobbes and with Huygens obviously made these troubles a matter of concern.[15] By August 1662 the

[13] Moray to Huygens, 6/16 March and 6/16 May 1662, in ibid., vol. IV, pp. 94, 130-132; Birch, *History*, vol. I, pp. 77-78.

[14] Birch, *History*, vol. I, pp. 75-78, 102, 106.

[15] Moray to Huygens, 3/13 March 1662 and 9/19 January 1663, in Huygens, *Oeuvres*, vol. IV, pp. 83-84, 297-298; Oldenburg to Huygens, 29 March/8 April 1662, ibid., p. 108 (and Oldenburg, *Correspondence*, vol. I, pp. 445-446); Birch, *History*, vol. I, pp. 77-78.

pump was working again, and on 5/15 November Robert Hooke was made Curator of experiments. From this moment, significantly, the pump in London worked better. Within six months it had been completely rebuilt along the lines of Boyle's machine at Oxford, a machine whose existence was partly due to Hooke himself. Thus it was not until Hooke was given responsibility for this pump, and until Huygens was present in person in London in summer 1663, that the Dutch experiments could be performed, or that the London pump, now reconstructed, could work reliably. Only with Hooke and Huygens as the specifically proficient operators could such performances be guaranteed. Under these circumstances, replication became a real issue: rival centres of work and different competences had developed. We turn to consider the network of replication that Huygens initiated in 1661.[16]

Replicating the Pump: London and Holland

The dissemination of air-pumps was a key aspect of the development of Boyle's pneumatic experiments. Hobbes was outside the experimental community and did not take part in this process. On the other hand, Christiaan Huygens was the only natural philosopher in the 1660s who built an air-pump that was outside the direct management of Boyle and Hooke. This was of immense significance for the production and career of pneumatic experiments. Huygens visited London in April 1661. On 1/11 April he attended a meeting of the Royal Society at Gresham College, where the air-pump was discussed and experiments performed. The following day, Huygens was visited by Boyle and they "discoursed for a long time." Other debates in which Huygens took part concerned the Society's work on the recoil of guns, on telescopic lenses, and on Boyle's work on siphons. Huygens told his brother Lodewijk that he had obtained a copy of *New Experiments*, and that he had seen "a quantity of beautiful experiments concerning the void, which they do not make with mercury in little pipes, but by extracting all the air from a large glass vessel by means of a certain pump."[17] Huygens was visited by Oldenburg when he returned to Holland in the summer, and remained in contact with Robert Mo-

[16] Birch, *History*, vol. I, pp. 102-106, 124, 139.

[17] Ibid., pp. 19-21; Huygens to Lodewijk Huygens, June 1661, in Huygens, *Oeuvres*, vol. III, p. 276; Huygens, journal, in ibid., vol. III, pp. 321-322, enclosing papers by Brouncker (pp. 323-328) and Boyle (pp. 328-331).

ray, who told him of the appearance of Hobbes's *Dialogus physicus* and of the work at the Royal Society that aimed to test the funicular hypothesis advanced by Linus. At the end of September 1661, Huygens wrote to Moray that he would soon begin to build his own machine, "to make some more new experiments in the void, and to have the pleasure of trying a part of those which are in [Boyle's] book." He also wrote to Paris that "the curious experiments of Mr. Boyle," which he had seen in London, "have given me the desire to built a machine like his . . . in the hope of trying yet more things which have not occurred to [Boyle]." Huygens had just obtained Sharrock's Latin translation of *New Experiments* and it was this text he cited when building his own pump. "When it is done," he now told Moray, "I will let you know what changes I have made, since it is necessary to see first how it will succeed." When Moray received this letter, in London, he replied that "since you are going to construct one in your fashion, I believe we will defer the reformation of ours until you have made yours. This is why you must let us know everything that concerns the design of the one you are going to make." The stages by which Huygens developed his own pump have now been impressively documented by Alice Stroup. She has also drawn attention to the manner in which Huygens acted as the centre of further dissemination of pump designs. Here, we concentrate on the two problems of the differences between pumps and the means by which information was exchanged between London and Holland. In these early announcements by Huygens, we can already detect the troubles that were to afflict his scheme for replication.[18]

During November and December 1661, Huygens kept notes on the process of constructing his pump. He informed Moray in London and Lodewijk Huygens in Paris of his progress. On 12/22 October, Huygens made a list of Boyle's tests for the goodness of a pump. Its integrity depended on the seals that linked the valve and the cylinder and which connected the glass receiver with the

[18] For Oldenburg's visit, see Oldenburg to Huygens, 24 July/3 August 1661, in Huygens, *Oeuvres*, vol. III, pp. 310-311, and in Oldenburg, *Correspondence*, vol. I, pp. 411-412, 412n-413n. For contacts with Moray, see Moray to Huygens, 9/19 August, 27 August/6 September and 18/28 September 1661, in Huygens, *Oeuvres*, vol. III, pp. 312, 317, 355; Huygens to Moray, 6/16 September 1661, ibid., p. 319. For news of the attempt to build a pump, see Huygens to Moray, 20/30 September 1661, ibid., vol. XXII, p. 72; Huygens to Montmor, ? October 1661, ibid., p. 76; Huygens to Montmor, 26 September/6 October 1661, ibid., vol. III, pp. 358-359; Moray to Huygens, 9/19 October 1661, ibid., pp. 368-370. For Huygens' role in disseminating air-pumps, see Stroup, "Huygens & the Air Pump," p. 138.

top of the pump. On 25 October/4 November, he told Moray that the machine was "not ready" because of "the cavity of the cylinder not being made exactly enough, which I will now have corrected." Huygens complained about the lack of "good workers," as Boyle also did repeatedly. He requested information from Boyle about valve design. Huygens was using a copper valve. He was critical of Boyle's use of wood, which Huygens claimed would warp. He told Moray that "since you wish to profit from my trials of what changes I make in my construction, it is right that I also profit from yours, and that you do not hide from me the faults and inconveniences which may be in Mr. Boyle's machine, and which he believes to be corrigible."[19] During the rest of November, Huygens attempted to finish his machine, and then to test it against Boyle's claims. He abandoned his first cylinder and replaced it with one of "massive copper." He made a basin to fit above the pump and below the receiver, allowing him to remove the glass bowl more easily. At this point, he wrote, his elder brother Constantijn decided to pull out of the project, "being afraid of the cost."[20]

By the end of November 1661, Huygens had a pump which he was satisfied was at least as good as that of Boyle. On 19/29 November, he possessed a receiver that was closed at the top and had to be removed from its basin to have apparatus placed inside. (Boyle's machine had an orifice in the top of the glass receiver.) This was one way in which Huygens' machine differed from those in England: it was a difference that had been crafted to improve the integrity of the air-pump. Other differences included a turpentine seal for the bottom of the receiver and a new recipe of yellow wax and resin to seal the valve and the connecting tube. Most importantly, Huygens claimed proof that this design (figure 11) was *better* than Boyle's. He told his brother as much on 20/30 November: "My pneumatic pump has begun to work since yesterday, and all that night a bladder stayed inflated within it . . . which Mr. Boyle was not able to effect." The use of an inflated bladder was a common means of calibrating the contents of the receiver. Boyle had said that in his pump such bladders did deflate very

[19] Huygens to Moray, 25 October/4 November 1661, in Huygens, *Oeuvres*, vol. III, pp. 383-385; Huygens, notebook, ibid., vol. XVII, pp. 306-312. For Huygens' troubled relationship with his instrument makers, see van Helden, "Eustachio Divini versus Huygens" and Leopold, "Huygens and His Instrument Makers."

[20] Huygens to Lodewijk Huygens, 13/23 November 1661, in Huygens, *Oeuvres*, vol. III, p. 389.

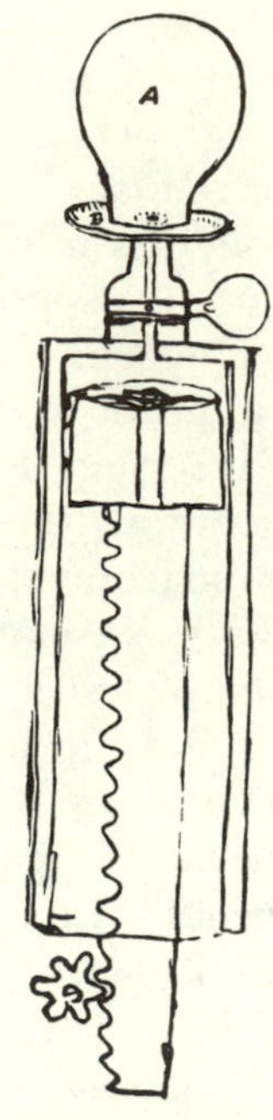

FIGURE 11

Huygens' first design for his air-pump (November 1661). Drawing from Huygens' working notebook, reproduced from Huygens, Oeuvres, *vol.* XVII, *p. 313 (figure 36). (Courtesy of Edinburgh University Library.)*

slowly.[21] Huygens now moved on to more complex means of calibrating his receiver and thus showing how much better his own pump was.

The process occupied Huygens throughout December 1661. By the middle of the month, he was claiming that he could take 99% of the air from the receiver. To test this claim, on 11/21 December Huygens attempted to repeat Boyle's nineteenth experiment, a version of the void-in-the-void phenomenon. This trial involved putting a barometer of water inside the receiver of the pump, and using the height of water as a measure of the air left in the receiver (figure 12). Huygens recorded that his new pump was "hermetically sealed, so that it took all the air from the receiver, which is shown by the following experiment." Huygens reported that "after five or six strokes of the pump" the water in the barometer fell to the level of the water in its basin. When he allowed air to reenter the receiver,

[21] Huygens to Lodewijk Huygens, 20/30 November 1661, in ibid., p. 395.

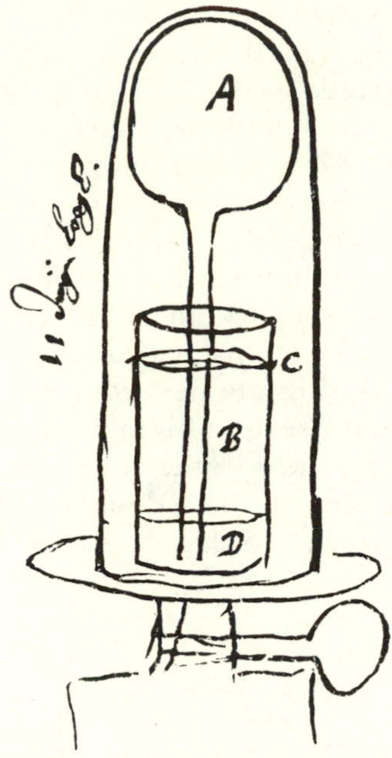

FIGURE 12
Huygens' diagram of his trial of the void-in-the-void experiment with his new pump (December 1661). A: flask full of water; D: water in outer vessel B; C: water level in both A and B after exsuction of air from receiver. From Huygens, Oeuvres, *vol.* XVII, *p. 317 (figure 39). (Courtesy of Edinburgh University Library.)*

the water rose again up the barometer, filling it completely "except that there remained a small air-bubble, a little larger than a hemp seed." This bubble eventually disappeared after a day and a night. By concentrating on the behaviour of this air-bubble, Huygens focused his attention on means by which he could ascertain the porosity of his receiver and the pump.[22]

In experiment 22 of *New Experiments*, Boyle discussed whether air could be contained in water, and therefore whether the receiver

[22] Huygens, notebook, in ibid., vol. VXII, pp. 316-322. Boyle's experiment is in "New Experiments," pp. 33-39. Huygens' claim to be able to extract 99% of air from the receiver can be compared with late eighteenth-century claims: Joseph Priestley to Richard Price, 27 September 1772, in Schofield, *Scientific Autobiography of Priestley*, p. 109, claimed "the best common pumps scarce go beyond 1/100. Mr. Canton, I remember, said that his went to 1/120 when it was in its best order." John Smeaton had built an air-pump which "has sometimes exhausted above 1/1000" but only Smeaton's and Priestley's were "in being."

could be made genuinely secure against air. We may recall that *air* was defined as the sole fluid that *could* be sucked from the receiver and whose presence could be detected there. So Boyle had decided that only a part of the bubble was such true air, and, because Huygens observed that this bubble disappeared after a day and a night, "on two successive occasions," Huygens also admitted that "I am not sure it was true air." To imply that it was all true air would prejudice the calibration and the integrity of the pump. Huygens immediately wrote to Lodewijk describing this set of trials, arguing that it showed conclusively that "the receiver must also be completely empty." During the next four days, Huygens was occupied in reinforcing the seals of his machine. He changed the tap by covering it with oil-soaked leather wrapped round with copper wire, and on 16/26 December he redesigned the top of the sucker and developed a new recipe for his cement (figure 13). At the same

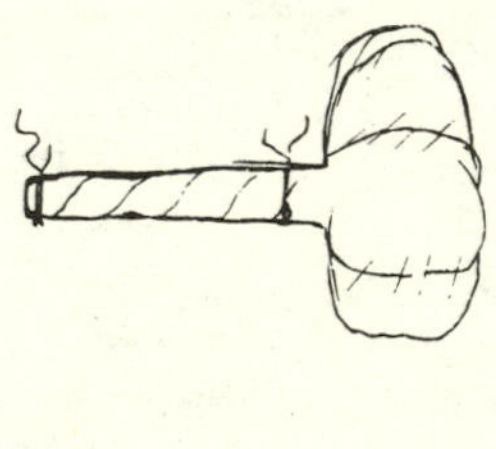

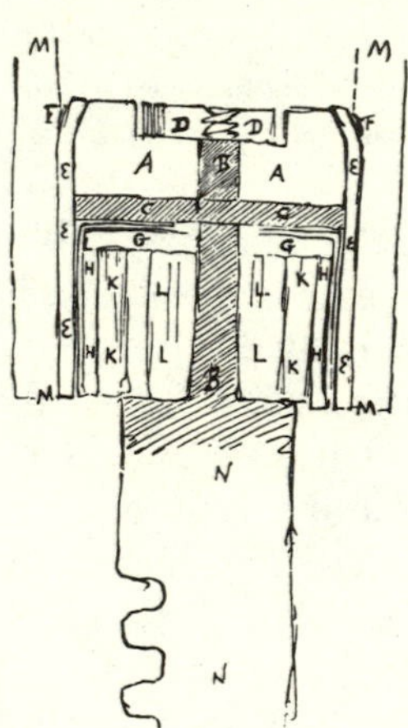

FIGURE 13

Huygens' diagrams of his changes in the stopcock (top) and sucker (December 1661). M: cylinder; N: piston rod; D: iron screw; A: wood; C: iron; E: stout leather attached by nails F; K, L: cork; H: wool inside pig's bladder. From Huygens, Oeuvres, *vol.* XVII, *p. 319 (figures 40 and 41). (Courtesy of Edinburgh University Library.)*

time, he received a letter from Moray in London that informed him that Boyle was making "another machine, even more exact than his first," to replace that given to the Society in May.[23]

It was now vital for Huygens that he establish for Moray and for Lodewijk that his pump was superior. He had tried Boyle's nineteenth and twenty-second experiments, and was perturbed by the presence of quantities of air-bubbles inside the water in the Torricellian tube. The character of the air in these bubbles would determine the porosity of his machine. So on 17/27 December he repeated these trials with water he had purged of air by leaving it many hours in the receiver of the air-pump. He then filled the barometer with this purged water and evacuated the pump. However, Huygens reported that "I established with astonishment that it did not wish to redescend even when I had evacuated the air as completely as possible." This failure of purged water (and later mercury) to descend in a barometer when the receiver was evacuated came to be labelled *anomalous suspension* (figure 14).[24] Huygens also noted that when he introduced an air-bubble into the Torricellian tube, the water did, at last, fall. The introduced air-bubble expanded at each stroke of the pump, until the water was driven down into the basin. He repeated this many times, investigating ever more closely the relation between these bubbles and the water. Huygens continued these experiments through January and February 1662, extending the length of the Torricellian tube until it projected above the receiver, constantly reinforcing the seals, and constantly checking the character of this air which could make the water fall, and whose absence seemed to allow the water to maintain its height. He wrote in his notebook that

> It seems, from these experiments, that air can expand itself one hundred thousand times, and yet still exercise a force when

[23] Huygens to Moray, 20/30 December 1661, in Huygens, *Oeuvres*, vol. III, pp. 439-440; Huygens to Lodewijk Huygens, 11/21 December 1661, ibid., p. 414; Moray to Huygens, 13/23 December 1661, ibid., pp. 426-427; Huygens, notebook, ibid., vol. XVII, pp. 318-320. Boyle's twenty-second experiment is in "New Experiments," pp. 47-55.

[24] Huygens, notebook, in Huygens, *Oeuvres*, vol. XVII, pp. 320-322. To realist readers still with us, we can offer little consolation. But here are some of the factors which would be considered relevant to a current scientific explanation of anomalous suspension: (1) short-range attractive forces between fluid and glass; (2) viscosity; (3) surface tension; and (4) the presence of residual air. The phenomenon of a "sticky vacuum" is still a trouble for experimenters. See DeKosky, "William Crookes and the Quest for Absolute Vacuum in the 1870s," p. 12.

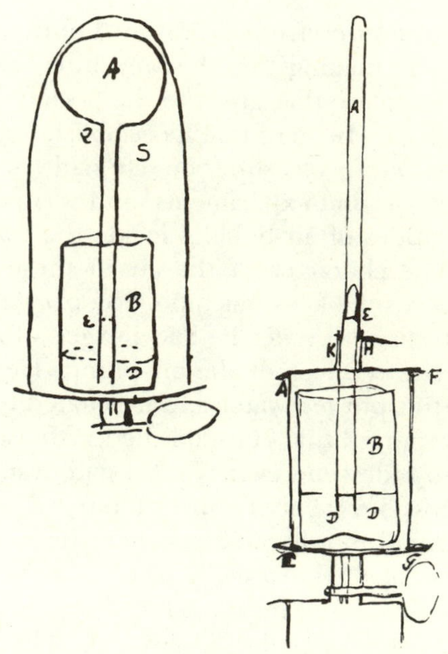

FIGURE 14
Huygens' diagram of the experiment that produced anomalous suspension (December 1661). A: glass tube full of water; D: water level in flask B; E: air-bubble in tube A. From Huygens, Oeuvres, *vol. XVII, p. 323 (figures 42 and 43). (Courtesy of Edinburgh University Library.)*

in this condition, because of its elasticity, unless there is something else to consider in this, as yet unknown, apart from the weight of the air and its elasticity.[25]

Huygens produced this phenomenon in the specific context of trials of the comparative excellence of his pump. By the end of 1661 his pump was already significantly different from those in London and Oxford. Moray had confirmed this in his letter of 13/23 December. Moray had mentioned Boyle's new and as yet unseen pump, and had reported Boyle's rejection of Huygens' plan to use a copper valve. Huygens, in turn, rejected Boyle's idea of placing the pump underwater. Appealing to the new phenomenon of anomalous suspension, Huygens argued that the new valve design and the new sucker were superior. His pump was evidently less porous. "It would need almost a whole letter to describe the architecture, but the main thing is that I do not put the piston into

[25] Huygens, notebook, in Huygens, *Oeuvres*, vol. XVII, pp. 320-330.

the copper cylinder after it is all finished but before, and when inside I fill it little by little with wool and other things until it can hold no more." In fact, Huygens did not describe all the changes he had made: the complex structure of the piston he recorded in his notebook is not treated fully in any of his letters to Moray, nor is the elegant device that enabled Huygens to isolate the receiver above the pump by means of three stacked copper plates with white iron and wax seals.[26] This meant that the new phenomenon had to be the vehicle used by both English and Dutch workers to compare their pumps.

By February 1662 Huygens took the anomalous suspension of water well purged of air, and the fall of that water when a bubble was introduced, to be marks of a good machine. Crucially, Huygens now denied that the air contained in this water was the same as common air. This strange air in the water was treated as the most important substance in pneumatics. Water descended in its tube when the pump was set to work not because of the absence of air in the receiver, but because of the necessary pressure of this new substance above the water. This substance had "a greater power of expansion than ordinary air," since if an extraneous bubble of common air were added to the substance above the water, the fall of the water was not proportionately greater.

On 30 January/9 February 1662, Huygens used this new substance to account for his failure to produce anomalous suspension with mercury. This failure was entirely due to the greater difficulty in removing the substance from mercury as compared with its removal from water. There would always be some of the substance remaining within the mercury, and the substance would emerge from the mercury in the tube and force it down: hence the difficulty of producing anomalous suspension with mercury.[27] Two results had emerged from Huygens' attempt to replicate Boyle's air-pump in autumn and winter 1661-1662. First, Huygens had established for himself a decisive phenomenon whose outcome measured the excellence of any air-pump. Second, to interpret this calibration-phenomenon, Huygens had summoned into existence a new fluid and challenged the sufficiency of the weight and spring of common air. The effects of this fluid were only visible in good pumps. This

[26] Huygens to Moray, 24 January/3 February 1662, in ibid., vol. IV, p. 24; Moray to Huygens, 13/23 December 1661, ibid., vol. III, pp. 426-427; Huygens to Moray, 20/30 December 1661, ibid., pp. 439-440; Huygens to Lodewijk Huygens, 25 December 1661/4 January 1662, ibid., vol. IV, p. 6.

[27] Huygens, notebook, in ibid., vol. XVII, pp. 326-328.

new fluid could *not* be common air, since that would show the pump to be porous. Decisions about the excellence of pumps would now be connected with decisions about the spring of the air and the existence of new kinds of fluid. For more than eighteen months neither of Huygens' claims was granted the status of matters of fact. We now consider the way in which the English experimenters dealt with these anomalies and how they negotiated with Huygens about these troubles of replication.

Calibration and Anomaly: Holland and London

Huygens told Moray about anomalous suspension and its central use in calibrating his pump on 24 January/3 February 1662. Huygens also wrote to his brother Lodewijk in Paris, telling him to arrange with the natural philosopher Jacques Rohault to construct a pump, and to test it by producing anomalous suspension, of which Huygens enclosed a diagram. In January 1662, Huygens was visited by Robert Southwell and in March by Johann Kohlhans, a friend of Henry Oldenburg, both of whom witnessed the pump experiments.[28] During March the natural philosophers in England began to respond to Huygens' claims. There were just two pumps available to them: the machine Boyle had given the Royal Society in May 1661, which had been in an almost permanent state of reconstruction under the impact of messages from Holland and Oxford; and the new design under Boyle's direct supervision at Oxford. Neither machine was like that of Huygens. On 3/13 March Moray confirmed that the London machine was "not in condition," and so "it will be necessary to ask you [Huygens] to continue to examine this matter." This problem was to continue throughout the year. No machine in England produced anomalous suspension during 1662. In the exchanges between Boyle and Hooke, on the one hand, and Huygens, on the other, the very character of the experiments in Holland in autumn and winter 1661 was challenged.[29]

The English natural philosophers used three resources against Huygens. First, Boyle had now completed his response to the attacks from Linus and Hobbes. In this answer, Boyle described his

[28] Huygens to Moray, 24 January/3 February 1662, in ibid., vol. IV, p. 24; Huygens to Lodewijk Huygens, 5/15 February 1662, ibid., p. 53; Moray to Huygens, 24 January/3 February 1662, ibid., pp. 27-28; Oldenburg to Huygens, March 1662, ibid., p. 108.

[29] Moray to Huygens, 3/13 and 4/14 March 1662, ibid., pp. 83-86.

reconstructed pump and included reports of the trials that displayed the new Law of the spring of the air. Boyle used this Law against Huygens. Second, Boyle claimed that his machine was better than that of Huygens, so that anomalous suspension was a mark of Huygens' *incapacity* to make a good pump. Finally, Boyle denied that anomalous suspension *could* be used as a calibration of the pump: Huygens was attacked for his failure to test for the contents of the receiver with means acceptable in England. During March and April 1662, Moray told Huygens about Boyle's *Examen* and the Law of the spring of the air. He also enclosed Boyle's denial that anomalous suspension was a matter of fact. Boyle told Huygens that "the non descent of the water" might be due to "this that the air was not sufficiently pumpt out." Boyle recommended "making use of a gage (if I may soe call itt) or stander or index within the Receiver by which he may know how far it is evacuated & how well it keeps out ye air." Boyle wrote that two kinds of gauges were used in England: either a small bladder whose expansion would measure the "degree of expansion of the air in the Receiver," or else a water manometer made from a J-tube with water and a small bubble, where "by ye shrinking of this bubble both the leaking of ye vessell may be concluded & the quantity of the admitted air may be ghest at." Boyle insisted that the best Huygens had ascertained was that his machine did not admit any new air, but *not* that it exhausted all its original air. Finally, Boyle referred Huygens to the void-in-the-void trials printed in *New Experiments*: in those trials enough air remained in the receiver to keep up 14 inches of water. Without these checks, Huygens could not claim he had replicated Boyle's pump, and, without replication, anomalous suspension was a real catastrophe and not a means of calibration.[30]

Huygens had been told that his pump leaked more than those in England. Anomalous suspension was not accepted as an authentic matter of fact. So on 30 May/9 June he sent Moray the fullest account yet of anomalous suspension, since no other worker had replicated the phenomenon. He demanded more details of Boyle's Law of the spring of the air, since he said that Moray's report was incomprehensible. Finally, he wrote that Boyle should be told that "I have made this experiment more than thirty times, and I knew very clearly that the receiver was emptied of air as much as it could be by means of my pump." Huygens had used

[30] Moray to Huygens, 6/16 March 1662, ibid., p. 94; Boyle to Moray, March 1662, ibid., vol. VI, pp. 581-582. See also Birch, *History*, vol. I, pp. 77-78.

small bladders in the receiver and had also compared tubes of water purged of air with tubes of water not so purged.[31] Moray sent Huygens a copy of Boyle's *Defence against Linus*, and Huygens answered again on 4/14 July in a letter read at the Royal Society some days later. Huygens repeated his description of anomalous suspension and reaffirmed the reality of the effect and of the subtle fluid in the water. Huygens also acknowledged that if Boyle's Law applied to *all* fluids, then it would be difficult to make use of a fluid which possessed a much greater spring than did common air. He had now challenged the virtue of English pumps and the applicability of Boyle's Law. Boyle's response was swift. He conceded that "by reason of the peculiar texture of some bodys, or some unheeded circumstances there may happen some odd phaenomenon or other, very difficult to be accounted for." Nevertheless, Boyle continued to urge that Huygens' pump still leaked, while Hooke argued that all fluids must obey the Law of spring, since their fluidity was due to their coiled structure. The arguments of Boyle and Hooke, they claimed, "question not [Huygens'] Ratiocination, but only the stanchness of his pump."[32]

Huygens' next move was to redesign his machine and then produce anomalous suspension with this new design. He told Lodewijk on 25 September/5 October 1662 that he had so many visitors demanding to see the machine that he was forced to pretend that the pump was not working: but "for the most part I am not lying, since it can scarcely ever remain in its perfection because of the piston which easily goes wrong."[33] His new pump was supposed to correct this problem and to respond to English criticisms. It was completed by 27 September/7 October. While it bears some similarities to Boyle's new pump of the previous year, it differs in several crucial respects. Huygens put the receiver on a separate plate, and inverted the pump so that it descended on suction into the cylinder. He inserted a water-oil mixture above the sucker, which was supposed to insulate the connection and prevent it from drying out. A basin was placed beneath the cylinder to collect the overflow of

[31] Huygens to Moray, 30 May/9 June 1662, in Huygens, *Oeuvres*, vol. IV, pp. 149-150.

[32] Huygens to Moray, 4/14 July 1662, in ibid., pp. 171-173; diagram, ibid., p. 174; Boyle and Hooke for Huygens, July 1662, ibid., pp. 217-222; Birch, *History*, vol. I, p. 102. Huygens' letter to Moray is also printed in Rigaud, *Correspondence of Scientific Men*, vol. I, pp. 92-95.

[33] Huygens to Lodewijk Huygens, 25 September/5 October 1662, in Huygens, *Oeuvres*, vol. IV, p. 245.

this liquid. But the most important difference between the new Dutch and English pumps lay in their manner of working. In Boyle's new pump, a very long stick (R) was attached to the top of the sucker as a valve. To evacuate the receiver it was necessary to push the piston to the bottom of the cylinder (NO) and then pull it back up the cylinder with the tap (G) open (figure 7). Air would rush out of the receiver and through the valve in the top of the sucker. In Huygens' new pump, however, a small hole (B) was made in the bottom of the cylinder and then sealed with wax or leather (figure 15). To evacuate the receiver, it was necessary to push the piston to the bottom of the cylinder with this small hole open, forcing all the air out of the cylinder. Then the sucker would be drawn back up to the top of the cylinder with the small hole and the tap (A) both shut. The tap would be opened, allowing air out of the receiver. Then the tap would be shut, the small hole opened again, and the sucker pushed down the cylinder once more, driving out the included air.

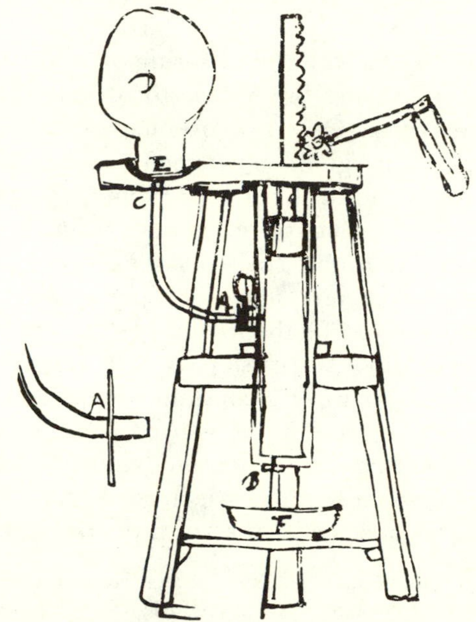

FIGURE 15

Huygens' revised air-pump (October 1662). D: receiver; A: connection of pipe from receiver to cylinder via stopcock; F: basin for collection of insulating liquid from cylinder overflow; B: valve. From Huygens, Oeuvres, *vol.* XVII, *p. 333 (figure 47). (Courtesy of Edinburgh University Library.)*

Huygens spelt out the superiority of this method in his notebook:

> Since following the original method of Boyle, the pump could only be rendered hermetically closed with difficulty, and that it soon lost this quality, it seemed good to me to invert the copper cylinder from below to above . . . so that the piston moved itself from down to up when the handle was turned. By this means the entry of air into the cylinder is rendered impossible, even if the body of the piston is not impermeable. . . . This method is much better than that of Boyle.[34]

Huygens sent descriptions to Paris and to Moray in London. He told Moray his pump was superior, that Hooke's account of spring attributed an inherent motion to matter, and that Boyle still ignored "certain very remarkable peculiarities which I have told you about in a full enough description." So it could not be "the air remaining in the receiver which prevents the descent," and with his new machine Huygens insisted that anomalous suspension must be accepted in England.[35]

However, during the winter and spring of 1663, the situation of the pump in London and the failure to produce anomalous suspension became critical. On 5/15 November 1662, as we have seen, Hooke was put in charge of experiments at the Society, and assumed direct command of the pump. From 3/13 December he set out to try "the experiment of purging water from air, to see whether it subsides, according to the Torricellian experiment." But repeatedly this trial was deferred, since the "engine was not tight." We can see how assessment of the virtue of the pump bore on the replication of anomalous suspension. In early January 1663, Moray again told Huygens of Boyle's continuing rejection of the matter of fact, because "you are not yet assured enough of the truth of the thing, since you have not used any kind of measure to know if the air is in the same state." Boyle argued that "before defining the cause let us be assured of the truth of the experience." Little

[34] Huygens, notebook, in ibid., vol. XVII, p. 332. For a comparison of means of evacuating the receiver through a valve, see Boyle, "Continuation of New Experiments," p. 180, and Huygens, *Oeuvres*, vol. VI, p. 586; vol. XVII, pp. 332-333, and fig. 47; vol. XIX pp. 204-205. These are also discussed in Stroup, "Huygens & the Air Pump," pp. 146-147.

[35] Huygens to Moray, 21 November/1 December 1662 and 23 January/2 February 1663, in Huygens, *Oeuvres*, vol. IV, pp. 275-276, 305; Huygens to Montmor, July 1663, ibid., vol. VI, pp. 586-587.

progress had been made since the previous spring.[36] But now Huygens had a new persuasive resource: he sent Moray a detailed diagram of his new pump, pointing out how it surpassed that of Boyle. Moray responded with news of Boyle's revised design at Oxford. He wrote that this machine was immersed in water and that Boyle continued to use his own recipe for the wax seal around the tap. Most importantly, Boyle had not yet produced anomalous suspension:

> That which Mr. Boyle has always used since he gave the first to the Society is scarcely different in shape, and it is submerged in the water like yours, but in a different manner. . . . [A]lthough my view is that the soft cement may be more proper than that Mr. Boyle uses, yet he always chooses his rather than yours. But up till now Mr. Boyle has never been able to do your experiment of the water which does not descend at all, even though he has taken all the care he could, without the air entering at all into the receiver, and the air having been well emptied from it, so that the mercury which he had put in a tube to be its measure, descended to the level of that which was in the little vessel beneath.[37]

So in March and April 1663 it became clear that unless the phenomenon could be produced in England with one of the two pumps available, then no one in England would accept the claims Huygens had made, or his competence in working the pump. On 19 February/1 March, Huygens was told that Hooke had been ordered to "accommodate our machine so that we can be enlightened by our own experience." Evidently, the problem here was that the pump in London could not reliably extract air from the water in the Torricellian tube. By 25 March/4 April, "the experiment of purging water from air" was still not working, "the engine not being tight." Moray even suggested that there might be "some difference between common water here and that of Holland." This might explain why air-bubbles were so much more difficult to remove from London water. However, the state of the pump was itself an almost permanent trouble because of its obvious leakage. On 1/11 April, for example, Hooke was ordered to bring a written account of the "construction of the pneumatic engine as it then was." Although

[36] Birch, *History*, vol. I, pp. 138-139, 212; Moray to Huygens, 19/29 January 1663, in Huygens, *Oeuvres*, vol. IV, pp. 297-298.

[37] Moray to Huygens, 19 February/1 March 1663, in ibid., p. 320; Huygens to Moray, 23 January/2 February 1663, ibid., p. 305.

he was instructed to try to purge the water and test anomalous suspension throughout April, Hooke repeatedly failed to do so. The troubles of replication were pressing: they were only to be resolved by Huygens' presence in London.[38]

Huygens travelled to Paris in March and reached London on 31 May/10 June. He remained in England until late September. He attended several meetings of the Royal Society, and on 22 June/2 July he was elected a Fellow. His presence at Gresham College was decisive for the career of anomalous suspension. At this stage, Boyle was staying out of London at the house of his sister, the Countess of Warwick, at Leighs in Essex. Boyle remained there until August and so did not witness the first replication of the vexed phenomenon. On 10/20 June, Oldenburg wrote a letter to Boyle announcing Huygens' arrival. On the same day, Hooke participated in an experiment to examine whether the bubbles which emerged from water in the tube were common air or some other fluid. Simultaneously, the Royal Society appointed a committee to examine the Torricellian phenomenon. Hooke then left London, carrying Oldenburg's letter to Boyle at Leighs, and stayed there for two weeks. Significantly, Oldenburg also told Boyle that it was important that Hooke return to London soon, since "ye abovementioned Strangers are like to continue here yet a while" and "ye Society shall much stand in need of a Curator of Experiments; wch, I hope, Sir, will ye sooner procure from yr obligingnes a dispensing with Mr. Hook for such a publick use." When Hooke returned, he soon set to work with Huygens on the air-pump.[39]

This programme began on 1/11 July. Initially, the phenomenon of anomalous suspension of water could not be produced, as Hooke told Boyle: Huygens "tried his own experiment, but it succeeded not, though he confessed the engine was very tight, and it will be tried again the next day." However, Huygens made notes on the significant differences between the Gresham pump and his own: "Their machine occupies a position opposite to mine, and is completely underwater. . . . The air hole is placed in the sucker, and carries a tube rising and falling with it." One important participant in the trials performed by Hooke and Huygens in London was the Halifax physician Henry Power. Power was in London to arrange

[38] Moray to Huygens, 19 February/1 March 1663, in ibid., p. 320; Birch, *History*, vol. I, pp. 212-214, 248.

[39] For Huygens' visit to England, see his journal, in Huygens, *Oeuvres*, vol. XXII, pp. 597-603; Oldenburg to Boyle, 10/20 June 1663, in Oldenburg, *Correspondence*, vol. II, pp. 65-67; Birch, *History*, vol. I, pp. 256-257. For another comparison of Huygens' and the Royal Society's pumps, see Monconys, *Journal*, vol. II, p. 73.

the publication of his book, *Experimental Philosophy*, which he showed to John Wilkins and to Hooke for comments. He also attended meetings at Gresham during June and July and was elected a Fellow on 1/11 July. On 11/21 July Power performed a series of experiments "in Boyles engine as it is now rectifyed & altered by Mr Boyle & ye Colledge." Power's collaborators included Walter Pope, Gresham astronomy professor and a member of the Society's Torricellian committee. The experiments they produced included the boiling of water in the receiver, and the trials with animals, including a sparrow ("shee did as Mr. Boyle says") and an eel. Most importantly, Power and his colleagues produced anomalous suspension of water in a tube in the receiver, noting that it required at least two days to extract bubbles from the water, and that after the third exsuction of air, "wee could not drawe it downe at all. Wt then was it suported ye cylinder of water[?]" By 16/26 July, Power had concluded that "in all water there is a competent proportion of aire." On the same day, Hooke told the Society of the experiments on anomalous suspension that he had performed between 6/16 and 8/18 July, and was able to confirm the success of trials like those of Power and Pope.[40]

With the replication of anomalous suspension of water in the London pump, it was necessary that Boyle be informed. Hooke immediately told Boyle of the events of early July, writing that "we made a trial of [Huygens'] experiment, where indeed it succeeded

[40] Huygens, journal, in Huygens, *Oeuvres*, vol. XXII, p. 599; Hooke to Boyle, 3/13 July 1663, in Boyle, *Works*, vol. VI, pp. 486-487, and in Huygens, *Oeuvres*, vol. IV, pp. 381-383; Birch, *History*, vol. I, p. 268 (for experiments on 1/11 July) and p. 275 (for experiments reported by Robert Hooke on 16/26 July). For the background to Power's pneumatics, see Webster, "Henry Power's Experimental Philosophy," and idem, "Discovery of Boyle's Law," pp. 472-479. Power's notes are in British Library Sloane MSS 1326 ff 46-47. In May 1661 Tillotson told Croune that Power had an air-pump; Croune to Power, 20/30 July 1661 (British Library Sloane MSS 1326 f 26^{v}): "A Gent: in the North (mentioning you) had made a great many experiments with Mr. Boyle's Engine not try'd by him . . . (thus Mr. Tillotson assures mee)." Boyle referred to Power's air-pump in "Defence against Linus," p. 155 (late summer 1662). We assume that Power's notes of 11/21 and 16/26 July 1663 record trials made in London. Power was certainly there on 24 June/4 July and 1/11 July. On 6/16 July it is suggested that Power was still "here" at Gresham; on 16/26 the Society was shown by Moray a stone found in the heart of a Scottish nobleman, and Power recorded on 16/26 July that "I saw a stone yt was generated in ye heart of a certain person of quality yt dyed in Scotland." (See Birch, *History*, vol. I, pp. 265, 268, 271, 276; British Library Sloane MSS 1326 f 47^{r}.) For Power's publication of *Experimental Philosophy*, see ibid., ff 39^{r}, 40^{v}. For Pope (tutor to Evelyn's nephew after summer 1663), see Frank, *Harvey and the Oxford Physiologists*, pp. 135, 326n.

so far, that with the pumping, that was used about it, the water would not descend, though I am very confident, that if the pump had been longer plied, the event would have been much otherwise." Evidently, it was still claimed that the phenomenon might be due to residual air in the receiver, rather than to some new subtle fluid. The phenomenon was produced again after Boyle returned from Essex in August. Huygens recorded that "I saw my experiment of purged water in the void in a pipe of 7 feet in height, where the water stayed up without falling, succeed 2 or 3 times, in the presence of Lord Brouncker, Mr. Boyle and of many other persons." Boyle himself conceded that the experiment "was try'd . . . wth very good successe" by Huygens and Brouncker, who was president of the Society. But he persisted in arguing that "in regard they had noe Gage to try how farre they had exhausted ye Aire in the Receiver it seem'd not absurd to coniecture that there might remaine in ye Receiver enough [air] to keep up in ye Tube 3 or 4 foot of Water." Thus, although anomalous suspension had now at last been made a matter of fact, it was very troubled and open to a range of competitive explanations.[41]

This was particularly disturbing, since Boyle was at the same period presented yet again with a rival model of the value of experiment and the possibility of a vacuum. At least one of the participants at the Royal Society in July 1663, Henry Power, was a declared plenist who announced his arguments against the Torricellian void in his new book *Experimental Philosophy*. His own notes on the July experiments concentrated on the permanent presence of air and aether in allegedly empty space and within the water. But Power did *not* indulge in an attack on the value of experiment. Such attacks were ruled out of debate within the experimental community. For example, Hobbes himself remained outside this debate on anomalous suspension. Huygens dined with Hobbes and his friend Samuel Sorbière at the house of the French ambassador, and also with Hobbes's patron, the Earl of Devonshire. Hooke also encountered Hobbes at Richard Reeves's instrument shop in London in early July. We have no evidence that the troubles of the air-pump were raised on any of these occasions.[42] This is important,

[41] Oldenburg to Boyle, 22 June/2 July 1663, in Oldenburg, *Correspondence*, vol. II, p. 75; Hooke to Boyle, July 1663, in Boyle, *Works*, vol. VI, p. 484; Birch, *History*, vol. I, pp. 275, 295; Huygens, *Oeuvres*, vol. XVII, p. 324n; Boyle to Oldenburg, 29 October/8 November 1663, in Oldenburg, *Correspondence*, vol. II, p. 124.

[42] Huygens to Lodewijk Huygens, 3/13 July 1663, in Huygens, *Oeuvres*, vol. IV, p. 375; Tönnies, *Hobbes*, p. 63; Hooke to Boyle, 3/13 July 1663, in Boyle, *Works*,

because anomalous suspension posed the issue of plenism in an acute form. Stroup perceptively argues that it showed that the hypotheses of the spring of the air and of the void were "mutually inconsistent." Huygens, "one of the earliest and most influential champions of the air-pump," was "taking the side of the plenists in the old debate." Soon after the successful trial at Gresham College, Oldenburg told his foreign correspondents that "it seems a cleer conclusion" that air pressure could *not* be the cause that kept up water or mercury in the tube. On 31 July/10 August, he also told Spinoza that this experiment "much troubles the vacuists and much pleases the plenists." Spinoza met Oldenburg in Holland in summer 1661 and got a copy of Boyle's *Certain Physiological Essays* in October. From spring 1662, he attacked Boyle's work on the redintegration of nitre and denied that certain knowledge could be generated by experimental work alone. On 17/27 July 1663, he wrote that such knowledge would only be achieved "when we have first learnt the mechanical principles of philosophy." Boyle responded via Oldenburg, insisting that "the doctrines of the new and more solid philosophy are elucidated by clear experiments," just as he had told Hobbes. Spinoza also shared with Hobbes a commitment to plenism. Boyle refused to debate this issue, since plenism was not "proved by any phenomenon; but . . . it is assumed only from the hypothesis that a vacuum is an impossibility."[43] But now the phenomenon of anomalous suspension *did* seem to offer resources for the plenists. Spinoza and Hobbes remained outside the experimental community. They showed that a rival and dangerous interpretative schema existed in which this new phenomenon might be not an anomaly but a matter of course. Moreover, it was not just their plenism that made Spinoza and Hobbes po-

vol. VI, pp. 486-487. Sorbière was in London posing as a delegate of the Montmor Academy: see Oldenburg, *Correspondence*, vol. II, pp. 115-118, 133-136, and our chapter 4, note 6. For Power's plenism, see Power, *Experimental Philosophy*, p. 169; Webster, "Discovery of Boyle's Law," p. 472. For Reeves's shop in Long Acre and its central place in experimental work at this period, see idem, "Henry Power's Experimental Philosophy," p. 158; E.G.R. Taylor, *Mathematical Practitioners of Tudor & Stuart England*, pp. 223-224.

43 Stroup, "Huygens & the Air Pump," pp. 136-137; Oldenburg to John Winthrop, 5/15 August 1663, in Oldenburg, *Correspondence*, vol. II, p. 106; A. R. Hall and M. B. Hall, "Philosophy and Natural Philosophy: Boyle and Spinoza"; Spinoza to Oldenburg, 17/27 July 1663, in Oldenburg, *Correspondence*, vol. II, p. 94 (translation); Oldenburg to Spinoza, 3/13 April, 31 July/10 August and 4/14 August 1663 (diagram of anomalous suspension; translations), ibid., pp. 41, 99-100, 103; R. McKeon, *Philosophy of Spinoza*, pp. 137-152.

tentially dangerous; it was the conjunction of plenism with a refusal to play by the rules of the experimental game.

Boyle's strategy, therefore, was to separate the troubles of anomalous suspension from the context of the air-pump. Its status had changed. It was no longer a reliable means of showing the inferiority of Huygens' pump, since with Huygens' help the phenomenon had been produced in London. It was no longer a plausible "gauge," since rival schemes existed that described the fluids inside the receiver. Huygens had now returned to France. Boyle suggested that anomalous suspension should now be tried with mercury rather than water. Huygens had never produced the anomalous suspension of mercury. Atmospheric air kept mercury at a height of 30 inches, so any extra height would be due to something other than "ye externall Aire alone." On 9/19 September, Brouncker and Boyle were both invited to try this experiment. Now Boyle made a further move. He suggested that the suspension of mercury in long tubes should be tried *without* using the air-pump. Boyle confessed that he had never been able to produce the anomalous suspension of water with his machines at Oxford, "my owne Engines being either out of ye way or out of order." Furthermore, "the sustentation of tall Cylinders of Mercury in the Engine seem'd to me to have too little Analogie wth all ye Experiments yt have been hitherto made about those of Torricellius." Anomalous suspension was therefore not to be connected with the behaviour of residual air in the air-pump. When Boyle returned to Oxford, he considered "yt twould be to litle purpose to make use of the Engine till we were first satisfy'd yt in ye open aire the Mercury might be kept suspended in a Tube longer than 30 inches." It took at least four days for Boyle and his assistant to purge the mercury in the open air and to produce anomalous suspension. On 23 September/3 October and on 7/17 October both Boyle and Brouncker reported that well-purged mercury would stand at a height of at least 52 inches without using an air-pump.[44] The outcome of these trials was now unrelated to the comparison of air-pumps. On 29 October/8 November, Boyle wrote out his comments on this work to Oldenburg. Moray translated the letter and sent it to Huygens. Boyle pointed out that the lower 30 inches of mercury would be due to atmospheric air; Brouncker pointed out that the mercury above this level was being sustained by some other fluid. So Boyle insisted that anomalous

[44] Boyle to Oldenburg, 29 October/8 November 1663 in Oldenburg, *Correspondence*, vol. II, pp. 123-124; Birch, *History*, vol. I, pp. 301, 305, 310.

suspension did not "overthrow our former Hypothesis" of the spring, but *supplemented* it. At this point, no one hazarded an account of what this other fluid might be. Just as the air-pump experiments showed that the air had a spring as well as a weight, so now some further property was required: "[U]pon the new Expts exhibited by our Engine we did not think fit to reiect the Hypothesis of ye Weight of ye Aire maintain'd by Torricellianists but added to it ye Spring of ye Aire to improve a Theory wch these new Discovery's shew'd to be not false but insufficient."[45] Anomalous suspension was thus an accepted but troubled fact in pneumatics. Its troubles would no longer threaten the integrity of Boyle's air-pumps and their place in the experimental programme.

From November 1663, Huygens responded gratifyingly to the work in England. Problems of communication between the experimenters remained. Huygens found it difficult to see how mercury could be purged without an air-pump. He repeatedly asked Moray how "they could purge the mercury so well of all the air," and whether "the 55 inches remained when the receiver was evacuated, or only beforehand, for this is already a miracle." By December, Huygens had accepted the extended accounts of the matter of fact produced in England. So for Huygens, too, the phenomenon was no longer a means of distinguishing his pump from those in England.[46] Boyle worked to separate the integrity of the pump and the treatment of anomalous suspension. Boyle's only published reference to any of the troubles posed by anomalous suspension was his comment, in experiment 14 of his *Continuation* (1669), that different heights of mercury and water in barometers in the exhausted receiver might be due to "some aerial corpuscles yet remaining, in spite of all we had done, in the water." In fact, the spring of the air was to retain its sole interpretative power. Correspondents such as John Beale suggested various supplementary accounts, including a diminution in magnetism which might explain the failure of mercury to descend. But Boyle never discussed any of these issues, and he never published any account of anomalous suspension throughout his career. It might challenge the air's spring and the worth of the air-pump. We consider the later career

[45] Boyle to Oldenburg, 29 October/8 November 1663 in Oldenburg, *Correspondence*, vol. II, pp. 125-126, and in Huygens, *Oeuvres*, vol. IV, pp. 437-440.

[46] Moray to Huygens, 29 October/8 November and 16/26 November 1663, in Huygens, *Oeuvres*, vol. IV, pp. 426, 436-440; Huygens to Moray, 8/18 November and 29 November/9 December 1663, ibid., pp. 432, 459.

of anomalous suspension at the end of this chapter. We now examine the way in which air-pumps themselves developed.[47]

Identifying the Air-Pump: London and Oxford

In the 1660s there was no definitive version of any air-pump. The flexibility of the machine was a powerful resource in the negotiations we have described. In chapter 5 we pointed out how Boyle used the leakage of the pump to defend his early failure to achieve the separation of marbles in the receiver. When Huygens built his pumps in autumn 1661 and autumn 1662, he immediately introduced a range of changes to the design. There were differences between the pumps in London and Oxford, and the pumps were very often failing to work or else being reconstructed: hence the need for Hooke's account of "the construction of the pump as it then was" in April 1663, and for Moray's reports to Huygens throughout the year.[48] We have also said that such changes were seen as significant by all experimenters. They explained why Huygens could produce the anomaly and why Boyle or Hooke could not. Any judgment about whether these differences were significant was actually a judgment about how air-pumps worked, what they contained and whether different experimenters were competent. But in principle the air-pump was supposed to be an easily identifiable material object: otherwise, the whole pattern of replication would collapse. So the troubles of the air-pump in the 1660s centred on the way in which rivals identified their pump and assessed which differences might *matter*.

This identity was reinforced through the use of the pump as the emblem of experimental philosophy. In chapter 2 we examined the iconography of the air-pump. In the later 1660s air-pump trials were shown to the Duchess of Newcastle and to the Grand Duke of Tuscany when they visited the Royal Society. In July 1663 Wren discussed the possible use of the pump if the King himself visited Gresham College. Both Faithorne's engraving of Boyle, commissioned in summer 1664, and John Evelyn's design, which appeared

[47] Beale to Boyle, 11/21 January 1664, in Boyle, *Works*, vol. VI, p. 378; Boyle, "Continuation of New Experiments," p. 204; idem, "New Pneumatical Experiments about Respiration," pp. 361-363.

[48] Huygens, notebook, in Huygens, *Oeuvres*, vol. XVII, pp. 312, 314, 316; Boyle to Moray, March 1662, in ibid., vol. VI, pp. 581-582; Birch, *History*, vol. I, p. 214; Moray to Huygens, 19 February/1 March 1663, in Huygens, *Oeuvres*, vol. IV, p. 320.

in Sprat's *History* in 1667, contained representations of an air-pump (figures 16b and 2). However, even here the identity of the air-pump to be displayed was a matter of debate. Faithorne's original design was purely conventional in its use of a background landscape (figure 16a). On 25 August/4 September 1664 Hooke asked Boyle whether "you will have any books, or mathematical, or chemical instruments, or such like," included in Faithorne's picture. Hooke and Boyle resolved that an air-pump should serve as emblem here. Hooke told Boyle in September that "I have made a little sketch, which represents your first engine placed on a table," and wondered whether Boyle might consider adding "your last emendation of the pneumatic engine." Boyle did not: Faithorne's ultimate engraving has the earliest form of the pump (figure 16b). On the other hand, Evelyn's design shows a later version of the machine, immersed in a water tank (figure 17).[49] The pump in London had already been reconstructed this way by summer 1663, and in March 1664 Jonathan Goddard performed an experiment "under a glass body cemented to the engine," indicating that at this stage a separate plate had been developed for the receiver. During the first half of 1665, and after the suspension of meetings during the Plague, the pump was used for trials of magnetism and the motion of pendulums, and was often redesigned.[50]

Defining a standard version of the pump depended on securing it against leakage and getting it to function at all. A further problem was the changing size of the receiver. This was fairly easily changed, provided the cost of a new glass bowl could be borne. There are varying accounts of the receiver's capacity: we noted that in spring 1662 the Society was planning a receiver large enough to contain a man. The Duchess of Newcastle was entertained in May 1667 with a machine of capacity "9 gallons and 3 pints": this was roughly that of the *first* design of the air-pump. The separate plate of Boyle's revised design made receiver-changing easier.[51] Hooke also de-

[49] Birch, *History*, vol. II, pp. 177-178 (and our chapter 2, note 15); Wren to Brouncker, 30 July/9 August 1663, British Library Sloane MSS 2903 f 105, and Birch, *History*, vol. I, p. 288; Oldenburg to Boyle, 2/12 July 1663, in Oldenburg, *Correspondence*, vol. II, pp. 78-79. On the Evelyn design, engraved by Hollar for Beale, but ultimately destined for Sprat, see Hunter, *Science and Society*, pp. 194-197. On the Faithorne engraving, see Hooke to Boyle, 25 August/4 September and 8/18 September 1664, in Boyle, *Works*, vol. VI, pp. 487-490; and Maddison, "The Portraiture of Boyle."

[50] For the use of the air-pump, see Birch, *History*, vol. I, p. 398; vol. II, pp. 17, 19-20, 25-26, 31, 46; Frank, *Harvey and the Oxford Physiologists*, pp. 161-162.

[51] Birch, *History*, vol. II, pp. 177-178.

FIGURE 16a
William Faithorne's original design for Boyle's portrait (summer 1664).

voted some attention to the enlargement of the receiver. In June 1667 "it was proposed by Mr. Hooke to have a rarefying engine made of wood big enough for a man to fit in. This was approved of by Mr. Boyle." The estimated cost was £5. This engine was produced on 11/21 July, but it proved not to be "sufficiently tight." Sustained discussion followed on means of sealing the engine: lead was preferred to cement, but in any case "the air (as Mr. Hooke supposed) getting in at the brass-sucker, he informed the Society, that he had since fitted it with a wooden-sucker instead." A further problem about the identity of these machines was their specific location. After the Great Fire of 1666, the Society moved to Arundel

FIGURE 16b
William Faithorne's portrait of Boyle with air-pump in background (1664). Figures 16a and 16b reproduced by permission of Ashmolean Museum, Oxford (Sutherland Collection).

House but the pump stayed at Gresham under Hooke's management. He complained about the effects of moving the machine, "since the cement about the engine was very subject to crack in the

FIGURE 17
Boyle's revised air-pump depicted in frontispiece to Sprat's History of the Royal Society *(1667). Enlargement of detail of figure 2. The bust is of Charles II. (Courtesy of Cambridge University Library.)*

carriage from Gresham-college to Arundel-house, [almost 1½ miles] where it became defective."[52] Thus, when Boyle left Oxford for London in April 1668, the pumps both at London and Oxford had changed considerably. They had worked only intermittently and had been beset by troubles such as Huygens' novel anomaly and the endemic problem of leakage. The problem of replication was made all the more difficult by the persistently changing state of the air-pumps.

At the start of February 1668, Boyle was preparing to leave Oxford. He told Oldenburg that he had collected "a pretty number of pneumatical experiments, made in order to the continuation of the *Engine Book*." These experiments were those performed under Boyle's direction through the 1660s, and they were to be collected in the *Continuation*, completed by March 1668 and published by December. A further collection appeared in Boyle's *New Experiments touching the Relation betwixt Flame and Air* (1672).[53] As was common

[52] Ibid., pp. 184-189, 464, 467-468, 472-473.

[53] Boyle to Oldenburg, 1/11 February 1668, in Oldenburg, *Correspondence*, vol. IV,

with his publications, Boyle presented experiments tried over a number of years, and with a variety of collaborators. His discussion of the development of the air-pump since 1659 was neither consistent nor exact. It was in *Continuation* that Boyle first printed an account of the revised pump, which, as we have seen, he had developed from winter 1661-1662. Boyle announced that when he gave his original to the Royal Society in 1661 he was "unable afterwards to procure another so good," and allegedly suspended work for a time. He then claimed that because of the difficulties in replicating this original, very few other pumps had been made, and so he built a new version. Though the new version was built *before* spring 1662, Boyle told a different story:

> . . . observing that the great difficulties men met with in making an engine that would . . . keep out a body so subtle as air, and so ponderous as the atmosphere (besides, perhaps, some other impediments) were such, that in five or six years I could hear but of one or two engines that were brought to be fit to work, and of but one or two new experiments that had been added by the ingenious owners of them; I began to listen to the persuasions of those that suggested that unless I resumed this work myself, there would scarce be much done in it. And therefore having (by the help of other . . . workmen than those I had unsuccessfully employed before) procured a new engine, less than the other, and differing in some circumstances from it, we did (though not without trouble enough) bring it to work as well as the other, and, as to some purposes, better.[54]

The *Continuation* presented motives for replicating his pump and his experiments, therefore, but it also spelt out the troubles that attended any attempt to accomplish replication.

Boyle pointed out three areas of difficulty for his readers in the experimental community. First, even though immersed underwater and with a receiver on a separate plate, the new pump was still notably deficient. "It may fall out . . . that the air will insinuate itself between the wooden board and the iron-plate, and so get up . . . into the cavity of the receiver": this was the kind of problem which Hobbes had exploited in the early 1660s. Similarly, Boyle acknowl-

---

p. 140; for the appearance of *Continuation*, see ibid., vol. v, pp. 36, 21, 240; for Boyle's work on fire and air at this time, see McKie, "Fire and the *Flamma vitalis*"; Frank, *Harvey and the Oxford Physiologists*, pp. 250-258; Boyle, "New Experiments touching . . . Flame and Air."

[54] Boyle, "Continuation of New Experiments," pp. 176-178. The text is dated as "24 March 1667" [= 1668] on p. 276.

edged that trouble which Huygens had indicated in December 1661: "If great care be not taken in turning the stop-cock, the water will be impelled into the receiver, and much prejudice sundry experiments."[55] Second, both in his *Hydrostatical Paradoxes* (1666) and in the *Continuation*, Boyle continued to complain about the difficulty of obtaining reliable workers. He had not been able to perform some trials "by the help of the bare spring of the air," because he "wanted dexterous artificers to work according to a contrivance I had designed." This also affected the troubles associated with the size of the receiver. Boyle argued that his new pump was less porous because its receiver was smaller, but he also encouraged "attempts to make receivers capacious enough to contain larger animals, and perhaps even a boy or a man." Yet when he attempted to do this with "an improvement made of our metalline cylinder by additional contrivances," Boyle "could not . . . get artificers that would perform what was directed." Such problems dominated Boyle's presentation of the career of his pump. Finally, Boyle drew attention to the difficulties of identifying the details of a standard or fixed pump design. This was partly due to the troubles in making pictorial representations of his experiments and his machines. He told his readers that he hoped "they who either were versed in such kind of studies, or have any peculiar facility of imagining, would well enough conceive my meaning only by words." The diagrams of the air-pump were unreliable: "Having occasion to alter the method of my experiments, when I began to foresee that I should be obliged to reserve divers things for another opportunity; and being myself absent from the engraver for a good part of the time he was at work, some of the cuts were misplaced, and not graven in the plates." Even with Boyle's full texts, however, we have already pointed out that no one managed to replicate his pump without seeing one at work. Furthermore, Boyle wrote that "some virtuosi may be furnished with the other [i.e., the earlier pump] already." Boyle advised that "they may make use of, or at least make a shift with the first engine, with a few alterations." These alterations principally involved the insertion of a plate between the tap and the receiver, "fastened to it by sodering or screwing," which would adequately insulate the base of the receiver in roughly the way it was now done in Boyle's new pump. Boyle's claim, therefore, was that it was possible to replicate his experiments with a different

[55] Ibid., pp. 181-182.

pump—and thus that these differences were not significant for the experiments.[56]

The assessment of the significance of *differences* between pumps directly depended on the form of gauge used by experimenters and their definition of what counted as *air*. For Boyle any elastic fluid was air, since it possessed a spring. In March 1665, the Royal Society witnessed experiments on the generation of an elastic fluid from "the dissolving of powdered oister-shells." Witnesses "inquired, how it was known, that what was supposed to be air . . . was true air." Brouncker said that "a body rarefied by heat, and condensed by cold, was true air." This definition prompted the common use of bladders as a gauge of the spring. The "air" generated was collected in a bladder that did expand when put over a fire. In May 1667, when Boyle commented on a visit of the Duchess of Newcastle, he suggested a "gauge" be used to judge the exhaustion of the pump. These measures were directly connected with the possibility of securing the matters of fact made with the pump. Thus, when discussing the respiration experiments of the early 1660s, Boyle explicitly linked the identity of his own pump (as opposed to any other) with the definition of "our vacuum" and the measure of air. The identity of the pump and the identity of its contents were closely connected: "By the *Vacuum Boylianum*, he means such a vacuity or absence of common air, as is wont to be effected or produced in the operations of the *Machina Boyliana*." Yet if the vacuum was as unique as the individual air-pump, how were comparisons to be effected?[57]

Comparison demanded gauging the contents of the receiver. In the *Continuation*, Boyle surveyed the means used to gauge the air-pump. Each measure depended on a different model of how the pump worked and what the air was. Boyle dismissed the use of bladders, since they were wont to become too large, and small mercury barometers or siphons were "shaken by the motion of the engine." He recommended the use of a mercury manometer: a small tube bent into a curve with one end containing an air-bubble and then sealed, filled with mercury and then opened at the other end into the receiver. This "standard-gage (if I may so call it)" was then to be calibrated with standard volumes of water. If "when the

[56] Boyle, "Hydrostatical Paradoxes," pp. 738-744; idem, "Continuation of New Experiments," pp. 245, 258-259, 178, 180-182.

[57] Birch, *History*, vol. II, pp. 26, 177-178; Boyle, "New Experiments touching . . . Flame and Air," pp. 564-566; idem, "Continuation of the Experiments concerning Respiration," p. 372.

quicksilver in the gage is depressed to such a mark you let in the water, and that liquor appears to fill a fourth part of the receiver, you may conclude that about a fourth part of the air was pumped out, or that a fourth part of the spring . . . was lost by the exhaustion." The scale was marked with glass balls or sealing wax, and coloured water in a longer tube could be used. But Boyle's recommended method was by no means acceptable to all experimenters. It was too dependent on a very specific model of the structure of the air. Boyle used water as a *surrogate* for air in this calibration; then the mercury gauge, not water itself, became the measure of spring. Hooke worked differently. In his own experiments on respiration and combustion, he sought the effects of fire and of animal life on the air contained in the receiver. He used water volume as a *direct* measure of spring. Hooke and Boyle had different models of the air: for Boyle, air was any elastic fluid and so it could be measured with a mercury manometer. For Hooke, air was one of a complex chemically active mixture of fluids. In the 1670s, both Hooke and Mayow observed that when respiration took place in the air in the receiver, the mercury gauge registered very little change. However, the volume of water would rise markedly into the receiver, so it seemed best to use water level as the measure of spring. The choice of measure followed from the choice of ontology. The same trouble arose in combustion trials, when the air grew warm. Boyle claimed that the mercury gauge could be used to measure the spring of compressed air; but he accepted that the gauge overestimated exhaustion when the included air was warm. This happened in combustion experiments. Then the pump should be worked till the gauge fell no more: "One that is versed in these trials, may well enough judge when he needs to pump no longer." To establish such craft-experience, of course, it was necessary to provide clear identifying marks for the air-pump, and it was also necessary to disseminate the air-pump and to compare claims to replication. We have already charted the troubles of replication; we now consider Huygens' work in disseminating his pump.[58]

[58] Boyle, "Continuation of New Experiments," pp. 211-214. For the work on air and spring by Hooke and Mayow, see Frank, *Harvey and the Oxford Physiologists*, pp. 256-263; Mayow, *Tractatus quinque* (1674), pp. 66-71; Hooke, *Diary*, pp. 32-35; Birch, *History*, vol. III, pp. 58-60, 78, 84, 89-90, 109, 143, 156-157, 177; Boyle, "New Experiments about the Weakened Spring," p. 218n; idem, "New Experiments about . . . Air and the Flamma Vitalis," pp. 586-587.

Disseminating the Pump: Holland and Paris

The only other air-pump of the early 1660s was that built for the group of natural philosophers in Paris known as the Montmor group. This machine was constructed under the direct supervision of Huygens; without his presence the machine did not work, nor did the members of the Paris group understand what Huygens had done until he demonstrated it for them in Paris. As Stroup points out, although the French developed the work on pneumatics pursued in the 1640s and 1650s, nevertheless it was Huygens who dominated air-pump work in Paris. This testifies to the importance of Huygens' transmission of skill in pump experiments and of his work to win acceptance elsewhere to the matter of fact of anomalous suspension. These factors made it possible and necessary to disseminate air-pumps in France. Huygens was in touch with the Montmor group from its formation in December 1657. Its members included Sorbière, Auzout, Montmor, Thévenot, Pecquet, Petit (who had collaborated with Pascal in Rouen), Roberval (who left the group after 1658), Rohault, and Chapelain. Much of their collective work debated the vacuum and capillary rise, and these ideas were discussed when Huygens visited Paris in 1660-1661. Petit had already read Boyle's reports on the air-pump, but told Oldenburg in October 1660 that he "could not get appropriate glass vessels made" in France. During autumn 1661, when both Boyle and Huygens were building air-pumps, Huygens wrote to Paris about his design and heard from Petit that glassware was being brought from Rouen to perform trials on the air. Huygens also told the Parisian natural philosophers about the appearance of two Latin translations of Boyle's book. In January 1662 Huygens wrote to his brother Lodewijk, then in Paris, to suggest that Rohault attempt to build an air-pump for himself in order to try anomalous suspension, but Rohault failed.[59] The result was a sustained correspondence with the Montmor members in which the difficulty of winning their understanding and assent was clearly demonstrated.

[59] H. Brown, *Scientific Organizations*, pp. 66-89, 107-134; Mesnard, "Les premières relations parisiennes de Huygens"; Brugmans, *Le séjour de Huygens à Paris*; Stroup, "Huygens & the Air Pump," p. 138. For contacts with the Montmor group: Petit to Oldenburg, 13/23 October 1660, in Oldenburg, *Correspondence*, vol. I, p. 398 (translation); Huygens to Thévenot, 26 September/6 October 1661, in Huygens, *Oeuvres*, vol. III, pp. 359-360; Petit to Huygens, 28 November/8 December 1661, ibid., p. 398; Huygens to Lodewijk Huygens, 11/21 December 1661, ibid., p. 414. See also McClaughlin, "Le concept de science chez Rohault."

During February and March 1662 Huygens sent the French increasingly detailed accounts of the structure of his pump, diagrams of its insulation, and reports on anomalous suspension. On 19/29 March he told Lodewijk that "it is necessary to look for some other principle than that of the spring of the air," and this was passed on to the Montmor group. Huygens soon realized that his correspondents "have not been well instructed on the fact," because their accounts mistook or ignored significant aspects of Huygens' procedures. For example, on 20/30 April Jean Chapelain sent an analysis of a phenomenon he took to be anomalous suspension. He understood Huygens to have inserted a tube of water in the receiver, exhausted the receiver, and prompted the fall of water into the basin below; readmitted air into the receiver, forcing water back up the tube; and, finally, observed that after a second exsuction the water did not descend. Huygens annotated Chapelain's letter with a series of comments that this was *not* the phenomenon he had produced. Huygens complained to Lodewijk about his brother's failure to explain this clearly to the French, and told Chapelain that his proffered theory was not necessarily false but certainly irrelevant. Chapelain replied on 5/15 June that "I would be more convinced if I was at the place where you are." Huygens repeated that he had used purged water admitted slowly into the tube, and that his own model used a highly elastic subtle fluid which would violate Boyle's Law of spring. The Montmor group failed to grasp any of these claims.[60]

So during the summer of 1662, Huygens considered the need to return to Paris and build a pump there to establish anomalous suspension in France. Thévenot told him that the French had now heard of the new pump Boyle had made at Oxford.[61] That autumn Huygens finished his own new pump and then set off for Paris in March 1663. On 31 March/10 April he visited Montmor and Sorbière to discuss plans to build the machine and to learn of the "new laws and ordinances" which they planned for a projected organization of natural philosophers. On 10/20 April Montmor sent Huygens "a mathematician and a worker in copper with the request that I instruct them in making a vacuum machine like the one I have made." This machine was "half finished" by late May, "that is

[60] Huygens to Lodewijk Huygens, 12/22 and 19/29 March, 9/19 and 16/26 April, 8/18 May 1662, in Huygens, *Oeuvres*, vol. IV, pp. 96-97, 111, 117, 133; Chapelain to Huygens, 14/24 and 20/30 April, 5/15 June 1662, ibid., pp. 112-124, 154-156 (Huygens' annotations are on pp. 123-124); Huygens to Moray and Chapelain, June 1662, ibid., pp. 174-175.

[61] Thévenot to Huygens, 12/22 June 1662, in ibid., p. 161.

to say the cylinder with the tap." But then Huygens left for London and the work had to stop.[62] Thévenot and Auzout were particularly keen to complete the machine, and in July they pressed Montmor and Petit to write to Huygens in London "to send an exact diagram of the joinery, whether of the height or the width or the quantity of supports and places where the pump, the handle and the receiver with the gear must be situated. The work having been interrupted by the absence of M. Huygens who is the inventor and consequently the promotor of this work." Petit also told Huygens that "we await your presence at our academy for the completion of the vacuum machine."[63]

Huygens responded to this request by sending an annotated diagram that corresponded in most details to the one he drew in Holland the previous autumn. The main change was that the pump was simplified: instead of the complex sucker which Huygens never successfully copied, it possessed a wooden shaft wrapped tightly with oil-soaked rope. Huygens also improved the solder of the pipe joining the receiver to the cylinder with a copper plate cemented onto the brass tap (figure 18). This diagram was not enough. Huygens returned to Paris from London in September 1663 and the pump was at last completed. It was working well by the end of November and Huygens began a set of trials that would assess its porosity. During December, Huygens exhibited this machine to notables in Paris, and told Moray that Rohault had offered a new explanation of anomalous suspension. But the pump in Paris could not produce anomalous suspension until March 1664. In the interim, Huygens was worried by the French theories of the phenomenon, since they depended on the size of the glass tube. As we have seen, Huygens had now learnt that the natural philosophers in London had produced the effect with mercury without using an air-pump, and so he asked Moray to tell him the exact thickness of the tubes used in London to do this. Finally, on 2/12 March he wrote that the pump had begun to work well in Auzout's rooms in Paris, and that he was now able to produce anomalous suspension for other eminent Parisians.[64] When Moray wrote to say that the

[62] Huygens to Lodewijk Huygens, 27 March/6 April, 10/20 April and 15/25 May 1663, ibid., pp. 325-329, 334, 345. For the "Project de la Compagnie des Sciences et des Arts," see Hahn, "Huygens and France," pp. 60-62.

[63] Montmor to Huygens, July 1663, in Huygens, *Oeuvres*, vol. IV, p. 365, and Petit to Huygens, 5/15 July 1663, ibid., p. 377.

[64] Huygens to Montmor, July 1663, ibid., vol. VI, pp. 586-587 (diagram on p. 587); ibid., vol. XVII, p. 258n; Petit to Oldenburg, 2/12 October 1663, in Oldenburg, *Correspondence*, vol. II, p. 117 (translation; Petit requested news of Boyle, "whose

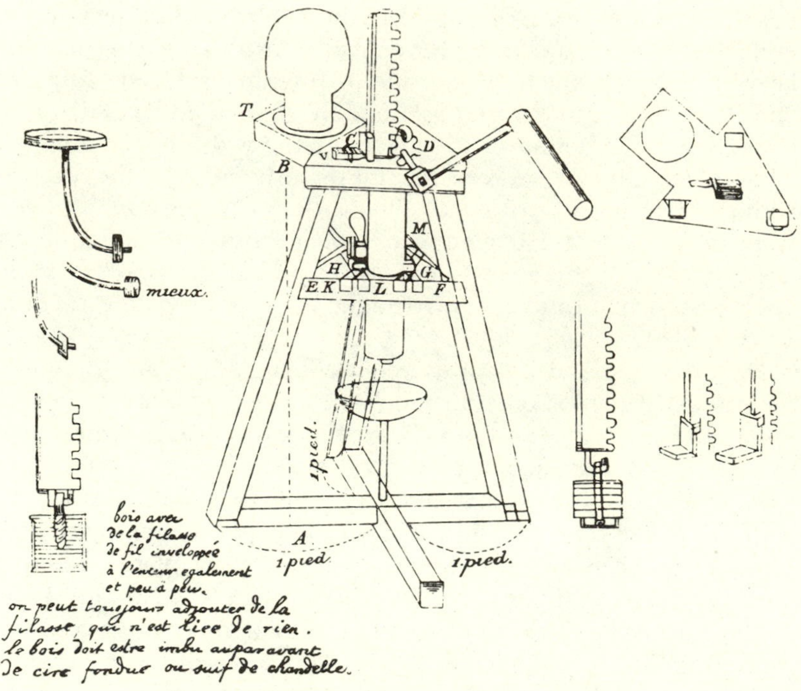

FIGURE 18

Design of Huygens' second air-pump; drawing sent from London to the Montmor group in Paris (July 1663). Top right is the base-plate for the receiver and the top of the piston; centre left is the connection of the incoming pipe from the receiver with the stopcock and cylinder; bottom left is the sucker and the head of the piston, rubbed with candle grease or melted wax and then wrapped with fine thread. From Huygens, Oeuvres, *vol.* VI, *opposite p. 586. (Courtesy of Edinburgh University Library.)*

London tubes were less than a finger's breadth wide, Auzout abandoned his account of the phenomenon, and Rohault also withdrew. By spring 1664, therefore, the Montmor group had accepted the reality of the matter of fact and of Huygens' subtle fluid. The future career of their pump is not clear. The Montmor group broke up

---

experiments we are about to verify with an engine similar to that of Mr. Huygens, and to make others upon which we will decide"); Huygens to Moray, 8/18 November and 29 November/9 December 1663, in Huygens, *Oeuvres*, vol. IV, pp. 433, 459; Auzout to Huygens, December 1663, ibid., pp. 433, 459; Auzout to Huygens, December 1663, ibid., p. 482; Huygens to Lodewijk Huygens, 5/15 December 1663, ibid., p. 472; Huygens to Moray, 9/19 December 1663, ibid., p. 474; Huygens to Lodewijk Huygens, 12/22 February 1664, ibid., vol. V, p. 31; Huygens to Moray, 2/12 March 1664, ibid., p. 41.

in May 1664 and Huygens returned to Holland in June. The Montmor machine then seems to have disappeared.[65]

Work on air-pumps only revived in France when Huygens returned to Paris in summer 1666. He had come to participate in the new Académie Royale des Sciences, which was initiated in the autumn. Once a member of the Académie, under Colbert's patronage, Huygens soon started planning the construction of an air-pump.[66] In early 1667 he drafted a scheme for a pump based on those he had built in Holland in 1662 and in Paris in 1663. By spring 1668 Huygens was able to present this new pump at the Académie. Huygens recommended his own pump since Guericke's "had been judged too inconvenient," and while Boyle "had perfected it or rather had made a new one, . . . nevertheless faults had still been found in it." So Huygens had built a pump that was "much more convenient" in order to perform experiments on the vacuum. Huygens made a set of revisions in this new machine: it possessed a much longer turning key, a redesigned plate to carry the receiver, and a piston made of a copper cylinder wrapped with fine string, rubbed with turpentine inside and candle grease outside. Huygens paid particular attention to the friction developed when the piston moved in the cylinder, but he maintained his criticisms of the English practice of immersing the pump in water. Huygens claimed credit for the idea of using water as an insulation of the piston, but insisted that his way of covering the top of the sucker with water and oil was definitely better. Furthermore, the new pump of 1668 was worked in the same manner as that of 1662: the air was extracted from the pump through a small hole in the bottom of the copper cylinder, which was then sealed either with a leather or wax cover or with one's finger (figure 19, above Z). Each of these provisions was claimed to render the machine both easier to use, and thus to copy, and also less porous, and thus more capable of displaying the effects of Huygens' subtle fluid.[67]

Between March and May 1668 Huygens tested the working of

[65] Huygens to Lodewijk Huygens, 5/15 June 1665, ibid., vol. v, p. 375; the end of the Montmor group is noted in ibid., p. 70. Neither Auzout, Petit, nor Thévenot became members of the new Académie. For the relations with this group, see McClaughlin, "Sur les rapports entre la Compagnie de Thévenot et l'Académie Royale," and Roger, "La politique intellectuelle de Colbert."

[66] "Biographie," in Huygens, *Oeuvres*, vol. XXII, pp. 625-626; Roger, "La politique intellectuelle de Colbert"; Hahn, "Huygens and France," pp. 62-66; idem, *Anatomy of a Scientific Institution*, chap. 1.

[67] Huygens, *Oeuvres*, vol. XIX, pp. 199, 201-202; see also the discussion on pp. 189-196 and p. 205n.

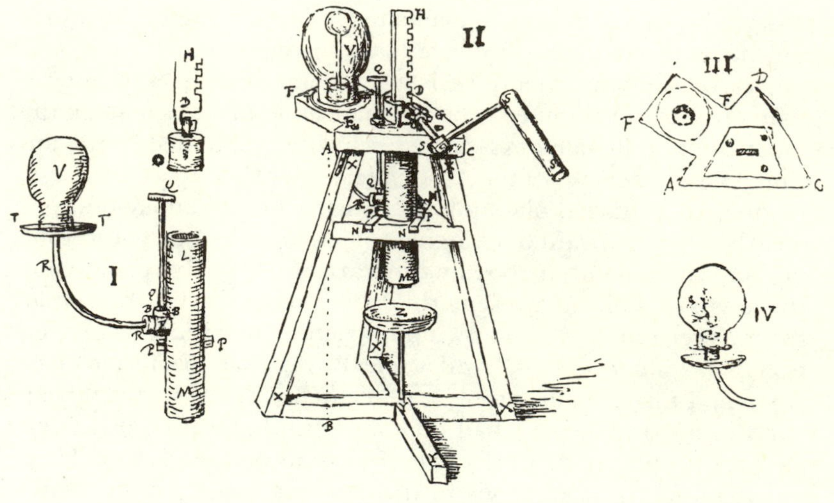

FIGURE 19

Design of Huygens' air-pump as demonstrated at the Académie Royale des Sciences in Paris (May 1668). Note the long key to the stopcock. From Huygens, Oeuvres, *vol.* XIX, *p. 202 (figure 95). (Courtesy of Edinburgh University Library.)*

this pump at the Académie. He used the range of standard gauges to assess its performance: bladders stayed inflated in the receiver, alarm clocks became inaudible when placed there, and alcohol visibly boiled. Moreover, Huygens and his colleagues outlined a new programme of research on plant growth inside the air-pump. These experiments would demand much longer periods of evacuation, and Huygens was confident his pump could stay in a working state over a greater length of time than those in England. The standard means of calibrating the machine, a small water barometer of six inches in length, seemed to confirm this claim.[68] Huygens also had his own theoretical interests which made it important to have access to a reliable pump in Paris. These interests centred on his model of the subtle fluid. In autumn 1667, soon after planning the construction of the pump, Huygens drafted a note on this "matière subtile," debating whether such fluids would behave differently with respect to weight and inertia in comparison with common matter. In spring 1668, at the moment when he presented his pump to the Académie, he wrote a text entitled "De gravitatione." The presentation of this text at the Académie then elicited the violent debate between Huygens, Roberval, and Mariotte on impact laws

[68] Ibid., pp. 200, 207-213.

and the character of subtle fluids that raged in the Académie during the autumn of 1669. On 18/28 August, Huygens stated his axiom that "to find an intelligible cause of weight, one must see how it can be done while supposing nothing in nature but bodies made of one common matter." These exchanges immediately prompted Huygens to return to his work on pneumatics to produce matters of fact that displayed the effects of a much wider range of subtle fluids, and here his new air-pump and anomalous suspension took pride of place.[69]

Anomalous suspension was no longer a means of showing the excellence of Huygens' pump. By assuming that the approval of the Académie established the worth of his machine, Huygens now used anomalous suspension to argue against critics of his notions of subtle fluids. In July 1672 Huygens published a letter in the influential *Journal des sçavans*, which made use of the air-pump work to this end. Huygens recapitulated the history of the air-pump and of anomalous suspension. By describing the Royal Society's resistance to the matter of fact, he made its ultimate assent all the more compelling. He reemphasized the superiority and uniqueness of his own pump, thus giving himself privileged access to the effects of subtle fluids. He appealed to a variety of relations between matter, including the concept of *liaison*, which violated his mechanical principles but made the action of fluids more plausible. He recalled that whereas in Boyle's pump, water only fell to one foot above the basin, in his own pump water fell all the way to the basin level on exhaustion of the receiver. This showed his pump was better: "I could scarcely suspect that there was any fault in my pump." His trial of anomalous suspension showed that "apart from the pressure of air, which sustains mercury to a height of 27 inches in the Torricellian experiment," there was "another pressure, much stronger than this, due to a matter more subtle than air, which penetrates glass, water, mercury and all other bodies which we see to be impenetrable to air."[70]

Huygens introduced some revisions in his account of the working

[69] For Huygens on subtle matter (1667): ibid., vol. XIX, p. 553; on gravitation (1668): ibid., pp. 625-627; Huygens' polemic with Roberval (1669): ibid., p. 631. For aspects of this work, see Dugas, "Sur le Cartésianisme de Huygens" (for an emphasis on the commitment to Descartes); Westfall, *Force in Newton's Physics*, chap. 4 (for kinematics); Snelders, "Huygens and the Concept of Matter"; Gabbey, "Huygens et Roberval"; Halleux, "Huygens et les théories de la matière."

[70] Huygens, "Lettre . . . touchant les phénomènes de l'eau purgée d'air," in *Oeuvres*, vol. VII, pp. 201-206.

of the pump and the matter of fact. In 1662 Huygens said the subtle fluid was a specific rare and elastic susbstance contained within the liquid. Its enormous spring would force down the liquid when it escaped above the water in the tube. But in 1672 Huygens had now to show the real presence of such fluids throughout space. So his fluid of 1672 was present in the atmosphere, and then penetrated the glass wall of the receiver. This fluid came from outside the receiver, not inside the water. Huygens used his concept of *liaison* to explain why it could not then get through the glass of the tube or the water itself. He also cited a set of further experiments, particularly those on siphons and the cohesion of marbles in the air-pump. We have noted that Huygens persisted in citing Boyle's reports of the failure of marbles to separate in the void, even after Boyle announced in 1669 that they had been made to separate. These were resources Huygens could not abandon. He produced detailed accounts of pneumatic phenomena made in the air-pump in his analysis of weight and gravity in 1678 and in his *Discours de la cause de la pesanteur* in 1686. The principal phenomena of the pump, the cohesion of marbles and the anomalous suspension of water, were always Huygens' best weapons in his effort to establish the reality of a range of space-filling subtle fluids: "I hold myself very assured of the new pressure I have supposed apart from that of air." To make this claim compelling, Huygens had to show that his pump was absolutely impermeable to air and absolutely permeable to subtle matter. The distinction between these two kinds of substance defined how the pump worked.[71]

In 1672-1673 Huygens' claims were discussed in detail in England and France. In summer 1672 Oldenburg printed a translation of Huygens' paper on anomalous suspension. This prompted hostile remarks from Sluse, who suggested the subtle fluid might be unnecessary, and from Towneley, who said he had failed to replicate the phenomenon. Hooke addressed the Royal Society on this issue in November 1672, and Wallis sent Oldenburg a series of letters which surveyed the variety of explanatory accounts and their different definitions of "air" and "subtle matter." We cited Wallis's letters in chapter 4, when discussing these different stipulations about the contents of the air-pump. The English response showed

[71] Ibid., pp. 204-206. For Huygens on magnetism and subtle fluids (1678), see ibid., vol. XIX, pp. 584-585; "Discours de la cause de la pesanteur," ibid., vol. XXI, p. 380 (text of 1686) and p. 474 (text of 1690). For the commitment to the range of fluids, see A. Shapiro, "Kinematic Optics," sect. 5; Rosmorduc, "Le modèle de l'éther lumineux"; Albury, "Halley and the *Traité de la lumière* of Huygens."

the relation between models of the air and rival interpretations of the use of the air-pump and its phenomena.[72] In France, Huygens was attacked by Pierre Huet, tutor to the Dauphin, and by Pierre Perrault, author of an authoritative work on hydrology. Huet denied the matter of fact. He said it was well known that air-pumps leaked. He distinguished between "the sensible demonstrations of Geometry" and "the clearer ones of Physics," where Huygens' grasp was allegedly less sure. Huet defended the Académie against embarrassing charges that it merely generated controversy: he avoided inviting Huygens to respond. Instead, he attacked the notion of *liaison*, arguing that any such link between water or mercury and glass could sustain long columns of liquid, without any need for Huygens' more dubious subtle fluid.[73] In summer 1672 Huygens also received a text from Perrault on the *horror vacui*. Huygens rejected this concept and made further notes on anomalous suspension and the subtle fluid contained in the receiver of the air-pump. In May 1673 Perrault turned Huygens' letter against him. Perrault advocated pugnacious scepticism: "Experiments do not give general decisions, and most often prove nothing." Perrault said that since "the principles of motion are not known," there was "no reason absolutely to reject attraction and only to admit impulse." His prime example against mechanism was, again, anomalous suspension. This showed that effects hitherto attributed to air pressure had been explained falsely. Huygens answered that "I have imagined causes for this which satisfy me well enough, without at all destroying that which depends on air pressure." Huygens compared the work of the natural philosopher to that of the code-breaker, and insisted that his conjectures were secure but always provisional.[74] The debate on the authority of experimental philos-

[72] Huygens, "An Extract of a Letter to the Author of the Journal des Sçavans," *Philosophical Transactions* 7 (1672), 5027-5030. For the English response, see Towneley to Oldenburg, 15/25 August and 30 September/10 October 1672, in Oldenburg, *Correspondence*, vol. IX, pp. 212, 267; Wallis to Oldenburg, 16/26 September, 26 September/6 October 1672 and 19 February/1 March 1673, in ibid., pp. 259, 279, 519-520 (the last addition was omitted from the published paper); Birch, *History*, vol. III, pp. 58-60. Compare Oldenburg to Sluse, 11/21 November and 16/26 December 1672, in Oldenburg, *Correspondence*, vol. IX, pp. 316-317, 363 (translations); Sluse to Oldenburg, 26 November/6 December 1672, ibid., p. 336; Sluse to Huygens, 26 September/6 October 1662 and 3/13 October 1664, in Huygens, *Oeuvres*, vol. IV,p. 248, and vol. V, p. 121.

[73] Huet, *Lettre touchant les expériences de l'eau purgée* (1673), discussed in Huygens, *Oeuvres*, vol. XIX, pp. 242-243.

[74] Perrault to Huygens, May 1673, ibid., vol. VII, pp. 287-298, and Huygens to

ophy was a direct consequence of the presentation of Huygens' work at the Académie. These exchanges were comparable with those of Boyle in his contest with Hobbes. The claims for experiment made by Boyle and Huygens were seen to rest on the integrity of the air-pump. That integrity was rejected by Hobbes, by Huet and by Perrault. These claims were also connected with attitudes to the factual status of anomalous suspension. Huygens put this fact at the centre of his matter-theory, and so published his report and alleged it was misused by his critics. This fact did not suit Boyle's purposes; he never published it, and Hobbes could never use it. It is likely that the career of anomalous suspension was affected by protagonists' consideration of Hobbes's reaction and uses of the phenomenon *had he known*. Thus, we have the apparently paradoxical situation in which Hobbes was a major actor on a stage where he never appeared.

During the 1670s air-pumps gradually changed from being the restricted property of a few privileged individuals to becoming commercially available articles. They were still displayed as emblems of experimental philosophy. Huygens' pump appeared in the frontispiece of a volume sponsored by the Académie Royale des Sciences in 1671 (figure 20). However, air-pumps were now an unproblematic resource and no longer an assemblage of controversial theoretical and practical components. Their calibration became a matter of routine rather than a matter of dispute. This was partly due to the work of Denis Papin, who began working with Huygens in Paris from summer 1673. The following year he published his *Nouvelles expériences du vuide*, which included Huygens' account of his new air-pump presented at the Académie in 1668. Papin also offered a discussion of the anomalous suspension of water, but now it merely served to show why a water barometer should not be used to calibrate the air-pump. Since water was prone to contain air-bubbles, its height was a bad measure of the exhaustion of the receiver, and when all the bubbles were removed, then the water would not fall at all. So Papin explained that "by the word *test* I will understand a bolthead or a tube which thus serves to measure the quantity of air in the receiver. And often I fill these tubes with mercury rather than with water." Anomalous

Perrault, 1673, ibid., pp. 298-301. On Perrault's scepticism, see Delorme, "Pierre Perrault," and on Huygens' probabilism, see Elzinga, *On a Research Program in Early Modern Physics*, pp. 36-44; idem, "Huygens' Theory of Research." Huygens' notes on anomalous suspension (1673) are printed in *Oeuvres*, vol. XIX, pp. 214-215.

FIGURE 20

Frontispiece of Claude Perrault, Mémoires pour servir à l'histoire naturelle des animaux *(Paris, 1671). Engraving by Sébastien Le Clerc (detail). Huygens' revised air-pump is on the left. The scene represents an imaginary visit in 1671 to the Académie by Louis* XIV *(centre) and Colbert (right). (Courtesy of Cambridge University Library.)*

suspension had now lost most of its interest for users of the air-pump.[75]

Papin also assisted the commercial dissemination of the air-pump. He wrote that "we are in a century where we are strongly attached to this kind of study: and having made the construction of vacuum machines so simple and so easy that everyone can have one at their own disposition, it very much looks as if we will, in the passage of time, experiment on more new things than we ever have before." Papin gave detailed descriptions of a very simplified version of the pump, designed after 1673. He referred his readers to the Paris clockmaker Gaudron, where one could find "machines ready-made."[76] There is also evidence that in Novemer 1673 Huy-

[75] Papin, *Nouvelles expériences du vuide* (1674), chap. 2 (printed in Huygens, *Oeuvres*, vol. XIX, pp. 217-218). For Papin's career with Huygens, see *Oeuvres*, vol. VII, p. 478; Cabanes, *Denys Papin*; Payen, "Huygens et Papin."

[76] Huygens, *Oeuvres*, vol. XIX, p. 216.

gens himself was acting as consultant on the despatch of air-pumps from Paris to Provence. In May 1674 Mariotte obtained a copy of Papin's book, and told a correspondent in Burgundy that Papin's machines were ten times cheaper and "more secure" than those of Huygens: he offered to send an example. A commercial market for air-pumps seems to have developed by the mid-1670s. In 1678 air-pumps in Paris were sold for the equivalent of four guineas by Hubin, "enameller to the King," and a former instrument-maker for Huygens. In Holland, Samuel van Musschenbroek made air-pumps at Leyden, and in England, Papin worked with Boyle on air-pump experiments with a new double-barrelled design from 1675. These experiments appeared in Boyle's second *Continuation* in 1680.[77] Air-pumps had now become cheaper and more widely available. Ultimately, so Guerlac has suggested, the air-pump techniques developed by Boyle and Papin were transmitted to Francis Hauksbee, who began his research at the Royal Society for Isaac Newton in 1703. Hauksbee's experiments, at least, do recall those of the 1660s: his work on capillarity in the void was to be profoundly influential in the matter-theory Newton developed after 1706, in the Queries to the *Opticks*. We can trace in this process the manner in which the closure of debate was accomplished: the air-pump as an instrument had been stripped of its contingent interest, and now appeared to be an unproblematic resource for the "business of experimental philosophy."[78]

The Limits of Replication: Germany and Florence

We now consider some sites at which experimenters did *not* attempt to replicate Boyle's engines. In this section, we look at two examples: the members of the Accademia del Cimento in Florence, and Otto von Guericke and his collaborators in Germany. The Accademia del Cimento was one of the most important centres of research on pneumatics in the 1650s and 1660s. Oldenburg sent the Florentines

[77] Ibid., p. 233; ibid., vol. VII, p. 412; Guisony to Huygens, 8/18 November 1673, ibid., p. 361; Gallon to de Puget, ibid., vol. X, pp. 730-732; Pelseneer, "Petite contribution"; Daumas, *Les instruments scientifiques*, pp. 115-117, 184.

[78] For Newton on "the business of experimental philosophy," see *Opticks*, p. 394; for Newton on anomalous suspension, see Westfall, *Force in Newton's Physics*, p. 412n. For Hauksbee, see Guerlac, *Essays and Papers*, pp. 107-119; Hawes, "Newton and the Electrical Attraction"; Home, "Hauksbee's Theory of Electricity"; Gad Freudenthal, "Early Electricity."

a copy of Sharrock's Latin version of Boyle's *New Experiments* in October 1661. In August 1662 the Accademia talked of several experiments reported by Boyle. They also compared the manufacture of the void with a Torricellian tube and that made by Boyle's pump. The comparison was based on the extinction of animals, which was faster in the barometer than in the air-pump. Boyle made the same comparison, but used different gauges. He insisted that he used small receivers which could be exhausted in less than thirty seconds; that very few experiments could be done "with any conveniency and some of them not at all" in the barometer; that the Torricellian space would inevitably be filled with "aerial particles lurking in the mercury" or between the mercury and surfaces "to which it does not closely adhere." Nevertheless, the Florentines concluded that "these two experiments, far from contradicting one another, agree marvellously well." They argued that if the barometer were slowly tilted, then "the air would have to pass through all degrees of rarity, successively greater and greater (much like what happens in the exhaustion of his [Boyle's] receiver.)" They were satisfied that nothing Boyle produced was a challenge to Torricellian pneumatics or demanded direct replication.[79]

Yet, as the leading historian of the Accademia has observed, despite their interest in this comparison and its uses the Florentines made no attempt to build an air-pump, nor to repeat Boyle's trials in a similar apparatus. They claimed to "verify all those [experiments] that Mr. Boyle has made in his instrument," but built no such instrument. The nearest equivalent was a machine formed of a copper box cemented to a pump like Guericke's machines. The Florentines reported that with it, they could not empty "vessels in this way as perfectly as with the quicksilver," and few trials seem to have involved this machine. Members of the Accademia, including the secretary Lorenzo Magalotti, visited England in 1668 and witnessed air-pump trials in Oxford with Boyle. Boyle apparently convinced them that his machine was superior, for while "the excellent Florentine academicians were pleased to confess to me" that they could not smash glass-bubbles in the Torricellian space, Boyle

[79] Middleton, *The Experimenters*, pp. 162, 263-270; Oldenburg to Boyle, October 1661, in Oldenburg, *Correspondence*, vol. I, pp. 440-442. The Oxford Latin edition, *Nova experimenta* (1661), is no. 19 of Fulton's Boyle bibliography. For other reports on the despatch of these editions, see Huygens to Montmor, 26 September/6 October 1661, in Huygens, *Oeuvres*, vol. III, p. 358, and Huygens to Moray, 25 October/4 November 1661, ibid., p. 384. For Boyle's comment, see "New Experiments touching . . . Flame and Air," p. 565.

had often done this "with the engine I employ, and convinced them that I could do so by doing it in their presence." This is a case of experimenters who possessed the resources to construct a version of Boyle's air-pump, but who did not do so, and who asserted the equivalence of their own machine.[80]

Otto von Guericke had started work on air-pumps in the 1650s with his design for what was called an *antlia pneumatica* at Regensburg, and his work was reported in Caspar Schott's *Mechanica hydraulico-pneumatica* in 1657. Boyle read this report and commented on the severe disadvantages of Guericke's initial design (see figure 22). In chapter 2 we listed those criticisms: Boyle argued that the pump needed to be immersed in water, that it did not offer a space accessible to experimentation, and that it was extremely difficult to work.[81] So it is very significant that Schott and Guericke *also* rejected Boyle's new design with equal vehemence. Nor did they attempt to replicate any of Boyle's air-pumps. Boyle and Schott corresponded during the 1660s, and Boyle sometimes commented favourably on his work along with that of other Jesuits. In February 1662 Guericke wrote to Schott telling him that news of Boyle's book had reached Magdeburg, though Guericke had not yet read it. He also told Schott of his plans for a revised pump design that was to occupy two storeys of his house (figure 21).

The pump was to be immersed underwater, and would need the labour of at least two men to operate it. On 30 April/10 May, Guericke wrote again to Schott, discussing those sections of the 1661 Latin translation of Boyle's book, which he had now read. The sections that attracted Guericke's attention were just those which showed that Boyle's pump did not extract all the air from the receiver. These were the failure of marbles to separate in the receiver and the failure of water to fall to the horizontal in the experiment of the void-in-the-void. Guericke wrote that

> . . . from these and other places, it appears that while a large part of the air is extracted, yet at the same time more air insinuates itself furtively round the sides of the sucker. Which by no means happens in my machine, whose picture you will

[80] Middleton, *The Experimenters*, pp. 152, 264; Boyle, "New Experiments," p. 64. For Magalotti and his visit to England, see Middleton, *The Experimenters*, p. 291; Waller, "Magalotti in England." Boyle reported a public comparison with the Florentines in "New Experiments touching . . . Flame and Air," p. 566; see also Oldenburg, *Correspondence*, vol. IV, pp. 193, 234.

[81] Boyle, "New Experiments," p. 6.

FIGURE 21

Otto von Guericke's second pump at his house in Magdeburg. From Schott's Technica curiosa *(Würzburg, 1664), p. 67. (Courtesy of Cambridge University Library.)*

> recall from what I recently sent you, Reverend Father. I have not rarely kept glasses, copper globes and other similar vessels evacuated for three months or longer.

He concluded that Boyle's machine was "in no way suitable for producing a vacuum, partly because the pump is not immersed in water, and partly because it cannot be kept going so quickly because of the shaking of the support." There was thus no reason to attempt the replication of the English machine.[82]

In 1664 Schott gave his most complete presentation of Guericke's pneumatics in his *Technica curiosa sive mirabilia artis*. The first volume of this massive work contained Guericke's reports; the second volume contained a Latin version of the whole of Boyle's *New Experiments* under the title "English marvels, or the pneumatic experiments displayed in England." Schott also printed Guericke's letters on Boyle and added his own comments on the failings of the English pump. He wrote that

> Guericke says well in affirming with certainty that [in Boyle's pump] much external air manages to get in between the sides of the cylinder and the sucker of the machine, since it is not immersed in water. The same could be said of the cover and the opening of the receiver. So internal air may be extracted, and external air may be excluded, much more perfectly and with no more labour in the Magdeburg machine than in this one.

Schott went further, and pointed out that Boyle's *provisional* definition of the vacuum indicated how porous was his pump:

> This author observes that by a vacuum he understands not some space in which there is no body at all, but such a space from which the air has been completely or incompletely removed. From this it is clear how much better the Magdeburg machine is than the English; since in the former the entry of any external air is excluded because of the water in which all those places by which it could enter are immersed; which is not the case in the English machine.[83]

[82] Boyle to Schott, n.d., in Boyle, *Works*, vol. VI, pp. 62-63 (on the forthcoming *Technica curiosa*); Guericke to Schott, 18/28 February 1662, in Schott, *Technica curiosa* (1664), pp. 54-58, and in Guericke, *Neue Magdeburger Versuche*, p. [33]; Guericke to Schott, 30 April/10 May 1662, in Schott, *Technica curiosa*, pp. 74-76. On Guericke's pneumatics, see Krafft, *Guericke*, pp. 98-108; Kauffeldt, *Guericke*.

[83] Schott, *Technica curiosa*, pp. 87-181, esp. pp. 97-98. For Guericke's notes on the void-in-the-void trial, see Guericke, *Neue Magdeburger Versuche*, pp. [70]-[71].

Of course, by 1664 Boyle's machine *was* underwater, and was insulated in just the way Schott described. Schott was also in touch with Huygens' work on anomalous suspension via Gottfried Kinner, a fellow Jesuit in Prague. In January 1665 Huygens sent Kinner a report of the anomalous suspension of mercury and this was transmitted to Schott. Despite these debates, Schott never reported any trials of anomalous suspension, nor did these natural philosophers ever attempt to build such a pump.[84]

We have given two cases where competent and informed natural philosophers did not attempt the replication of Boyle's pumps. In both cases they asserted the equivalence or superiority of their own machines. They appealed to specific models of how air-pumps worked in order to justify this claim. The case of Guericke is particularly revealing. The German attack on Boyle was like that mounted by Hobbes on the integrity of the air-pump. Ironically, they were also much the same as those which Boyle himself made on Guericke's *antlia pneumatica.* The career of the air-pumps in the 1660s shows how experimenters made matters of fact. Two points can be made: (1) the accomplishment of replication was dependent on contingent acts of judgment. One cannot write down a formula saying when replication was or was not achieved. The construction of any device which could be taken as a successful copy of an existing pump was entirely dependent on direct witnessing. No one built a pump from written instructions alone; the transmission of pump-building and pump-operating skills required the transfer of people (figure 10).[85] Moreover, as the Florentine case illustrates, even the notion of *verification* itself is profoundly problematic. The Florentines announced that they had verified Boyle's results without needing Boyle's machine. (2) Thus, if replication is the technology which turns belief into knowledge, then knowledge-production depends not just on the abstract exchange of paper and ideas but on the practical social regulation of men and machines.

[84] Huygens to Kinner, 26 December 1664/5 January 1665, in Huygens, *Oeuvres*, vol. v, p. 221; Kinner to Huygens, 25 January/4 February 1665, ibid., pp. 217-219; Kinner to Schott, 25 January/4 February 1665, ibid., pp. 219-221; Schott to Kinner, February 1665, ibid., pp. 253-254; Kinner to Schott, 11/21 March 1665, ibid., pp. 272-274. For details of Guericke's programme, see Heathcote, "Guericke's Sulphur Globe."

[85] Cf. Collins, "The TEA Set"; idem and Harrison, "Building a TEA Laser" for the transmission of skills generally. Collins argues against the "algorithmic model" of skill transmission, in which written instructions are accorded efficacy, in favour of an "enculturation model," in which the transfer of skills is assimilated to a craft pattern.

The establishment of a set of accepted matters of fact about pneumatics required the establishment and definition of a community of experimenters who worked with shared social conventions: that is to say, the effective solution to the problem of knowledge was predicated upon a solution to the problem of social order. Hobbes's criticism was that no matter of fact made by experiment was indefeasible, since it was always possible to display the labour expended on making it and so give a rival account of the matter of fact itself. The decision to display or to mask that labour was a decision to destroy or to protect a form of life.

· VII ·

Natural Philosophy and the Restoration: Interests in Dispute

... kindred intellects evoke
allegiance per blunt instruments
e e cummings, *i sing of olaf*

HOBBES and Boyle used the work of the 1640s and 1650s to give rival accounts of the right way to conduct natural philosophy. We have examined the way in which the experimental philosophers sustained Boyle's programme against adversaries and how they dealt with trouble within their community. What hinged on the acceptance of such a programme? We now consider the issues that bore on the way Hobbes's and Boyle's schemes were assessed in the 1660s. This demands an outline of the political and ecclesiastical context of the Restoration. The crisis of the Restoration settlement made proposals for a means of guaranteeing assent extremely urgent. We explore the importance of conscience and belief in the intellectual politics of the 1660s. The experience of the War and the Republic showed that disputed knowledge produced civil strife. It did not seem at all clear that *any* form of knowledge could produce social harmony. Yet this was just what the experimenters and their propagandists did claim. Furthermore, the restored régime concentrated upon means of preventing a relapse into anarchy through the discipline it attempted to exercise over the production and dissemination of knowledge. These political considerations were constituents of the evaluation of rival natural philosophical programmes.

The link between the means of guaranteeing assent and the establishment of indefeasible civil order was obvious both to the experimentalists and to Hobbes. In chapter 2 we argued that Boyle's technologies could only gain assent within a secure social space for experimental practice. Hobbes assaulted the security of that space because it was yet one more case of divided power, of double vision in political allegiance. Thus the disputes between Boyle and Hobbes became an issue of the security of certain social boundaries and the interests they expressed. For Boyle this would inevitably involve the connection between the work of the experimental philosopher and that of the priest as Christian apologist. Their functions reinforced each other and Hobbes was their common enemy. But for Hobbes any profession that claimed such a segregated area of competence, whether priestly, legal, or natural philosophical, was thereby subverting the authority of the undivided state. The events of the Restoration made that authority a vital concern for philosophy.

"Tender Consciences" and the Restoration Settlement

In May 1659 the fragile government of the Protectorate, of which Boyle's brother Roger, Lord Broghill, was a leading member, collapsed. After nine months of protracted dispute, between the army, Parliament, and a variety of contending political factions, the army commander in Scotland, George Monck, established contact with the exiled Charles II. By May 1660 the monarch had returned from Holland and a Convention met at Westminster. Hobbes travelled from Derbyshire to witness the King's return and was presented to Charles.[1] Boyle was also in London. In June his *New Experiments Physico-Mechanical* was published. The events of 1660 began a long drawn-out search for stability by the restored régime. By the mid-1660s the Restoration settlement came to be embodied in the harsh legislation of the Clarendon Code. It represented the result of a set of attempts to control the beliefs and behaviour of unruly subjects through forms of discipline. These attempts gave immediate point to the models of knowledge and social organization offered by men like Hobbes and Boyle. It was held that strong links connected subjects' vulnerable beliefs and their assent to Church and state. It was in this context that any bold proposal for winning assent to right knowledge and to communal behaviour would now be assessed. Any such proposal must be shown to be possible, effective and safe. That is to say, it had to be shown how knowledge was connected with public peace; it had to be shown how such knowledge might be produced; and it had to be shown that such communities would not threaten existing authorities such as the clergy or the power of the restored régime.

Opponents of the régime were commonly branded as sectarian

[1] Woolrych, "Last Quests for a Settlement"; Davies, *The Restoration of Charles II*. For Hobbes in London in 1660, see Aubrey, "Life of Hobbes," p. 340.

and connected with the subversives of the Interregnum. These sectaries were seen as the major threat to public order. Their various professions of belief were the principal target against which discipline should be directed. Indeed, the term "fanatic" entered common parlance at this moment following its use by Monck in a speech in February 1660 on "military or civil power." During 1659 Broghill had gone on campaign against such "fanatics" in Ireland to clear himself of the charge of being "half a Presbyter."[2] Plots and rumours of plots fomented by the government and allegedly planned by the sects enabled harsh legislation to be enacted with even less resistance, while measures designed to police meetings of the sects were also put into effect.[3] These actions defined the task for the proponents of settlement: their job was to outline ways in which the sects could be controlled and sectarian knowledge contested. This was a powerful constraint, since it was widely argued that knowledge itself was a source of sectarian conflict. So the proponent of any successful model of pacific knowledge must both deny its tendency to promote dissension and also deny the basis of the sects' own forms of belief.

The two areas in which these models of knowledge were debated were the proposals for Church settlement and the proposals for control of the distribution of information. The reestablishment of the Church clearly involved these issues of conscience and sectarian dissent. The imposition of censorship and licensing also raised the problem of the character and the effects of public knowledge. On 4/14 April 1660, with the counsel of his chief adviser Edward Hyde, the King issued the Declaration of Breda, which explicitly declared a "liberty to tender consciences," since, so it was argued, "the passion and the uncharitableness of the times have produced several opinions in religion by which men are engaged in parties and animosities against each other; which, when they shall hereafter unite in a freedom of conversation, will be composed, or better understood." The Declaration endorsed a move towards a public resolution of dispute based on the free play of differing consciences. This manifesto made the source of compulsion the key political issue.[4] The leader of the moderate Presbyterians, Richard Baxter, preached

[2] A. Wood, *Life and Times*, vol. 1, p. 303 (on Monck); Green, *Re-establishment of the Church of England*, p. 18 (on Broghill).

[3] Abbott, "English Conspiracy and Dissent"; compare Green, *Re-establishment of the Church of England*, p. 182.

[4] The Declaration of Breda is printed in Kenyon, ed., *The Stuart Constitution*, p. 357; Clarendon's role is indicated in Abernathy, "Clarendon and the Declaration of Indulgence," p. 56n; compare Clarendon, *History of the Rebellion*, pp. 898-902.

the message of discipline over animosity in a sermon at the end of the month: "Unhappily, there hath been a difference among us, *which is the higher Power*. . . . The question is not, *whether Bishops or no*? but *whether Discipline or none*? and *whether enow to use it*?"[5] From summer 1660 the issue of settlement was far from clear, nor was that settlement seen as secure. For Edward Hyde, now Earl of Clarendon, the key lay in a form of discipline which would then allow the peaceable resolution of conflicting interests. His role here was highly ambivalent. Historians have seen him as an authoritarian, aiming at the extinction of dissent, or as a moderate proponent of toleration forced into repression by expediency and the extremism of the Commons. In his apologetic autobiography, written after his fall in 1667, he was keen to emphasize the providential basis of the Restoration and yet the inherent instability of the restored régime: "The King was not yet master of his kingdom, nor his security such as the general noise and acclamation, the bells and bonfires, proclaimed it to be."[6]

Government policy from summer 1660 thus involved the installation of favoured candidates in vacant ecclesiastical offices, the control of the Commons through the use of rumours of sectarian revolt, and, finally, the continued negotiation with the leaders of the principal Dissenting groups. In October 1660 negotiations with Baxter and his colleagues at Worcester House resulted in a Declaration on ecclesiastical settlement that pointed towards a measure of toleration while emphasizing the disasters of dispute. Clarendon was also prepared to discuss proposals for moderate episcopacy, as suggested by Baxter and Ussher in 1655 and promoted by Boyle and his allies under the Protectorate. Boyle himself now encouraged colleagues such as Thomas Barlow, Peter Pett, and John Dury to propagandize for the virtues of toleration and such a model of settlement. But in the following month a narrow majority in the Convention rejected the Worcester House Declaration, and, after a spate of sectarian risings during the winter, Parliament was dissolved and a massively royalist and Anglican Commons was elected

[5] Baxter, *Sermon of Repentance*, pp. 44-45, quoted in Lamont, *Baxter and the Millennium*, p. 201.

[6] Bate, *The Declaration of Indulgence*, p. 55; Bosher, *Making of the Restoration Settlement*, pp. 107, 211-224 (for Laudian policies); Green, *Re-establishment of the Church of England*, pp. 213-229, and Abernathy, "Clarendon and the Declaration of Indulgence" (for toleration and a state religion); Whiteman, "Restoration of the Church of England" (for expediency). Compare Clarendon, *History of the Rebellion*, p. 994.

in March 1661. Moves towards discipline and repression followed swiftly.[7]

Throughout these months the government monitored threats to the settlement and the activities of dissenting sects. A plot of disbanded soldiers was reported in December 1660, and when the King was out of London in early January 1661 a group of Fifth-Monarchy Men staged a riot in the capital. This rising, labelled "Venner's Plot," prompted a swift response. Secretary Nicholas wrote that though "the Fanaticks had laid their plot throughout the kingdom and actually broke forth in this city," yet "we at present (thanks be to God) enjoy a perfect quiet, for the preservation whereof his Majesty hath set forth a proclamation forbidding all private meetings and assemblies of the Fanaticks and Sectaries." This proclamation was issued on 10/20 January 1661: while acknowledging that it violated the Declaration of Breda by "restraining some part of that liberty, which was indulged to tender consciences," the proclamation nevertheless banned meetings of "divers persons (known by the name of Anabaptists, Quakers, and Fifth-monarchy men, or some such-like appellation)," who "under pretence of serving God, do daily meet in great numbers, in secret places, and at unusual times."[8] During the spring, most elections returned high Anglican members, though there were indications from the City of London and elsewhere that not all were "friends to bishops turning out godly ministers." Clarendon initiated further discussions with the Presbyterians at Savoy House: among the Presbyterian delegation were Richard Baxter and Hobbes's enemy John Wallis. The meetings there continued until July, but proved abortive.[9]

[7] For Dissent at the Restoration, see L. Brown, "Religious Factors in the Convention Parliament" (on the strength of the Presbyterians); J. R. Jones, "Political Groups and Tactics in the Convention" (on the campaign in spring 1660); Abernathy, "English Presbyterians and the Stuart Restoration"; Lacey, *Dissent and Parliamentary Politics*, chaps. 1, 3. For moderate episcopacy and Boyle's role, see Lamont, *Baxter and the Millennium*, pp. 153, 212-214; Spalding and Brown, "Reduction of Episcopacy"; J. Jacob, *Boyle*, pp. 133-144. The Worcester House debate is in *Journals of the House of Lords* (hereafter *L.J.*), vol. XI, p. 179; the debate on the Uniformity Act is in *Journals of the House of Commons* (hereafter *C.J.*), vol. VIII, pp. 442-443.

[8] *Calendar of State Papers (Domestic)* (hereafter *C.S.P.D.*) (1660-1661), pp. 515, 561; Abbott, "English Conspiracy and Dissent," pp. 503-529; Gee, "The Derwentdale Plot"; Nicholas, *Mr. Secretary Nicholas*, p. 302; Bosher, *Making of the Restoration Settlement*, pp. 204-205; Ashley, *John Wildman*, pp. 161-165. For Fifth Monarchism, see P. Rogers, *The Fifth Monarchy Men*, esp. pp. 112-122; Capp, *The Fifth Monarchy Men*; Clarendon, *History of the Rebellion*, p. 1033; C. Hill, *The Experience of Defeat*, pp. 62-66; idem, *World Turned Upside Down*, pp. 72, 97, 171-173, 347.

[9] *C.S.P.D.* (1660-1661), p. 541; on Savoy House, see Green, *Re-establishment of the Church of England*, p. 200; Bosher, *Making of the Restoration Settlement*, p. 210.

The government now began the preparation of a bill to compel uniformity in the Church, while Secretary Nicholas and Clarendon both reported further sectarian revolts. In September 1661 Nicholas told Clarendon that Presbyterians, Anabaptists, and Fifth Monarchists were preaching in the same churches, and persuading their congregations to "fight perpetually against their Sovereign." Two weeks later Nicholas heard that "two late prisoners on Venner's business" had high hopes of further revolt, while the government produced evidence of a series of conspiracies in London and the provinces. By the end of the year acts were passed banning Dissenters from office and against treasonable utterances.[10] By February 1662 Parliament had approved the Uniformity Act against Dissent in religion, and despite misgivings at Court it received royal assent in May. On St. Bartholomew's Day 1662 hundreds of Dissenting ministers were formally ejected from their posts, while legislation against refusal to take the oaths and against sedition in print was also enacted. On 14/24 October 1662 the King issued instructions to preachers "tuning the pulpits" against "the extravagance of preachers," which "has much heightened the disorders and still continues to do so by the diligence of factious spirits who dispose them to jealousy of the government." The instructions advised preachers against using sermons "to bound the authority of sovereigns, or determine the difference between them and the people, nor to argue the deep points of election, reprobation, free will &c. They are to abstain as much as possible from controversies."[11] In December 1662 the King attempted a Declaration of Indulgence to lessen the effects of the Uniformity Act, but Parliament resisted the Declaration the following spring, and during the next eighteen months passed legislation which strengthened the campaign against Dissent. The Triennial Act governing elections was repealed, and conventicles were banned. Most of the measures which had justified resistance in the 1640s were now removed, and Anglican and royal power was extended. "We cannot forget the

[10] *C.S.P.D.* (1661-1662), pp. 97-98; Ashley, *John Wildman*, pp. 166-181; Bosher, *Making of the Restoration Settlement*, p. 238; Sacret, "The Restoration Government and Municipal Corporations." For Nicholas and the collection of news, see Fraser, *The Intelligence of the Secretaries of State.*

[11] On the Uniformity Act and the expulsions, see Feiling, "Clarendon and the Act of Uniformity"; Beddard, "The Restoration Church"; Abernathy, "Clarendon and the Declaration of Indulgence"; Bate, *The Declaration of Indulgence*, pp. 25-35; Bosher, *Making of the Restoration Settlement*, chap. 5; *C.S.P.D.* (1661-1662), p. 517; Clarendon, *History of the Rebellion*, pp. 1077-1080.

late disputing Age, wherein most Persons took a Liberty, and some men made it their Delight, to trample upon the Discipline and Government of the Church," declaimed the Speaker of the Commons on the introduction of the Uniformity Bill. Similarly, the Commons resisted the Declaration of Indulgence in spring 1663 by arguing that any toleration of sectarian Dissent would lead to a return to the age of civil war: "The variety of professions in religion, when openly indulged, doth directly distinguish men into parties, and, withal, gives them opportunity to count their numbers."[12]

The Clarendon Code, perhaps inaccurately attributed to Clarendon's own proposals, now formalized a specific view of the means of discipline and the way to prevent reversion to civil chaos. By 1665 legislation against any who "at any time endeavour an alteration of Government either in Church or State" had been directed at most areas of civil life. The target of this legislation was subjects' consciences and their assent to the legitimate restored order. In his speech at the opening of Parliament in May 1661, Clarendon had spelt out the link between subjects' beliefs and the communal agreement and assent to the Restoration settlement:

> If the present oaths have any terms or expressions in them that a tender conscience honestly makes scruple of submitting to, in God's name let other oaths be formed in their places, as comprehensive of all those obligations which the policy of government must exact, but still let there be a yoke; let there be an oath; let there be some law, that may be the rule to that Indulgence, that under pretence of liberty of conscience, men may not be absolved from all the obligations of law and conscience.[13]

Such speeches summed up the means by which the boundaries of debate might be securely established. Dispute might be tolerated within such boundaries, and communities of believers might publicly discuss forms of knowledge. However, without the commitments and compulsions which invested such disciplined boundaries

[12] Bate, *The Declaration of Indulgence*, pp. 36-40; Lacey, *Dissent and Parliamentary Politics*, pp. 47-56; Beddard, "The Restoration Church," p. 168; *C.J.*, vol. VIII, pp. 442-443; *L.J.*, vol. XI. p. 470. Compare Seth Ward's comments as Bishop of Exeter, printed in Bosher, *Making of the Restoration Settlement*, p. 266.

[13] *L.J.*, vol. XI, p. 243; *C.J.*, vol. VIII, pp. 172-174; on the Clarendon Code, see Kenyon, ed., *The Stuart Constitution*, chap. 10; Beddard, "The Restoration Church," pp. 161-170; Bosher, *Making of the Restoration Settlement*, chap. 5; Western, *Monarchy and Revolution*, chap. 3; J. R. Jones, *Country and Court*, pp. 136-139.

with potency, any discussion of this type would at once generate dissension. Because the régime portrayed itself as vulnerable, and because dissension challenged peaceful assent and divided subjects into rival communities, it would collapse into civil war. As Hobbes's patron, the Earl of Newcastle, had reminded the King, "controversye Is a Civill Warr with the Pen which pulls out the sorde soone afterwards."[14]

Discipline and "Coffee-house Philosophies"

At the Restoration it seemed clear that all free debate bred civil strife. It seemed less plausible that *some* forms of free debate might produce knowledge which could prevent that strife. In one fundamental text on this issue, Milton's *Areopagitica* of 1644, the case for safe and effective knowledge reached by open dispute had been argued at length: "Where there is much desire to learn, there of necessity will be much arguing, much writing, many opinions; for opinion in good men is but knowledge in the making. Under these fantastic terrors of sect and schism, we wrong the earnest and zealous thirst after knowledge and understanding which God hath stirr'd up." Yet Milton's vaunted "Nation of Prophets, of Sages, and of Worthies" had soon decayed to that anarchy decried in Thomas Edwards' *Gangraena* (1646). Even Milton listed Hobbes as one author who should not be tolerated. During the 1650s radicals still argued that "the mystery of iniquity . . . is the magistrate's intermeddling with Christ's power over the judgments of men." They were reacting against the attempts of men like Richard Baxter to advocate fresh forms of discipline over debate. It was reported that Charles II "always lamented" popular access even to the Bible: "This liberty was the rise of all our sects, each interpreting according to their vile notions and to accomplish their horrid wickednesses."[15]

The ordinances against printing and publication enacted from autumn 1649 and renewed intermittently under the Protectorate were reconfirmed in June 1660. The Licensing Act of 1662 con-

[14] Newcastle's manuscript "Of Government" is printed in Strong, *Catalogue of Letters*, pp. 173-236 (cited in Turberville, *History of Welbeck Abbey*, vol. I, p. 174).

[15] Milton, "Areopagitica," in *Prose Works*, vol. II, pp. 554; Ailesbury, *Memoirs*, vol. I, p. 93; Stubbe, *Malice Rebuked* (1659), pp. 7-8; Nicolson, "Milton and Hobbes"; C. Hill, *Milton and the English Revolution*, pp. 149-160; J. Jacob, *Stubbe*, p. 31. Compare the King's sentiments on the accessibility of the Bible with those of Hobbes, "Behemoth," p. 190 (noted in chapter 3 above).

siderably tightened this control. Monopoly of the press was granted to the Stationers' Company and the universities. The government was now to license any works on politics or history; the number of printers was to be reduced from sixty to twenty, and then controlled by the Archbishop of Canterbury and the Bishop of London. Roger L'Estrange, an aggressively high Anglican journalist, was given responsibility for licensing the press. In 1663 he wrote that "the spirit of hypocrisy, scandal, malice, error and illusion that achieved the late rebellion was reigning still." The following year it was argued that any liberty of the press automatically led to war: a royal monopoly should therefore replace the work of the stationers.[16]

Typical casualties of the new policy were the astrologer and Rosicrucian John Heydon (imprisoned in 1663, and again in 1667 after casting the King's horoscope) and Giles Calvert, the most important radical publisher of the Interregnum. Calvert published works by Thomas Vaughan, John Webster, Boehme, Dell, Winstanley, and Leveller and Ranter writers. His shop served as a centre for such activity. In 1654 one defender of Scholastic learning complained of "Lame Giles Calvers shop, that forge of the Devil, from whence so many blasphemous, lying scandalous Pamphlets . . . have spread over the Land . . . to the provocation of Gods wrath against us." Samuel Hartlib worked with Calvert in 1658. But Calvert was imprisoned in 1661 for publishing the seditious *Phoenix of the Solemn League and Covenant.* He fled the country and his wife was jailed for producing an almanac, "instilling into the hearts of subjects a superstitious belief thereof and a dislike and hatred of his Majesty's person and government." Calvert's work was tried again in 1664 and he died in Newgate jail. These measures went some way to regulate the dissemination of ideas developed before 1660: men like Harrington, Neville, or Vane were in jail or dead. Furthermore, the appearance of a tighter control affected a wider group of authors: they were prudent in their expressions of dissent.[17]

[16] For the imposition of censorship and L'Estrange, see L'Estrange, *Considerations and Proposals,* p. 8; Muddiman, *The King's Journalist,* pp. 150-167; Bourne, *English Newspapers,* vol. I, p. 32; Fraser, *The Intelligence of the Secretaries of State.* For the effects of censorship, see Weston and Greenberg, *Subjects and Sovereigns,* p. 158; Western, *Monarchy and Revolution,* pp. 61-64; McLachlan, *Socinianism,* pp. 327-331, 338; Redwood, *Reason, Ridicule and Religion,* pp. 79-81; C. Hill, *Milton and the English Revolution,* pp. 64-66, 217-218; idem, *Some Intellectual Consequences,* pp. 46-52. See also Zwicker, "Language as Disguise"; idem, *Politics and Language,* chap. 1.

[17] John Heydon and his prosecution are discussed in *C.S.P.D.* (1666-1667), pp. 428-431, 490, 541; Debus, *The Chemical Philosophy,* vol. II, p. 387n; Capp, *Astrology and the Popular Press,* p. 48. Giles Calvert is mentioned in *C.S.P.D.* (1661-1662), pp.

Licensing of the press became a common problem once again. The right to issue books under its own *imprimatur* was a valuable privilege for the Royal Society. In this context, debates among the learned must not seem to be too polemical: in December 1669, Peter du Moulin told Boyle that the high Anglican bishop Peter Gunning had refused to license his "heroics in commendation of the Royal Society." In 1668 an opponent of Joseph Glanvill, Robert Crosse, was refused a licence to print his attack on the Society. It has recently been shown that Glanvill's own *Plus ultra* was linked with the sectarian disturbances in Somerset. Crosse accused Glanvill of atheism, and Glanvill was attacked by Dissenters who "say they shall have liberty of conscience, and that the government, which cannot stand much longer, durst not do otherwise than permit their freedom." The practice of licensing closely connected the religious import and the political security of Restoration debates. Glanvill's ally, Henry More, travelled to Lambeth in July 1670 to obtain a licence for his *Enchiridion metaphysicum,* in which Boyle's "monstrous spring of the Ayre" was challenged. The book's title announced the "Vanity and Falsehood" of all "who suppose the Phenomena of the World can be explained by purely Mechanical Causes." The licenser, Samuel Parker, granted a licence "at first sight, seeing from the very title which way it tended."[18]

Public meeting-places were also treated with suspicion. However ineffectively, surveillance included the control of the new coffee-houses. The coffee-houses began in London at the lapsing of the licensing acts in 1652, and soon became linked with the spread of Dissent and the new philosophy. Tillyard's at Oxford opened in 1656 and was the venue for the experimental philosophy group. It was there, too, that in 1661 Peter Stahl, invited to Oxford by Boyle and Hartlib, began chemistry lectures. In 1662 a pamphlet arguing for the *fuga vacui* as the cause of the Torricellian phenomenon was published "at the Coffee House in Wilde Street, where the dispute was held; for every man that hath eyes in his head to

---

23, 572, 592; T. Hall, *Histrio-mastix* (1654), p. 215; Clarkson, *The Lost Sheep Found* (1660); Hartlib to Boyle, 16/26 December 1658, in Boyle, *Works*, vol. VI, p. 115; Ashley, *John Wildman*, pp. 194-195, 204-209; Muddiman, *The King's Journalist*, pp. 142-143, 169. Compare the arrest of Henry Oldenburg, possibly because of his millenarian sympathies, in Webster, *From Paracelsus to Newton*, p. 38.

[18] Steneck, " 'The Ballad of Robert Crosse and Joseph Glanvill' "; for du Moulin, see du Moulin to Boyle, 28 December/7 January 1670 and 23 February/5 March 1674, in Boyle, *Works,* vol. VI, pp. 579, 581; for More, see More to Anne Conway, 6/16 August 1670, in *Conway Letters*, p. 303.

read and to judge himself, on which side lies the Non-sense."[19] The new régime suspected such freedom. The arch-conspirator and ex-Leveller John Wildman bought his own coffee-house in summer 1656 and the Commonwealth Club met there during summer 1659. Harrington's Rota Club was a very similar group. John Aubrey was a member, and noted that after 1660 it became "treason" for the Rota to meet. It was even claimed by some clergymen that the King established the Royal Society to contest the Rota's influence. Hobbism was also branded a "coffee-house philosophy." In 1673 Glanvill wrote that at such places "each man seems a leveller." Lord Keeper Guilford considered that coffee-houses should be suppressed, since "if the opportunities of promiscuous and numerous assemblies of idle spenders of time were removed, ill men would not be able to make such broad impressions on people's minds as they did." In the terrible year of 1666, Clarendon contemplated suppressing such meeting-places. Their importance for political debate was a further illustration of the suspicion of public dispute. England's "natural governors" attributed great power to any rival form of knowledge.[20]

The Restoration crisis and these measures of discipline forced those who sought to produce such knowledge to adapt or to keep silent. Hobbes's "coffee-house philosophy" enjoyed a measure of fashionable support at Court; but it was also charged with subversion by the "hunters of Leviathan." Hobbes's experience in the 1660s illuminated his analysis of the right relation of legal power and the public declaration of philosophy. In 1661 he was prosecuted in Convocation: Oldenburg reported the "asserting of Hobbes's principles in parlement" and Aubrey wrote that the bishops had "made a motion to have the good old Gentleman burnt for a heretique." In 1666 the Commons ordered the inspection of "Hobbe's book called *Leviathan* and examination into abuses in printing."[21] In 1668 Hobbes was ordered not to publish any work

[19] For coffee-houses, see Robinson, *Early History of Coffee Houses*, pp. 77-79; G. H. Turnbull, "Peter Stahl"; compare Anon., *An Excerpt of a Book shewing that Fluids Rise not in the Pump . . . at the Occasion of a Dispute in a Coffee-House* (1662), p. 8.

[20] North, *Lives of the Norths*, vol. 1, pp. 316-317; Ashley, *John Wildman*, pp. 103, 119, 142-148; Robinson, *Early History of Coffee Houses*, p. 167; Reiser, "The Coffee-Houses of Mid-Seventeenth Century London." For Hobbism as "coffee-house philosophy," see Mintz, *Hunting of Leviathan*, p. 137; for Glanvill, see J. Jacob, *Stubbe*, p. 84; for the Royal Society and the Rota, see C. Hill, *The Experience of Defeat*, p. 191.

[21] Oldenburg to Beale, 30 May/9 June 1661, in Oldenburg, *Correspondence*, vol. 1, p. 410; Tönnies, *Hobbes*, pp. 59-60; *C.S.P.D.* (1666), p. 209.

on politics or religion in English—a significant limitation—and in the same year Daniel Scargill, a Cambridge Fellow, was forced publicly to recant his "Hobbist" beliefs. Hobbes himself found that his history of the Civil War, *Behemoth*, composed in English, had been suppressed. It did not appear in print until 1679. In June 1668 the Latin collection of Hobbes's works, including *Leviathan* and the *Dialogus physicus*, was issued at Amsterdam. In September, Pepys wrote that an English *Leviathan* was "mightily called for" and very expensive, since "the Bishops will not let [it] be printed again." Hobbes also added an appendix on heresy and persecution to this Latin *Leviathan*. He based the principal argument on the drafts which were originally composed in 1662, at the time of the ecclesiastical charges against him. Hobbes composed an "Historical Narration concerning Heresy." He wrote to the Secretary of State, Joseph Williamson, including "the words concerning heresy which you mistake and which may be left out without trouble, but I see no cause of exception against them, and desire they may stand, unless the rest of the book cannot be licensed without them." This appeal failed: his work on heresy was delayed until 1680. Its argument responded directly to the conditions of the censorship imposed after the Restoration. Hobbes illustrated the relation between the debate on public knowledge and the attempts at discipline of belief. He specified the measures which could be taken against dangerous knowledge, and described how such knowledge should be defined.[22]

In his original drafts on persecution and heresy, Hobbes had already rehearsed his claim that there was no current law in force which authorized the persecution of belief by an independent priestly power. Even if a belief were repugnant to the churchmen, no purely ecclesiastical power could compel the subject to "accuse, or to purge him or her selfe of any criminall matter or thing." In the "Historical Narration," Hobbes extended this account. He linked persecution with forms of knowledge, and traced the genealogy of heresy to philosophical dispute:

> After the study of philosophy began in Greece, and the philosophers, disagreeing amongst themselves, had started many

[22] For Scargill, see Linnell, "Daniel Scargill"; Axtell, "The Mechanics of Opposition"; for Hobbes and heresy, see *C.S.P.D.* (1667-1668), p. 466; Mintz, "Hobbes on the Law of Heresy" (and the correction of the date of this document in Willman, "Hobbes on the Law of Heresy"); for Pepys and *Leviathan*, see Pepys, *Diary*, vol. IX, p. 298 (3/13 September 1668).

> questions, not only about things natural, but also moral and civil; because every man took what opinion he pleased, each several opinion was called a *heresy*; which signified no more than a private opinion, without reference to truth or falsehood.

Hobbes's account insisted on the detailed analysis of the language of the Creed and the provisions of statute law: the label of heresy as "ignominy" did not license persecution, for only the enforcement of civil law could do that. Thus, under the Commonwealth "men obeyed not out of duty, but for fear; nor were there any human laws left in force to restrain any man from preaching or writing any doctrine concerning religion that he pleased. And in this heat of the war, it was impossible to disturb the peace of the state, which then was none." It was in this context that Hobbes published *Leviathan*, "in defence of the King's Power," he claimed; and he pointed to bishops and to Presbyterian divines such as John Wallis who would now enforce law against belief illegitimately, since they had no legal warrant for their action: "So fierce are men, for the most part, in dispute, where either their learning or power is debated, that they never think of the laws, but as soon as they are offended, they cry out, *crucifige*."[23]

Hobbes's analysis of the legal basis of persecution was consistent with his declaration on assent to the state and on the unreliable character of private belief. Such belief was beyond control; those who claimed to sway subjects' beliefs or to base a system of knowledge upon belief were both dishonest (since they claimed to do what could not be done) and dangerous (since they claimed power separate from that of the civil authority). The response of churchmen and experimental philosophers such as John Wallis or Walter Pope was to point out Hobbes's suspect political record and his alleged support for the Republic: they claimed Hobbes came back from Paris in 1651 "to print his *Leviathan* at London to curry favour with the Government."[24] One more radical exponent of toleration,

[23] Mintz, "Hobbes on the Law of Heresy," p. 414; Hobbes, "Historical Narration concerning Heresy," pp. 387, 407. Compare Hobbes, "Concerning Body," p. ix: "There walked in old Greece a certain phantasm, for superficial gravity, though full within of fraud and filth, a little like philosophy; which unwary men, thinking to be it, adhered to the professors of it, some to one, some to another, though they disagreed among themselves." See also Sprat, *History*, pp. 11-12.

[24] Pope, *The Life of Seth*, p. 125. Compare Hobbes, "Considerations on the Reputation of Hobbes," pp. 416-420. In July 1651 Hartlib was told that *Leviathan* was written by "a man passionately addicted to ye royall interest" (see Skinner, "Conquest and Consent," p. 94n).

Roger Coke, printed an attack on Hobbes in 1660 which linked him with the equally suspect political doctrines of Thomas White. White's *Grounds of Obedience and Government* (1655) was also examined by Parliament in 1666. White and Hobbes were both assailed by the propagandists of experimental philosophy for their dogmatism, principally by Glanvill in *Scepsis scientifica.* They were both attacked for authoritarian politics by writers like Peter du Moulin, an ally of Boyle, who charged White with the same crime of support for Cromwell "in the height of Oliver's Tyranny." Hobbes responded by aiming at royal patronage. Both his *Problemata physica* (1662) and some of his papers in geometry were dedicated to the King: "Making the authority of the Church wholly upon regal power . . . , I hope your Majesty will think is neither atheism nor heresy." But Hobbes's ecclesiastical and natural philosophical opponents did claim as much. They were also concerned by any attempt to gain support for Hobbes at Court. John Worthington wrote to More in June 1668, when Hobbes's works appeared at Amsterdam, that they were "dedicated to the King, and if so, I told Dr. Cudworth that it might be well that his [attack on Hobbes] should be so dedicated."[25]

Hobbes's critique pointed to the connection between the settlement of civil authority and the social organization of knowledge-production. Hobbes suggested that clergymen and experimental philosophers were dangerous because they claimed an independent competence in disseminating their views. For many of the Anglican divines, however, it was a commonplace that in the 1660s the Church itself was an inadequate instrument for the policing of consciences. We have indicated that Glanvill and his colleagues in Somerset had encountered threats to established religion, and the incompetence of the unsupported Church. Glanvill told Beale in 1667 that the Dissenters "still grow in numbers and insolence." In 1663 Seth Ward, now Bishop of Exeter, wrote to his Archbishop of the troublesome effects of the Uniformity Act. An "outed presbyter" in Devon had been forced to return to preaching there, since "no man was put in his stead and . . . the people went off, some to

[25] Coke, *Justice Vindicated*; J. Jacob, *Stubbe*, p. 114 (for Coke as member of the Green Ribbon Club); Glanvill, *Scire/i tuum nihil est* (1665), "To the Learned Thomas Albius"; du Moulin, *Vindication of the Sincerity of the Protestant Religion* (1664), pp. 61-63; Sylvester, *Reliquiae Baxterianae*, p. 118. For White, see Henry, "Atomism and Eschatology"; for Hobbes's dedication, see Hobbes, "Seven Philosophical Problems," pp. 5-6; John Worthington to Henry More, December 1667 and June 1668, in Worthington, *Diary and Correspondence*, vol. III, pp. 288, 293.

atheism and debauchery, others to sectarism." After preaching to flocks of more than 1500, the man had been arrested and imprisoned.[26] Forms of "atheism" were often linked with the influence of Hobbes and with the current weakness of the Church. John Eachard wrote of the contempt suffered by the clergy and followed this with a bitter attack on Hobbes's "state of nature." Bishop Lucy of St. David's republished his *Observations* on Hobbes in 1663. He also had direct experience of the problems of priestly power over Dissent or atheism:

> These fanatics . . . fear as little our excommunication as the Papists, and indeed I find no sect much dreading it; but although I doubt every diocese . . . hath all sects in Amsterdam, and more by the Papists; yet I fear a secret Atheism more than all of them, for I hope in time by degrees they will wear away with the reviving of ecclesiastic discipline, but Atheism will not be overcome but by apostolical men . . . we must act what we can with counsel, with menace, with deeds.[27]

The Church needed "apostolical men" and "counsel" to combat and control private belief, but few churchmen expressed the hope of unaided success in this task. In 1660 Henry More deplored the condition of ecclesiastical order: it was "a wilderness of Atheism and Profaneness; in a manner wholly inhabited by Satyrs and Savage Beasts." In his *History*, More's admirer Gilbert Burnet recalled that "the clergy themselves became lazy and negligent of their proper duties, leaving preaching and writing to others, and buried their parts in ease and sloth." Burnet indicated two threats to the Restoration settlement: the condition of the Church itself, and the appeal of *Leviathan*, "a very wicked book with a very strange title." He also pointed out two moves made against these threats. One was the reform campaign mounted by More and his allies in Cambridge. These men "studied to assert and examine the principles of religion and morality on clear grounds and in a philosophic

[26] Steneck, " 'The Ballad of Robert Crosse and Joseph Glanvill'," p. 62; J. Jacob, *Stubbe*, pp. 78-81; Seth Ward to Gilbert Sheldon, 19/29 December 1663, in Thirsk, ed., *The Restoration*, p. 38.

[27] Eachard, *Hobbes's State of Nature Considered* (1672), sig A6^v (for the attack on the clergy); idem, *Grounds and Occasions of the Contempt of the Clergy* (1670); Lucy, *Observations . . . of Notorious Errours in Leviathan* (1663); Lucy to Isaac Basire, 1661, quoted in Bosher, *Making of the Restoration Settlement*, p. 233. See also Bowle, *Hobbes and His Critics*, pp. 135-137 (for Eachard), and pp. 75-85 (for Lucy); Mintz, *Hunting of Leviathan*, pp. 55-65, 65ff.

way." The other was the emergence of experimental philosophy led by Robert Boyle.[28] So producers of knowledge could find a place in Restoration society if they could supply weapons for the weakened churchmen. Boyle reckoned that experimental philosophy did provide such weapons. During the 1660s he instructed others on the way these weapons should be made and the right means of using them.

"The Contention of Hands and Eyes": Experiment and Coercion

In the 1660s the religious and political authorities identified subjects' beliefs as the source of danger to the Restoration settlement. The experimenters then offered a solution to the problem of settlement: they presented their own community as an *ideal society* where dispute could occur safely and where subversive errors were quickly corrected. Their ideal society was distinguished by the source of authority the experimenters recommended. The experimental philosophers warned against tyranny and dogmatism in their work. No isolated powerful individual authority should impose belief. The potency of knowledge came from nature, not from privileged persons. Matters of fact were made when the community freely displayed its joint assent. Three features of the social condition of experiment emerged in this context: (1) propagandists for experimental work argued that if it was properly made and used then sound knowledge could have valuable political effects; (2) the harsh imposition of uniformity could not establish a secure settlement of subjects' beliefs, whereas the free play of rival opinion could lead to social stability; (3) this free play would only be safe and effective if the boundaries within which dispute was allowed were carefully defined and defended. If it met these conditions, then the activity of the experimental group could aid the political and ecclesiastical settlement. Boyle and his allies made two things available to Restoration society: the form of life practised within the experimental space, and the matters of fact which experimenters helped make.

First it was necessary to show that there *were* forms of knowledge that did not tend to strife. Under the Republic, men like Petty,

[28] R. Ward, *Life of Henry More*, p. 178, cited in Duffy, "Primitive Christianity Revived"; Burnet, *History of His Own Time*, vol. 1, pp. 323, 333, 321.

Boyle and Dury wrote of the appalling condition of civil society and of established philosophy in much the same terms. They were "good for nothing but to feed ravens and infect the air," and needed "a soul to quicken and enliven them."[29] Right knowledge and an effective social organization would solve this problem. In June 1655 Oldenburg told Hobbes that secure knowledge could do political work:

> [T]his demonstratif knowledge stayeth and satisfieth the mind as much as food doth an hungry stomach: and the same diffusing itself through and to ye good of all ye parts of ye body politique, as good meat well concocted doth to all ye limmes of ye body naturall, & must needs beget ye greatest contentment yt any sublunary thing can doe.[30]

This image was a commonplace. In 1655-1657 Oldenburg wrote to John Milton and others about the work of the experimenters at Oxford, who included Boyle and Hooke. Oldenburg urged that "that is to be judged *knowledge*, as I see it, which does not disquiet the mind, but settles it." The imagery of civic harmony, and the constitutional and humoral balance of the body politic, was now put to use to give an opportunity for the ministration of curative knowledge—"aliments of the Politick Body." The proponents of experiment developed a prescription for the cure of that body. Specifically, they developed a different account of how assent should be won from unruly subjects.[31]

In this account, the rhetoric of *limited toleration* was set against that of legal coercion. Stable assent was won because believers organized themselves in a defined and bounded society that excluded those who did not accept the fundamentals of good order. Uniformity would then emerge *as an accomplishment*: it was not to be imposed upon believers who were members of the community.

[29] Boyle to John Dury, 3/13 May 1647, in Boyle, *Works*, vol. 1, pp. xxxix-xl; Petty, *The Advice of W.P. to Hartlib* (1648); Hunter, *Science and Society*, pp. 27-28. Compare Power, *Experimental Philosophy*, p. 187, and Charleton, *Physiologia*, p. 2.

[30] Oldenburg to Hobbes, 6/16 June 1655, in Oldenburg, *Correspondence*, vol. 1, pp. 74-75.

[31] Oldenburg to Thomas Coxe, 24 January/3 February 1657, in ibid., pp. 113-114; compare Oldenburg to Adam Boreel and to John Milton, April and June 1656, ibid, pp. 91, 100; J. Rogers, *A Christian Concertation* (1659), p. 92; Harrington, *A System of Politicks* (1658); and Hooke, *Micrographia*, "The Preface," sig b2ʳ. The image of the body politic at this time is discussed in Daly, "Cosmic Harmony and Political Thinking," pp. 17-20; also Diamond, "Natural Philosophy in Harrington's Political Thought," p. 391.

Boyle and his political allies denied that such externally imposed discipline worked. Harsh discipline, whether Presbyterian or Anglican, was seen as tyrannic and wrong. In 1646-1647 Boyle wrote from London about the "impostures of the sectaries, which have made this distracted city their general rendezvous." Presbyterians and their allies in Parliament planned laws against the blasphemies of the sects. Boyle argued instead that many "impostures" might well be "manifestations of obscure or formerly concealed truths." He had been some time at Geneva, and he told his friend John Dury that the Calvinist régime there was admirable: but no such ideal, however perfect, could justify the imposition of Presbyterian discipline by legal means. It was wrong "to think by a halter to let new light into the understanding, or by the tortures of the body to heal the errors of the mind." Such measures did "not work upon the seat of the disease."[32] Throughout the 1650s, the grounds of secure assent to the régime remained a fundamental political and ecclesiastical issue. Dury played a central role in debates with Hobbes and others on loyalty to the Commonwealth in 1649-1651. Dury consistently argued for an irenic reconciliation of parties in the Church and for a form of knowledge that would escape from political dispute. Boyle's allies in Ireland such as Archbishop Ussher developed the plans for ecclesiastical settlement and limited toleration that Broghill revived in 1658-1659. All these proposals represented unrealized ideals of government in Church and state: they were to be founded upon the right relation of subjects' assent and discipline. The schemes that Hartlib and Boyle proposed to the Protectorate government for "countenancing and advancing universal learning" also aimed at this form of settlement.[33]

The problem of toleration and coercion became even more acute at the Restoration. The move towards imposed uniformity that characterized the policies of the early 1660s posed severe difficulties for the exponents of toleration. Broghill quickly contacted the King when Restoration seemed imminent, but in late April 1660 he wrote

[32] Boyle to Isaac Marcombes, 22 October/1 November 1646, and Boyle to John Dury, 3/13 May 1647, in Boyle, *Works*, vol. I, pp. xxxii-xxxiii, xxxix-xl; J. Jacob, *Boyle*, p. 22.

[33] For Dury and Hobbes in the Engagement controversy, see Skinner, "Conquest and Consent," pp. 81-82, and Judson, *From Tradition to Political Reality*, pp. 60-65; for Ussher and Boyle on limited toleration and moderate episcopacy, see J. Jacob, "Boyle's Circle in the Protectorate"; Lamont, *Baxter and the Millennium*, p. 165; Ashley, *John Wildman*, pp. 121-130; for the reform of learning, see Hartlib to Boyle, 16/26 December 1658, in Boyle, *Works*, vol. VI, p. 115.

that he feared an extreme régime: "I do monstrously dread the Cavalier party, and if the Parliament should be of such, God only knows what will be the evills." In June, Oldenburg, now a close colleague of Boyle and tutor to his nephew, wrote of his hope that since the Restoration had been "bloodless," the new régime would be "equable and clement." During the summer, Boyle and his Oxford friend Peter Pett feared "the restored clergy might be tempted by their late sufferings to such a vindictive retaliation, as would be contrary to the true measures of Christianity and politics."[34] The relation between Christianity and politics divided the churchmen. Wilkins and Ward were ejected from the universities, yet soon acquired bishoprics. They argued against each other about the virtues of toleration or suppression of Dissent. Wilkins attacked the Uniformity Act as too coercive: he would have preferred that the Church "stand without whipping." At Cambridge both Henry More and Simon Patrick came under attack from the restored churchmen; yet Worthington pointed out the similarities between the proposals for toleration of conscience in the Declaration of Breda and in More's works of 1660. In 1662 Patrick produced his account of the "Latitude-Men" and their support for experimental philosophy: these priests were "so merciful as not to think it fit to knock people on the head because they are not of our Church."[35] These exchanges gave considerable point to the proposals that Boyle and his allies produced for the establishment of a social space in which dissent would be safe and tolerable.

The ideal community of the experimenters was described in the apologetic texts that appeared in the 1660s. In chapter 2 we indicated that Sprat's *History of the Royal Society* (1667) labelled Hobbesian dogmatism as tyranny and uncontrolled private judgment as enthusiasm. Such dangers were to be excluded from the community: otherwise debate would not be safe. This ideal matched demands of the Restoration settlement. Those outside the experimental group were also to be barred from the provisions of tol-

[34] Broghill to Thurloe, 24 April/4 May 1660, in Davies, *The Restoration of Charles II*, pp. 251-254; Oldenburg to de la Rivière, 11/21 June 1660, in Oldenburg, *Correspondence*, vol. 1, p. 373; for Boyle and Pett, see Boyle, *Works*, vol, 1, p. cxli.

[35] Wright-Henderson, *Life of Wilkins*, p. 115; Patrick, *Brief Account of the Latitude-Men*, p. 12; Worthington to Hartlib, 29 November/9 December 1660, in Worthington, *Diary and Correspondence*, vol. 1, pp. 233-234; Gabbey, "Philosophia Cartesiana Triumphata," p. 228 (for More on toleration); for Ward's views on comprehension rather than toleration of Dissent, see Simon, "Comprehension in the Age of Charles II."

eration. In *Plus ultra*, Glanvill argued that those who resisted experimental philosophy also resisted established religion: "Philosophical Men are usually dealt with by the zealous as the greatest Patrons of the Protestant Cause are by the Sects." In his analysis of religious and political toleration and its limits in 1660, Henry More declared that he would exclude both atheists and enthusiasts from the political nation. The criterion was whether open debate was safe: atheism was "the very plague of Human Politicks" and enthusiasm could not be made "accountable and intelligible to others."[36] Similar arguments were developed by Boyle's collaborator, Thomas Barlow, librarian at Oxford and future bishop. Soon after the Restoration, Boyle and Barlow worked with Pett and Dury on a set of texts that discussed toleration and the settlement. Barlow did not dare publish his analysis; it appeared posthumously under Pett's editorship. He conceded there were so many sectaries and Papists at the Restoration that it might "be more safe for the publick to pardon than punish." But some must be rigorously excluded: those who denied "all Magistracy" or believed "all oaths unlawful," the Catholics who "acknowledg a Power which can absolve them from that oath," and those like Quakers or Adamite nudists who violated natural law. For these churchmen, men who could not be swayed by evidence or free debate must be denied entry to the space in which that debate took place.[37]

Within this space conscience was allowed free play. The works of Barlow, Pett, and Dury argued that the balance of disputing sects was better than a state that included a cowed and disaffected party coerced into silence. Toleration would be stable because it would encourage labour instead of wasteful sectarian dispute. Assent to the political settlement would emerge from true principles, not from imposed force. Barlow argued for the political work of conscience and against such coercion. In 1656 he defended the universities against Hobbes, while as an opponent of Presbyterian or of Papist discipline he conceded that "the Civill Magistrate" should be "jealous of any power superior to his owne." Two years later he promoted Boyle's scheme to republish the lectures on conscience and political obligation of the Oxford casuist Robert Sand-

[36] Sprat, *History*, pp. 28-34, 360-362; Glanvill, *Plus ultra* (1668), p. 138; More, *Explanation of the Grand Mystery of Godliness*, pp. 516, 527.

[37] For Pett, see Pett, *Discourse concerning Liberty of Conscience* (1661); Pett to Bramhall, 8/18 February 1661, in Bosher, *Making of the Restoration Settlement*, p. 242; J. Jacob, "Restoration, Reformation and the Royal Society." For Barlow, see Barlow, "The Case of Toleration," pp. 15-16, 22-36.

erson. In 1660 Barlow urged that citizens should only be disciplined for "*matters of fact*" such as sorcery or sacrilege, and not for "faith or opinion." The civil power could only compel believers to try their belief, not to accept the result of any trial: "That *faith comes by hearing*, we read, and know, but that men are or can be beaten into a belief of Truth, we read not." Barlow condemned coercion of belief as tantamount to Papism, and insisted that only through a "trial of the truth" could subjects come to "Religion by choice, and not only by chance."[38] Boyle used the same strategy in arguing for limited toleration in the experimental form of life. In the debates on the effects of toleration and the achievement of assent, the experimenters in the 1660s showed how their community acted as just such an ideal and stable society.

The establishment of a space which was so securely bounded that dispute could occur safely within it was a difficult accomplishment in social cartography. The technologies Boyle developed were designed to sustain the integrity of this space. We have examined the debates between Boyle and his adversaries on the proper character of this boundary. The security of the matters of fact that experimenters made, and the character of the debates they conducted, depended on the exclusion of work that was vulnerable to "passion or interest, faction or party." Petty used this argument in commending political arithmetic to the Royal Society.[39] Experimenters were to be marked out by their membership of the community. Participation morally distinguished them from those outside the experimental form of life. There were massive rewards from a successful exercise in boundary maintenance. Within this community debate was free: hence the remarkable authority experimenters claimed. In 1666 Boyle wrote that he was "wont to judge of opinions, as of coins: . . . if I find it counterfeit, neither the prince's image or inscription, nor its date (how antient soever,) nor the multitude of hands, through which it has passed unsuspected, will engage me to receive it."[40] This was an extremely influential argument for a form of liberty. We have discussed the program-

[38] Barlow to Hobbes, 23 December/2 January 1657, in British Library MSS Add 32553, ff 22-23; Barlow to Izaak Walton, 10/20 May 1678, in Walton, *Lives of Donne, Wotton . . .*, vol. II, pp. 317-320 (on Boyle's patronage of Sanderson); J. Jacob, *Boyle*, pp. 130-132; Barlow, "The Case of Toleration," pp. 45, 52-53, 92; R. Sanderson, *Several Cases of Conscience* (1660).

[39] Hunter, *Science and Society*, pp. 121-123; Buck, "Seventeenth-Century Political Arithmetic" (for Petty and the search for state support for natural philosophy).

[40] Boyle, "Free Inquiry," p. 159.

matic recommendations in Boyle's *Proëmial Essay* of the 1650s, published in 1661. Henry Power quoted this text at length in the preface of his own *Experimental Philosophy.* If a writer provided no clear experimental warrant for a claim to knowledge, then "if he be mistaken in his Ratiocination, I am in some danger of erring with him." But within the rules of the collaborative experimental community, and through the use of the technology of witnessing, it was possible to act freely and yet grant unconditional assent: "I am left at liberty to benefit myself," even if the author's *opinions* were "never so false."[41] The rules of the experimental community offered this solution to the fundamental political problem of liberty and coercion.

However, just as toleration was to be limited in order to be secure, so experimental liberty was distinguished from individual antinomianism and from uncontrollable private judgment. Experimenters gained their authority from the balance of free judgment and communal discipline. Petty answered More's dismissal of "slibber sauce experiments" by valuing "the sweetness of experimental knowledge" above the "Vaporous garlick & onions of phantasmaticall seeming philosophy." For Henry Power, experimental work was a necessary "Rational Sacrifice," the homage men owed to nature's God. Historians have recently pointed out how these claims for experimental liberty and devotion were connected with Restoration millenarianism. Sprat's *History* was but one of a number of texts which commented on or responded to the providential events of the period between 1660 and 1666. Controlled rather than radical millenarianism was the only safe option for experimentalist propaganda. Publishers of directly radical eschatology, such as Calvert, suffered severely for their views. It was important to break with such radicalism and with the implications of sectarian views of the restoration of all things.[42]

"Restoration" was now past. Many apologists envisaged a distant but safe space for experimental philosophers. Thus Henry Power addressed the experimenters as "well placed in a rank specifically different from the rest of grovelling humanity." He wrote that "this is the Age when all men's souls are in a kind of fermentation," and

[41] Power, *Experimental Philosophy*, "Preface," sig c3^{v}, citing Boyle, "Proëmial Essay," p. 303.

[42] Petty to More, December 1648 and January 1649, in Webster, "Henry More and Descartes," pp. 365, 368; Power, *Experimental Philosophy*, p. 183. For Restoration millenarianism, see Webster, *From Paracelsus to Newton*, pp. 32-33, 67; M. McKeon, *Politics and Poetry in Restoration England*, chap. 8.

he referred to "so powerful an Inundation" that would see "the Rotten Buildings overthrown." But such enthusiast liberty must be balanced. Sprat wrote that experiment would "secure all the Ancient *Proprietors* in their *Rights*: A work as necessary to be done, in raysing a *new Philosophy* as we see it is in building a *new London*." It was safest to model the future state in the image of the idealized present, with a cosmic experimental philosophy at its centre. In his *Christian Virtuoso*, Boyle announced that "it is likely, that as all our faculties will, in the future blessed state, be enlarged and heightened; so will our knowledge also be, of all things that will continue worth it." It was commonly argued that "Natural Philosophy . . . shall then be improved to the utmost, and a Vertuoso shall be no rarity."[43]

Such visions were politically useful in the programmatic texts that Glanvill and Sprat produced in the 1660s. Their efforts to capture support were also answers to enemies and prescriptions of how experimenters should behave. Glanvill argued the experimental "Spirit" made men "so just, as to allow that liberty of judgment to others, which themselves desire, and so prevents all imperious Dictates and Imposings, all captious Quarrels and Notional Wars." Glanvill's target was tyrannic dogmatism, so he emphasized experimental liberty and doubt. But unconstrained Pyrrhonism was not sanctioned by the experimental community either: Boyle said it was "little less prejudicial to natural philosophy than to divinity itself." With Hobbes in view, however, Glanvill insisted that "*Dogmatizing* is the great disturber both of our *selves* and the *world* without us: for while we wed an *opinion*, we resolvedly ingage against every one that opposeth it . . . hence grow *Schisms, Heresies*, and *anomalies* beyond *Arithmetick*." By 1665 Glanvill began to mount a campaign against "extreme Confidence on the one hand, and Diffidence on the other." Diffidence was "more seasonable," but it was now necessary to show why experimental *confidence* was safe and justified.[44]

[43] For millenarian views of natural philosophy, see Power, *Experimental Philosophy*, pp. 191-192; Sprat, *History*, pp. 323, 352; Boyle, "The Christian Virtuoso. Second Part," pp. 776, 789; J. Edwards, *Compleat History of all the Dispensations and Methods of Religion* (1699), p. 745. See discussions in M. Jacob, *The Newtonians*, chap. 3, and Webster, *From Paracelsus to Newton*, p. 68.

[44] Glanvill, *Plus ultra*, pp. 147-148; idem, *Vanity of Dogmatizing* (1661), pp. 228-229; idem, *Scire/i tuum nihil est*, sig Al[v]; Boyle, "Experiments and Notes about the Producibleness of Chymical Principles," p. 591. For the background to Glanvill's scepticism, see Cope, *Glanvill*; for Sprat's comments, see Sprat, *History*, p. 107: "They are therefore as farr from being *Scepticks*, as the greatest *Dogmatists* themselves."

Sprat's *History* attempted this task by giving weight to the dangers of dogmatism and of private belief. The work was begun in 1663 and was sponsored by Wilkins and others at the Royal Society. It responded to the specific political demands of the 1660s as they impinged on the perceived place of experiment. As P. B. Wood has shown, Sprat's *History* served a very particular apologetic function. Millennial and political rhetoric was used to sell the work of the experimenters. Thus Sprat argued that the year 1666 was "now the fittest season for *Experiments* to arise." Errors in epistemology were also dangers in politics. "One of the principal Causes" of "*disobedience*" was "a misguided *Conscience* . . . opposing the pretended Dictates of *God* against the Commands of the *Sovereign*." This was sectarian enthusiasm. Alternatively, "the most fruitful Parent of *Sedition* is *Pride*, and a lofty conceit of mens own *wisdom*." Sprat attacked the overweening ambition of those who claimed private knowledge, and satirized "whoever shall impiously attempt to subvert the Authority of the *Divine Power*, on false pretences to better *Knowledge*."[45] But no enforced coercion could or should stifle such threats to civil peace. Experimental labour destroyed these sources of sedition by submitting private opinion to the judgment of others. This led to Sprat's dramatic assessment of the ambitions of experiment itself. Experiment was directly comparable with the works of Christ himself. It "gives us room to differ, without animosity; and permits us, to raise contrary imaginations upon it, without any danger of a *Civil War*." Dispute *within* the experimental space was possible, even necessary. While the variety of professions and beliefs of the experimental philosophers was a virtue, that variety *outside* this space had been disastrous. Inside this boundary, Sprat claimed, the *exact* equivalent of a civil war could be staged, as in a theatre, with no harmful result. If subjects did not know how safely to dispute, they should go and watch the experimenters: "There we behold an unusual sight to the *English Nation*, that men of disagreeing parties, and ways of life, have forgotten to hate, and have met in the unanimous advancement of the same *Works*."[46]

The manifestos of experimental philosophy made bold political and religious claims. They argued that a new and exclusive way of behaving could now resolve contentions safely. Sprat said the experimenters avoided "convers about affairs of state, or spiritual

[45] Sprat, *History*, pp. 362, 428-430, 346; see P. Wood, "Methodology and Apologetics"; Oldenburg to Boyle, 24 November/4 December 1664, in Boyle, *Works*, vol. VI, p. 180 (on the consultation with Wilkins and his colleagues).

[46] Sprat, *History*, pp. 352, 56, 427.

controversies" because "*Civil differences* and *Religious distractions*" were "the first cause of our *animosities*, and the more they are rubb'd, the rawer they will prove." This self-denying ordinance, however, did not deny the religious and political significance of experimental philosophy. In the Restoration context of licensing and uniformity the safe functioning of experimental philosophy must be secured. This security was conditional on the system of exclusions that experimenters declared they would obey. Then contention would be safe and it would be desirable: "The *Truth* will be obtain'd between them; which may be as much promoted by the *contention* of hands, and eyes; as it is commonly injur'd by those of *Tongues*."[47] In the 1660s critics attacked the discipline that experimenters claimed made their own work safe. Power announced that natural knowledge "must needs be the Office of onely the Experimental and Mechanical Philosopher." We have analyzed the way Boyle used this exclusive boundary. In his *Free Inquiry* (1665-1666) he distinguished between competent and incompetent critics. His adversaries were labelled "naturists: which appellation I rather chuse than that of naturalists, because many, even of the learned among them, as logicians, orators, lawyers, arithmeticians, &c., are not physiologers." Experimenters alone could decide a "physiological question": the rest was metaphysics.[48]

Most of those who criticized the claims of experimental philosophy at the Restoration pointed out the political and religious implications of the experimenters' set of practical refusals. The boundary around the experimental community was now challenged. First, adversaries satirized the low status of experimental labour, and so scorned any attempt to set up some newly autonomous discipline. For Hobbes, the best experimenters were quacks, and the air-pump was "of the nature of a pop-gun which children use, but great, costly and more ingenious." When Thomas White answered Glanvill, he argued that the works of experiment "belong to Artificers and Handy-Craft-Men, not Philosophers, whose office 'tis to *make use of* Experiments for Science, not to make them." In 1666 the Duchess of Newcastle also observed that "the speculative part of Philosophy" was "more Noble than the Mechanical." Experimenters' work was *banausic*: it could not mobilize the moral authority that was needed to allow the establishment of the exper-

[47] Ibid., pp. 426, 100.

[48] Power, *Experimental Philosophy*, p. 184; Boyle, "Free Inquiry," p. 168; idem, "New Experiments," p. 38; compare idem, "Origin of Forms and Qualities," pp. 7-8.

imental community. Experimenters' claims looked ludicrously ambitious without such authority. The Duchess of Newcastle attacked those for whom "the bare authority of an Experimental Philosopher is sufficient . . . to decide all Controversies and to pronounce the Truth without any appeal to Reason." To its critics, experimental liberty seemed like a new form of coercion.[49]

Second, adversaries argued that if experimenters did not engage in matters that touched religious controversy of "humane learning" then they would *weaken*, rather than strengthen, the fortunes of the Church. This exclusion, it was claimed, would subvert true religion. Henry Stubbe's broad swathe of charges used this rhetoric to catalogue the experimenters' imbecilities. No experimental philosopher could "come to be more acceptable to God than another." In 1669 Meric Casaubon also argued that historical and classical scholarship were key weapons in the defence of the faith. Experimenters seemed to underestimate, if not to discourage, this defence. Stubbe put it more bluntly the following year: "Where is the authority of the Church, in controversies of faith," he asked Sprat, "if a common apprehension be that according to which controversies of faith must be decided?" One skilled religious disputant would "be more serviceable unto Monarchy than a Fleet of Ships, Thirty thousand horse and foot, or Three hundred thousand Virtuosi."[50] Since experimental skill was ineffective, to set it up as the chief weapon of the Church and the state was trivial and dangerous.

This claim was very commonly made by churchmen in the 1660s and 1670s. It was made by Stubbe's former patron, John Owen, and by Boyle's erstwhile ally, Thomas Barlow. Owen was vice-chan-

[49] Hobbes, "Mathematicae hodiernae," p. 229; idem, "Seven Philosophical Problems," p. 19; White, *Exclusion of Scepticks* (1665), p. 73; Cavendish, *Observations upon Experimental Philosophy*, "Further Observations," pp. 1-4 (cf. chapter 2, note 14, above). For further evidence of Margaret Cavendish's critical attitude towards low-level experimentation, see her *Description of a New World*, added to the second (1668) edition of *Observations*, esp. pp. 28-32 (on telescopes, microscopes, and their failings); see also the comments of Thomas Wharton, cited in Hunter, *Science and Society*, p. 138.

[50] For Stubbe's attack on the Royal Society, see Syfret, "Some Early Critics of the Royal Society"; R. F. Jones, *Ancients and Moderns*, pp. 244-262; H. W. Jones, "Mid-Seventeenth-Century Science: Some Polemics" (for bibliographic details); J. Jacob, *Stubbe*, pp. 84-108. For Casaubon, see Spiller, "*Concerning Natural Experimental Philosophy*", and Hunter, "Ancients, Moderns, Philologists, and Scientists." Compare Stubbe, *Censure upon Certaine Passages in a History of the Royal Society* (1670), pp. 38-42, and idem, *Lord Bacons Relation of the Sweating-Sickness* (1671), "Preface to the Reader," pp. 9, 23; see also the comments on this polemic in Stubbe to Boyle, 4/14 June 1670, in Boyle, *Works*, vol. 1, pp. xc-xcvii.

cellor of Oxford during 1652-1658. He sponsored attacks on Hobbes's enemy John Wallis. He read *Leviathan* attentively, but considered that Hobbes deified "the magistrate and spoyled all by ye Kingdome of Darknesse." After 1660 Owen was forced into retirement. Hobbes now found it useful to dissociate himself from the former leader of the Independents, "Cromwell's archbishop." Hobbes cited Owen's dramatic condemnation of the whole of Hobbesian philosophy when illustrating the *odium Hobbii* in his *Dialogus physicus*. But Owen himself was not an ally of the experimental philosophers. They would, apparently, tolerate Papists but they declared that they opposed the Protestant sectaries. In 1663 Owen wrote satirically that no doubt the Royal Society were "upon some serious consultations for the benefit of mankind . . . how his Majestys bears may be taught to bite none but fanaticks and that without hurting their teeth."[51] Barlow expressed the same suspicions of experimental philosophy. In 1674 he was sent a copy of William Petty's *Discourse concerning the Use of Duplicate Proportion*, which combined an attempt to bring mechanical principles to a wide public with a remarkable and idiosyncratic matter-theory. Barlow responded with worries that such work was both atheistical (since it seemed to exclude God from nature) and Jesuitical (since its "novel Whimsies" distracted subjects from "the severer studies of . . . Divinity"). Thus the experimental philosophers met challenges to their attempt to win authority and autonomy. Suspicious churchmen and satirical publicists denied that the resources they offered were safe for true religion or effective against religion's enemies.[52]

The polemics of the 1660s affected the way the experimenters presented their programme. They had to seek a conciliation with the churchmen. Clerics made stringent demands on the work of Boyle and his collaborators. We have already examined the ne-

[51] For Owen, Stubbe and Hobbes, see Nicastro, *Lettere di Stubbe a Hobbes*, pp. 27-28 (cf. Stubbe to Hobbes, 11/21 April 1657, British Library MSS Add 32553 f 32); J. Jacob, *Stubbe*, pp. 18-23; C. Hill, *The Experience of Defeat*, pp. 252-254. Owen tried to dissuade Stubbe from translating *Leviathan* (Nicastro, *Lettere*, p. 28). For Hobbes on Owen in 1661, see "Dialogus physicus," p. 274; for Owen on experiment, see Owen to Thornton, ? autumn 1663, in Owen, *Correspondence*, p. 132; for Owen as "Cromwell's Archbishop," see C. Hill, *The Experience of Defeat*, pp. 170-178; idem, *God's Englishman*, pp. 184, 188, 197; Lamont, *Baxter and the Millennium*, pp. 220-224.

[52] Thomas Barlow to John Berkenhead?, 1674, in Pett, ed., *Genuine Remains of Thomas Barlow*, pp. 151-159; P. W. Thomas, *Sir John Berkenhead*, p. 234; Hunter, *Science and Society*, p. 138n.

gotiations between Boyle and More about these demands. More described the work that experimenters should do against political and religious enemies; if they did not, then they would be guilty too. Worthington told More in 1668 that he was worried by certain materialist passages on blood and the soul in Sprat's *History*. More became increasingly concerned by "Cartesian" exclusion of spirit from the world. Culverwell's *Discourse of the Light of Nature*, reissued in 1669, stated that this exclusion was "a meer Arbitrary determination and a Philosophical kinde of Tyranny." The next year More completed his *Enchiridion metaphysicum*, in which experimenters were instructed on the role they should play in producing evidence of spirit.[53] Boyle answered these demands. His debate with Hobbes, we have suggested, involved a contest for the title to mechanical philosophy; his debate with More involved the title to piety. Boyle portrayed the work of experiment as distinct from that of the Church. Yet its work was also valuable for the churchmen. If the rules of the experimental game were obeyed, then that game would work well for the godly. These were the aspects of experimental philosophy that More and his allies found useful at the Restoration. The same resources let Hobbes argue that experiment shared the characteristics of priestcraft: Boyle made experimenters a new kind of clergy.

Experimental Philosophy and the City of God

For Hobbes, the clergy were the first and best example of the disastrous effects of a divided power. Restoration experimenters who sought an alliance with the churchmen and their training ground, the universities, and who claimed an autonomous godly authority for themselves, were easily convicted of all the sins that Hobbes claimed the clergy had committed. In chapter 3 we indicated how Hobbes revealed the interests at play in the false metaphysics that fortified priestly authority. Any group of intellectuals that wilfully established itself in a position of independent power within the state would be attacked in this way. Such independence

[53] Worthington to More, 5/15 February 1668, in Worthington, *Diary and Correspondence*, vol. III, p. 265; for More's concerns about the Dutch Cartesians, see Gabbey, "Philosophia Cartesiana Triumphata," pp. 239-250; Sprat, *History*, p. 348: "What the *Scripture* relates of . . . the *Souls* of men, cannot seem incredible to him, when he perceives the numberless particles that move in every mans *Blood*"; Culverwell, *An Elegant and Learned Discourse* (1652; composed 1646), p. 15; More, *Enchiridion metaphysicum*, pp. 138-140.

made men, "see double, and mistake their *lawful sovereign*." We cited Hobbes's account in the historical narrative *Behemoth* of the way intellectual faction bred war. As early as July 1641, soon after leaving England, Hobbes told Devonshire of his broad approval of attacks on the bishops, since "I am of the opinion, that Ministers ought to minister rather than governe." Devonshire might "perhaps thinke this opinion but a fancy of Philosophy," but Hobbes insisted that dispute "betwene the *spirituall* and *civill* power, has of late more then any other thing in the world, bene the cause of civill warres, in all places of Christendome."[54]

The issue of priestly power over consciences was raised again in the 1650s; Hobbes used this in dispute with Wallis in *Six Lessons to the Professors of Mathematics* (1656) and in *Stigmai* (1657). Hobbes told Wallis that "you know how to trouble and sometimes undo a slack government, and had need to be looked to, but are not fit to hold the reins." Wallis had been secretary to the Westminster Assembly of Divines in the 1640s. Hobbes now revived charges of Presbyterian conspiracy against him. The assault on an autonomous and potent clergy was always an assault on the doctrines they taught at the universities: "Divinity may go on in Oxford and Cambridge to furnish the pulpit with men to cry down the civil power, if they continue to do as they did."[55] Hobbes's texts took their place in the radical campaigns for academic and political reform. The divines and natural philosophers responded in kind by pointing out that the new experimental philosophy was now an integral part of the universities' role as servants of Church and state. Seth Ward accused Hobbes of "the delation of us to the Civill Magistrate, and the Endeavours for our Extirpation"; John Wallis told Huygens that "our Leviathan is furiously attacking and destroying our universities . . . and especially ministers and the clergy and all religion." This made the response to Hobbes's natural philosophy directly relevant to the defence of the authority of clergymen and experimenters alike.[56]

Hobbes persisted in these attacks at the Restoration. In his *Ex-*

[54] Hobbes, "Leviathan," p. 460; idem, "Behemoth," pp. 346-348; Hobbes to William Cavendish, Earl of Devonshire, 23 July/2 August 1641, in Tönnies, *Studien*, pp. 100-101.

[55] Hobbes, "Six Lessons," p. 345; idem, "Stigmai," p. 398. For Wallis at the Westminster Assembly, see Scriba, "Autobiography of Wallis," p. 35: "When as they were called *Presbyterians*; it was not in the Sense of *Anti-Episcopal*, but *Anti-Independants*."

[56] Ward, *Vindiciae academiarum* (1654), p. 61; Wallis to Huygens, 22 December 1658/1 January 1659, in Huygens, *Oeuvres*, vol. 1, p. 296, and Scott, *Mathematical Work of Wallis*, pp. 170-171.

aminatio et emendatio mathematicae hodiernae, completed in summer 1660, he concluded a fresh assault on Wallis with the reaffirmation of the need for undivided state power, and he expressed his hopes of the settlement from those "who now debate about power in England." In the *Dialogus physicus*, Hobbes went on to attribute the resistance he encountered from both experimenters and priests to the fact that he had "freely written the truth about Academies."[57] He used the catalogue of crimes that experimenters and mathematicians had committed against him to show that the same interests were at play in both the priesthood and the experimental community. Priestcraft masqueraded as philosophy and did so incompetently. In his own answer to *Dialogus physicus*, Wallis pointed out that Hobbes's argument that natural philosophy should not be infected by priestcraft matched Hobbes's assault on the "absurd talk" of priestly metaphysicians about spirits: "He would not have it thought that a *Divine* can be a *Philosopher*, any more than that a *Substance* can be *Incorporeal*." Henry More reissued his *Antidote against Atheism* in the same year as Wallis's answer to Hobbes. More hammered home the message that the experimental proof of spirit was the prime political task for natural philosophers and clergy alike. "[A]ssuredly that Saying is not more true in Politicks, *No Bishop, No King*; then this is in Metaphysicks, *No Spirit, No God*." This defined the basis on which collaboration between priests and experimenters would be possible. During the 1660s Boyle and his colleagues explored the ways in which this task might be sustained. Hobbes pointed out the collaboration and challenged its political effects.[58]

Sermons against Hobbes were often larded with arguments drawn from Boyle. Moderate clergymen at Cambridge under attack from high Anglicans sought Boyle's aid. Simon Patrick wrote in summer 1662 that the true Church should "choose her servants where she best likes." His brother made extensive notes on Boyle's air-pump experiments. Simon Patrick used them against "the open violence of *Atheisme*" and "the secret treachery of *Enthusiasme* and *Superstition*." This invitation suggested means by which the stringent demands churchmen made of experimenters could be satisfied.[59]

[57] Hobbes, "Mathematicae hodiernae," p. 232; idem, "Dialogus physicus," p. 274; cf. idem, "Decameron physiologicum," pp. 73-78.

[58] Wallis, *Hobbius heauton-timorumenos*, p. 6; More, *Antidote against Atheism* (1662), p. 142.

[59] Patrick, *Brief Account of the Latitude-Men*, pp. 8, 21, 24; John Patrick's notes on Boyle's pneumatics, Cambridge University Library MSS Add 77 ff 11-32; More to

Boyle's reply to Hobbes appeared just after Patrick's book. We have pointed out that Boyle censured Hobbes for his comments on "things said in the books of naturalists concerning immaterial substances." Boyle now used More against Hobbesian materialism. He also referred to his own work on theology, which Boyle said he would publish in the near future. In the following October, Edward Stillingfleet wrote to Boyle asking him to "communicate to the world those papers . . . in behalf of Christianity." Stillingfleet also made plentiful use of Boyle's work in his own *Origines sacrae*, published in the same year. He told Boyle that it would be "seen yet further, that those great personages, who have courted nature so highly, that her cabinets are open to them, are far from looking on religion as mean and contemptible." During 1662, therefore, works by Boyle and Wallis, More, Patrick, and Stillingfleet all attacked Hobbism and all pointed to an alliance between the "bare authority" of experiment and the use of experiment in Christian apologetics.[60]

However, the condition of this utility was the achievement of disinterested autonomy by the experimental community. The *independence* attributed to Boyle's testimony was what made it valuable. Thus Stillingfleet encouraged Boyle to "employ your excellent pen in a further discovery of those rich mines of experimental philosophy." Simon Patrick aimed to win over "the Ingenuous Gentry who begin generally to be acquainted with the *Atomical Hypothesis*." Only the *independent* voice of God's Book of Nature and its disinterested interpreters could sway them.[61] Glanvill had direct experience of the weakness of the Church in gaining assent from believers. He reckoned that the Royal Society was a better defender of religion than "the *profest Servants* of the *Altar*" whose work was "interpreted by such as are not willing to be convinced, as the products of *interest*, or *ignorance*." Boyle agreed. He answered the charge of interest made against clergy and experimenters alike.

Boyle, 27 November /7 December 1665, in Boyle, *Works*, vol. VI, p. 513. Patrick wrote the *Brief Account* after a contest with Anthony Sparrow and his patron Clarendon for the presidency of Queens' College: see Cambridge University Library MSS Add 20 f 6, and Nicolson, "Christ's College and the Latitude-Men," p. 48. Compare chapter 5, note 117, above.

60 Boyle, "Examen of Hobbes," p. 187; Stillingfleet to Boyle, 6/16 October 1662, in Boyle, *Works*, vol. VI, p. 462; Stillingfleet, *Origines sacrae*, pp. 466-470 (for comments on the mechanical philosophy). For Boyle on Hobbes and "orthodox Christian naturalists," see "Animadversions on Hobbes," pp. 104-105, and "Examen of Hobbes," p. 187.

61 Stillingfleet to Boyle, 6/16 October 1662, in Boyle, *Works*, vol. VI, p. 462; Patrick, *Brief Account of the Latitude-Men* p. 24.

When he was offered an ecclesiastical living in the 1660s, he expressed the view that "the irreligious fortified themselves against all that was said by the clergy, with this, that *it was their trade*, and that *they were paid for it*." Hence it was best for the Christian naturalist to work in a space impervious to ecclesiastical control: "Having no other interests, with relation to religion, besides those of saving his soul, gave him, as he thought, a more unsuspected authority in writing or acting on that side." Most importantly, the experimental form of life was presented as a means of winning assent from otherwise unruly subjects. The autonomy of that form of life was necessary for the authority claimed by the experimenters. The godly could use this authority if they respected the integrity of experiment.[62]

A range of strategies linked experiment with the priestly efforts to win assent from unruly believers. Churchmen like More and Glanvill used experimental technologies to make ghost stories and witch testimonies into matters of fact and so convince men that extended spirits were real. This was a political task: sectarian enthusiasts saw spirits everywhere and materialist atheists saw them nowhere. It was also a juridical problem: lawyers such as Matthew Hale or John Selden allowed witch trials but denied demonic potency, while critics and radicals such as John Webster and John Wagstaffe affirmed "many thousands of spirits made of an incorporeal matter too fine to be perceived," yet denied that spirits ever made "contracts . . . with any man or woman." The experimental form of life showed how to tell between reliable and unreliable testimony; it showed how witnesses should be judged and how evidence could be made convincing.[63] In chapter 5 we indicated that Boyle collaborated in this attempt to make spirit testimonies compelling. More said that "Hobbians" could be won over by appearances of the "Drummer of Tedworth" in 1663; in 1665 he

[62] Birch, "Life of Boyle," p. lx; Glanvill, *Scepsis scientifica*, "To the Royal Society," sig bl^{v}.

[63] For More and spirits, see Burnham, "The More-Vaughan Controversy"; Guinsburg, "More, Vaughan and the Late Renaissance Magical Tradition"; Heyd, "The Reaction to Enthusiasm in the Seventeenth Century." For witchcraft as an issue of legal and disciplinary competence, see Macfarlane, *Witchcraft in Tudor and Stuart England*; Mandrou, *Magistrats et sorciers*; Ginzburg, *The Night Battles*, esp. pp. 125-129; Hirst, "Witchcraft Today and Yesterday." For Matthew Hale, see B. Shapiro, *Probability and Certainty*, pp. 206-208; for Selden, see K. Thomas, *Religion and the Decline of Magic*, p. 625; for Webster, see Jobe, "The Devil in Restoration Science"; for Wagstaffe, see Wagstaffe, *The Question of Witchcraft Debated* (1671), pp. 112-113, and Webster, *From Paracelsus to Newton*, p. 65 (for Wagstaffe's use of Hobbes).

talked with Boyle in a London bookshop about these appearances and about the faith-healers Matthew Coker and Valentine Greatrakes. Boyle and du Moulin published stories about spirits, and in 1668 Glanvill compiled an array of stories which he processed so that they might become authoritative experimental testimonies: "We know not anything of the world we live in but by *experiment*, and the *Phaenomena*; and there is the same way of *speculating immaterial* nature."[64]

Experimental matters of fact became the prized possessions of adversaries in these debates. The literary and social technologies Boyle advocated could distinguish between rational and obdurate participants so that too much confidence would look like enthusiasm. Experiment ought to sway "the mighty confidence grounded upon nothing, that *swaggers*, and *Huffs*, and *swears*, there are no *Witches*." Glanvill reckoned that those who denied "Matters of Fact well grounded" must be "hardn'd against Conviction." More told Glanvill that "such fresh examples of Apparitions and Witchcrafts," when subjected to proper trial, must win over "benummed and lethargick Mindes."[65] However, victory needed a definition of those who could count as reliable witnesses. The experimental form of life must be accepted. The controversies between More, Boyle, and Henry Stubbe about the interpretation of Greatrakes' cures in the mid-1660s showed that it was difficult to achieve such acceptance. Boyle wrote that "I hold it not unlawful to endeavour to give a physical account of his cures," but many churchmen held differently. Webster challenged Glanvill's stories in the 1670s and claimed that Boyle now doubted their veracity. Boyle assured Glanvill he remained convinced: good evidence of "intelligent beings, that are not ordinarily visible" would help in "the reclaiming" of atheists. The standards of experiment must be met: "Any one relation of a supernatural phaenomenon being fully proved and duly verified,

[64] For More and Boyle on spiritual cures, see Worthington, *Diary and Correspondence*, vol. II, pp. 216-217; for More and the "Hobbians" at Tedworth, see More to Anne Conway, 31 March/10 April 1663, in *Conway Letters*, p. 216, and Cope, *Glanvill*, p. 15n; for Glanvill, see Glanvill, *A Blow at Modern Sadducism* (1668), p. 116; idem, *Saducismus triumphatus* (1681), pp. 89-118; Prior, "Glanvill, Witchcraft and Seventeenth-Century Science."

[65] Glanvill, *Philosophia pia* (1671), pp. 25-34; idem, "Against Modern Sadducism," pp. 3-4, 58-60; More to Glanvill, in Glanvill, *A Praefatory Answer to Henry Stubbe*, p. 155. For More on the testing of witnesses in the case of Matthew Coker, see More to Anne Conway, 7/17 June 1654, in *Conway Letters*, pp. 101-102, and Kaplan, "Greatrakes the Stroker," pp. 182-183. For Glanvill and the witnesses at Tedworth, see Cope, *Glanvill*, p. 102.

suffices to evince the thing contended for." Well-made matters of fact would then "enlarge the somewhat too narrow conceptions men wont to have of the amplitude and variety of the works of God."[66]

Experimenters and clergymen said that witnessed and testified matters of fact were compelling. Their use of these techniques defined the community of believers and witnesses to which they could appeal. All those who were invulnerable were outside the political and religious nation. So Boyle could stipulate that when Hobbes withheld assent to experimental facts he was a fanatic; More argued that anyone who did not grant assent to spirit testimony was beyond reasoned debate. Furthermore, these techniques were used to make invisible and incorporeal substances into palpable realities. These realities were supposed to rebut Hobbists and their allies. Hobbes reckoned that his adversaries attributed false potency to such experienced matters of fact. We examined this issue of political epistemology in chapter 3. It was another example of the alliance of priestcraft and experimental philosophy. Boyle and his allies alleged that all matters of fact carried a badge of adequate "moral certainty": "A rational assent may be founded upon proofs that . . . are strong enough to deserve a wise man's assent in them."[67] This limited authority was sufficient to sway any member of the community. No higher certainty should be demanded from knowledge of experience or belief. English philosophers and churchmen at Great Tew in the 1630s had used this strategy against the Jesuits: they had argued then that some limited moral certainty was both adequate and compelling. The experimenters drew on these arguments in defending their own power, and churchmen such as Wilkins and Glanvill used the appeal to limited certainty when defending a rational religion in the 1660s. Such techniques were

[66] For Greatrakes, see Maddison, *Life of Boyle*, pp. 123-127; Boyle to Stubbe, 9/19 March 1666; in Boyle, *Works*, vol. I, p. lxxxi; J. Jacob, *Boyle*, pp. 164-176; idem, *Stubbe*, pp. 50-63, 164-174; Steneck, "Greatrakes the Stroker." For the broader context of healing and its uses, see Macdonald, "Religion, Social Change and Psychological Healing"; for Boyle in this connection, see B. Shapiro, *Probability and Certainty*, pp. 216-217; Boyle to Glanvill, 18/28 September 1677 and 10/20 February 1678, in Boyle, *Works*, vol. VI, pp. 57-60. (The work in question included the case of the "Demon of Mâçon": see Boyle to du Moulin, in ibid., vol. I, pp. ccxxi-ccxxii, and Labrousse, "Le démon de Mâçon.")

[67] Boyle, "Discourse of Things above Reason," p. 450. Compare More, "How a Man is to Behave Himself in this Rational and Philosophical Age for the Gaining of Men to . . . the Christian Faith," in idem, *Modest Enquiry*, pp. 483-489; Stillingfleet, *Origines sacrae*, pp. 171-176.

grounded in an appeal to subjects' beliefs, and they worked to defend the authority of those specially competent in the production of matters of fact within the community of reliable believers.[68]

The work to produce spirit testimonies and to win their acceptance showed the issues that bore upon the presentation of experiment as a vehicle for churchmen. Hobbes had a different answer to the disciplining of the enthusiasts and the interpretation of spirit stories. The key tactic, of course, was to locate the political interest:

> A private man has always the liberty, because thought is free, to believe or not believe in his heart those acts that have been given out for miracles, according as he shall see what benefit can accrue by men's belief, to those that pretend or countenance them, and thereby conjecture whether they be miracles or lies.

Witches and ghosts were only real because of legal and civil powers. Thus, while there was no physical reality in witchcraft, it was needful to prosecute those who claimed such powers, "for the false belief they have that they can do such mischief, joined with their purpose to do it if they can; their trade being nearer to a new religion than to a craft or science." Hobbes claimed that the experimenters presented themselves as ambitious masters of a power they did not have and should not claim. When used by priests, "the ghostly men," experimental technologies would be suspect too. The same interest was at work in these ways of going on. We have argued that when Hobbes inspected the language of those who talked of incorporeal bodies he saw the absurdity and danger in their work. "If this superstitious fear of spirits were taken away . . . by which crafty ambitious persons abuse the simple people, men would be much more fitted than they are for civil obedience." The ingenuity of godly men made subjects rebellious: it could never produce a settled peace.[69]

[68] The key texts of the Great Tew Circle include: Falkland, *A Discourse of Infallibility* (1645); Hales, *A Tract concerning Schism and Schismaticks* (1642); Chillingworth, *The Religion of Protestants* (1638). For connections with Boyle, see Canny, *The Upstart Earl*, p. 147; and with Hobbes, see K. Thomas, "Social Origins of Hobbes's Political Thought." For discussions of constructive scepticism, see McAdoo, *The Spirit of Anglicanism*, pp. 1-23; van Leeuwen, *The Problem of Certainty*; Orr, *Reason and Authority*; Popkin, *History of Scepticism*, chap. 7; B. Shapiro, *Probability and Certainty*, chap. 3.

[69] Hobbes, "Leviathan," pp. 9-10, 436-437, and see our discussion in chapter 3, above. Hobbes told Newcastle "that he could not rationally believe there were witches"; see Cavendish, *The Cavalier in Exile* (1667), pp. 142-143.

The experimenters consistently displayed themselves as a godly community. Boyle sponsored texts on Restoration politics that were couched in the language of casuistry and conscience: he wrote of his vocation as an experimental philosopher and believer by describing the "conversion experience" he had undergone at Geneva. These patterns aped the stories of spiritual awakening reported by Augustine and Luther, Bacon and Descartes, Herbert and van Helmont.[70] Contrast Boyle's account of an encounter with the voice of God in a thunderstorm with the story of Hobbes's awakening when reading Euclid. Conversion accounts showed how men were called to their destiny. They were the key moments at which assent to visible truths was won and retained. Thus the details of a conversion reported by an author would exemplify the author's ideal of how conviction was attained. Boyle's autobiography was composed in the late 1640s. It described at least two such conversion experiences. A storm reminded Boyle of the "consideration of his unpreparedness" for judgment. The next year, probably 1640, Boyle once again revived his doubts of "some of the fundamentals of Christianity" and he contemplated suicide. Some months later, while at Mass, it "pleased God . . . to restore unto him the withdrawn sense of his favour." It was from this period that Boyle "dated his conversion." Only after this did he begin to examine the rational grounds of faith and to compare the customs of other religious communities. Divine power summoned men to labour: thus work through experience was effective and necessary.[71]

Hobbes, by contrast, was summoned to geometry during his second journey to Europe in the late 1620s. Aubrey testified that Hobbes's call followed an exemplary experience of the power of demonstrative logic over impatient and recalcitrant refusal to accept the truth:

> He was 40 yeares old before he looked on geometry; which happened accidentally. Being in a gentleman's library . . . , Euclid's Elements lay open, and 'twas the *47 El. libri I.* He read the proposition. By G__, sayd he "this is impossible!" So he reads the demonstration of it, which referred him back to such a proposition; which proposition he read. That referred him

[70] For cases of conscience, see McAdoo, *The Spirit of Anglicanism*, pp. 24-80; Klaaren, *Religious Origins of Modern Science*, pp. 100-108, and, for conversion experiences, pp. 72-76; Farrington, *The Philosophy of Bacon*, pp. 59-72 (translating Bacon's *Masculine Birth of Time*); Shea, "Descartes and the Rosicrucians," pp. 42-46.

[71] Boyle, "Account of Philaretus," pp. xxii-xxiii; J. Jacob, *Boyle*, pp. 38-42.

back to another, which he also read. Et sic deinceps, that at last he was demonstratively convinced of that trueth. This made him in love with geometry.[72]

For Hobbes, philosophy was not a vocation to which men were to be summoned by divine grace. For Boyle, by contrast, the experimental philosophers were to be called the "priests of nature." Their doctrines could be traced back to Moses and they were charged with the production of "successful arguments to convince men there is a God." In 1662 More recommended that every priest should be a rational man or philosopher. Boyle used the same sources to argue that "if the world be a temple, man sure must be the priest, ordained (by being qualified) to celebrate divine service not only in it, but for it."[73]

The presentation of experimenters as the "priests of nature" was extremely influential: their work was held to have direct effects in the establishment of religion and their laboratories acquired a sacred status. Contemporaries recognized Boyle's hieratic vocation. Beale told him in October 1663 that the cure for dispute in matters of belief would be the construction of an "operative, practical, and experimental" religion: "Sir, you are by divine endowments consecrated a chief in that priesthood." These features linked the experimental form of life with priestcraft. Boyle suggested that experimental trials should best be performed on Sundays as part of the worship of God. Laboratories were to be contrasted with the private shrines of "secretist" philosophers and Hermetics whom Boyle criticized for their refusal to communicate in public. Boyle described his own laboratory as "a kind of *Elysium*, so as if the threshold of it possessed that quality the poets ascribed to that *Lethe*, their fictions made men taste of before their entrance into those seats of bliss." Hobbes challenged the privileged boundary that Boyle erected around this space. In so doing, he also pointed out the alliance experimenters aimed to forge with the priests, and the interests that alliance would serve.[74]

[72] Hobbes, "Vita," p. xiv; Aubrey, "Life of Hobbes," p. 332; de Beer, "Some Letters of Hobbes," p. 205; and note the conversion experience dramatized in the "Dialogus physicus," p. 271.

[73] Boyle, "Disquisition on Final Causes," p. 401; idem, "Usefulness of Experimental Natural Philosophy," p. 32. See also More, *Collection*, "Preface General," p. v; Fisch, "The Scientist as Priest."

[74] Boyle, "Usefulness of Experimental Natural Philosophy," essay III; Beale to Boyle, 17/27 October 1663, in Boyle, *Works*, vol. VI, pp. 341-342; Boyle to Katherine

The Kingdom of Darkness

Hobbes and Boyle set out two different ways of organizing the production of knowledge. They put forward two different images of the ideal community. The charges Hobbes levelled against the experimenters were charges against the political effects of their form of life. The issues of intellectual authority and autonomy were central components of that political critique. The attack on priest-craft and its natural philosophy made by Hobbes in his discussion "Of the Kingdom of Darkness" could be made on any independent intellectual coterie. We have read Hobbes's analysis at the end of *Leviathan* as an attack on a corrupt philosophy and those who peddled it. Scripture suggested the Kingdom of Darkness was ruled by Satan, "*the prince of the power of the air*," and filled with "phantasms" and "illusions." Hobbes decoded the allegory to show that the Kingdom was "a *confederacy of deceivers . . . to obtain dominion over men in this present world . . . by dark and erroneous doctrines*."[75] At the Restoration, Hobbes claimed the social reality of experimental philosophy was also that of a confederacy. As such, it could not present itself as an ideal community. Hobbes had his own ideal commonwealth, and in the realm of learning that was geometry.[76]

The disputing parties portrayed each other as gang leaders and their ideal communities as little better than a rabble. Hobbes argued that the experimental confederacy was both too exclusive and too open. First, it was private. It refused entry to Hobbes and Hobbesian philosophy. What were proclaimed as public truths were in reality the private judgments of a select few. Second, there was nothing special about experimenters and their practices. They were just as politically motivated as any confederacy. They were no more sophisticated than children or artisans.[77] The experimenters were just another conspiratorial group whose interests were in obtaining power over citizens, and whose devious confederacy sought an illegitimate autonomy from the state.

Hobbes claimed that the privacy of experimental space did its own political work. In *Leviathan* he attributed resistance to sound

---

Boyle, Lady Ranelagh, 6/16 March 1647, ibid., vol. I, pp. xxxvi-xxxvii; Boyle to Lady Ranelagh, 31 August/ 10 September 1649, ibid., vol. VI, pp. 49-50. For another attack on Hermetic secrecy, see Stillingfleet, *Origines sacrae*, pp. 103-104.

[75] Hobbes, "Leviathan," pp. 603-604.

[76] For geometry as Hobbes's ideal community, see Buck, "Seventeenth-Century Political Arithmetic," p. 82.

[77] Hobbes, "Dialogus physicus," pp. 240, 278.

political doctrine to the interests of "Learned Men" who would "digest hardly any thing . . . that discovereth their errors, and thereby lesseneth their authority." In *Dialogus physicus* he attributed resistance to sound natural philosophy to the fact that "there are very few of those who profess the sciences who are not pained by the discovery of difficult truths by others rather than themselves."[78] In his survey of the history of natural philosophy, Hobbes spelt out the ways in which such philosophers organized themselves to preserve their intellectual authority. The Papacy sustained Scholastic physics to "command men" because the Roman Church lacked troops. The exclusive confederacy of natural philosophers had a strong link with charlatanism: "When many of them are once engaged in the maintenance of an error, they will join together for the saving of their authority to decry the truth." We have seen how this autonomy and privacy was held to tend to civil war. This linked the strategy Hobbes suggested that the Royal Society should follow and the political security of the Restoration settlement. The "gentlemen of Gresham College" should "apply themselves to the doctrine of motion (as Mr. Hobbes has done, and will be ready to help them in it, if they please, and so long as they use him civilly)."[79]

Boyle and his allies did not invite Hobbes's help. In fact, they used the confederacy of which they were members as a powerful weapon against him. Boyle could portray himself as a loyal member of the experimental community. Then Boyle could read Hobbes as though he were attacking the Society as a whole. Alternatively, Hobbes could be read as criticizing views that were properly the opinions of Boyle alone. This was where matters of fact were crucially differentiated from metaphysical theses or bold conjectures. There was a social boundary between matters of fact and conjectures, a boundary that was highly functional in these disputes. Matters of fact were the property of the whole community that accepted them: hence it was possible for Boyle to point out the irony that Hobbes's patron, the Earl of Devonshire, was a Fellow of the Royal Society.[80] In *Micrographia* (1665) Hooke distinguished between his methodical accomplishments, which belonged to the

[78] Hobbes, "Leviathan," p. 325; idem, "Dialogus physicus," p. 274.

[79] Hobbes, "Six Lessons," pp. 344-348; idem, "Decameron physiologicum," pp. 73-78; idem, "Considerations on the Reputation of Hobbes," p. 437.

[80] For Boyle as individual or as spokesman for the Royal Society, see Boyle, "Examen of Hobbes," pp. 188, 190-191; idem, "Animadversions on Hobbes," p. 112; Wallis, *Hobbius heauton-timorumenos*, pp. 148-152; and our discussion in chapter 5, above.

community, and his conjectures, which were his own. Any reader should be instructed in this boundary, and was told that "I have produced nothing here, with intent to bind his understanding to an *implicit* consent." In 1663 Hooke's draft rules for the Society proposed giving this distinction a legislative force. These rules would ban hypothesizing within the Royal Society, and would stipulate the private character of such speculation. According to Oldenburg in 1663, "These are the bounds to which the Royal Charter limits this British assembly of philosophers, which they think it would be improper to transgress."[81] Experimenters presented their communities as bounded and disciplined and safe. Hobbes used the rhetoric of this exclusive group to sustain a charge of conspiracy against them.

Experimental discipline produced complex relations between individuals and their community. Sprat insisted that the experimental community should not be "a Company all of *one mind*."[82] Intellectual tyranny within the community was both a sin and an error. Boyle pointed out that intellectual sects which based their power on the authority of individuals were thereby made vulnerable to attack by an experimental and communal strategy: "*Aristotle* being himself a dark and dubious writer, and his followers being on that account divided into sects and parties, which for the most part had nothing to alledge but his single authority, it was not difficult to answer the arguments drawn from the Peripatetick philosophy."[83] Yet the experimenters did use the authority and power of individuals to sustain their claims to success. We showed in chapter 2 how the air-pump and Boyle himself became emblematic of experimental progress. Hooke described Boyle as "the *Patron* of *Philosophy* it self; which he every day *increases* by his *Labours* and *adorns* by his *Example*." In summer 1666 John Beale proposed to issue Boyle's complete works as a means of securing the products of the experimental philosophy: he told Boyle that "you will conduct the two rivulets of mechanism and chemistry into the ocean of theology." In his history of the sciences since Adam, *Plus ultra*, Glanvill devoted two entire chapters to Boyle's publications and his future plans: "Had this *great Person* lived in those days, when men *Godded* their

[81] Hooke, *Micrographia*, "The Preface," sig bl^r; Oldenburg to Leichner, April 1663, in Oldenburg, *Correspondence*, vol. II, pp. 110-111; Weld, *History of the Royal Society*, vol. I, pp. 146-148; M. B. Hall, "Science in the Early Royal Society," pp. 60-61.

[82] Sprat, *History*, p. 73.

[83] Boyle, "Some Considerations about Reason and Religion," p. 152; cf. Glanvill, *Scepsis scientifica*, "To the Royal Society," sig a1^r.

Benefactors, he could not have miss'd one of the first places among their *deified Mortals*.[84]

Hobbes argued that any confederacy of interested subjects would need such a powerful and sole authority. The illusion of communal decision reached freely was just that: a dangerous myth. He argued that

> no one man's reason, nor the reason of any one number of men, makes the certainty. . . . And therefore, as when there is a controversy in an account, the parties must by their own accord, set up, for right reason, the reason of some arbitrator, or judge, to whose sentence they will both stand, or their controversy must either come to blows, or be undecided, for want of a right reason constituted by nature; so is it also in all debates of what kind soever.[85]

Hobbes claimed that the debates of the experimenters illustrated this principle. They pretended that *nature* could set up some right reason, and that their private community of interpreters then spoke with the authority of that reason. Yet they excluded those who offered the path to certain knowledge of nature and they failed to exclude sectarian interest from their confederacy.

Hobbes's critics turned this argument back upon him. For Hobbes the mark of the ideal community was its certainty: hence the role of the geometers as the social ideal. From this characteristic of Hobbes's ideal flowed the salient features of the experimenters' response. First, they could claim that Hobbes aimed for a monopoly of learning. Wallis wrote in 1662 that Hobbes believed the experimenters' "*Pretenses* he would have it thought are such as His are wont to be." Hobbes allegedly considered "His Doctrine should be made the Standard for *Schools* and Pulpits. . . . But that any such have been the Pretensions of those at *Gresham-College*; as that none shall ever be able to adde to what They shall do: I have not yet heard." Seth Ward wrote that Hobbes's wish was that "his *Leviathan* be *by entire soveraignty imposed upon the Universities*, there to be read and publickly taught." Hobbes responded in 1656 that he could

[84] Hooke, *Micrographia*, "The Preface," sig dl^v; Beale to Boyle, 18/28 April, 13/23 July and 10/20 August 1666, in Boyle, *Works*, vol. VI, pp. 399, also pp. 405-407, 416-417; Glanvill, *Plus ultra*, p. 93. For Boyle as representative of experimental philosophy, see Klaaren, *Religious Origins of Modern Science*, p. 19; M. B. Hall, "Science in the Early Royal Society," pp. 72-73; Westfall, "Unpublished Boyle Papers," p. 64.

[85] Hobbes, "Leviathan," p. 31.

well have argued for a "lay-university" wherein "lay-men should have the reading of physics, mathematics, moral philosophy and politics, as the clergy have now the sole teaching of divinity."[86] This was entirely consistent with the Hobbesian suspicion of all autonomous professions, and, in the Restoration context of imposed uniformity and the debate on church discipline, his argument carried greater threat. It followed from the ideal of the geometrical and nonsectarian community. Because of this social monism, critics could claim that Hobbes's geometry was directly connected with his political campaign. To refute one was to refute all. Hobbes told Wallis that "the doctrine of the duty of private men in a commonwealth" was "much more difficult" even than geometry. If Wallis erred in geometry, "How then do you think, . . . you should be fit to govern so great nations as England, Ireland, and Scotland, or so much as to teach them?" Wallis answered in 1662 that "his *Geometry* was to have given credit to all the rest, and is it not able to support itself?" He told Christiaan Huygens that it was necessary to refute Hobbes's geometry and show "how little he understands of the mathematics from which he takes his courage."[87]

Finally, just as Hobbes charged the experimenters with the crime of forming a private confederacy, so the experimentalists saw Hobbes as the spokesman for an influential but secret party, powerful among the courtiers and threatening to the toleration of experiment. He was "the great Leviathan, the very Dagon of many young squires or squirrels." Quentin Skinner has argued that both Seth Ward and Abraham Cowley had been influenced by Hobbes, and that William Petty was decisively concerned with Hobbes's scheme for settlement.[88] In the 1660s, however, it was necessary to dissociate work from Hobbesian principles, and to identify the covert Hobbists who sought to satirize or subvert experimental philosophy. Thus, when Thomas White brought forward similar schemes in geometry and similar assaults on the dubious claims of the experimenters, Glanvill immediately accused him of being "in

[86] Wallis, *Hobbius heauton-timorumenos*, p. 149; Ward, *Vindiciae academiarum*, p. 52; Hobbes, "Six Lessons," p. 345.

[87] Hobbes, "Stigmai," p. 399; Wallis, *Hobbius heauton-timorumenos*, p. 6; Wallis to Huygens, 22 December 1658/1 January 1659, in Scott, *Mathematical Work of Wallis*, p. 170; Wallis to Owen, October 1655, in Owen, *Correspondence*, p. 86.

[88] Kendall, *Sancti sanciti* (1654), p. 153. For Hobbes's allies, see comments in Hobbes to Aubrey, 24 February/6 March 1675, in Tönnies, *Studien*, p. 112. Compare Skinner, "Ideological Context of Hobbes's Political Thought"; idem, "Hobbes and His Disciples in France and England"; Buck, "Seventeenth-Century Political Arithmetic," pp. 77-78.

the very rode of the Hobbian hypothesis." Some "Hobbians" could be swayed by the provisions of reliable spirit experiences, as both More and Glanvill testified. The best remedy, however, was the experimental community itself: "Divers of the brisker *Geniusses*, who desire rather to be accounted *Witts*, then endeavour to *be so*, have been willing to accept *Mechanism* upon *Hobbian* conditions," Glanvill told the Royal Society in 1665. " 'Tis not conceivable how a more suitable *remedy* could have been produced against the *deadly influence* of that *Contagion*, then your Honourable Society."[89]

Hobbes said that no independent group of intellectuals could avoid constituting a threat to civil society. On the contrary, such groups were themselves a danger. This was a general account of the link between civil strife and the implications of privileged disciplinary skills. Clerics and lawyers were no better than the radical sects. We have cited *Behemoth* to illustrate Hobbes's argument that the Protestant sects were pernicious because they, too, claimed private judgment and a right to personal interpretation. Only the civil power could act as "judge" and "interpreter." However much skill and experience were accumulated by any individual, however much illumination was claimed from practice or from inspiration, that individual would still gain no further competence in civic philosophy. For Hobbes, success in the construction of commonwealths, as of all artifacts, was a matter of rational rule-following, not of practice.[90] In *De corpore politico*, Hobbes also pointed out the political effects of the distinction between the pursuit of deductive rules and the attempts of the ingenious to secure belief. The former tended to peace, the latter to rebellion. In this classical condemnation of oratory, subversive eloquence was set against pacific demonstration: "To demonstration and the teaching of the truth there are required long deductions and great attention, which is unpleasant to the hearer." The demagogues who sought credence took another path: "by aggravations and extenuations" they "make good and bad, right and wrong, appear great or less, as shall serve their turn." Thus Hobbes set up the disinterested teacher as an ideal contrasted to the rabble-rouser. Walter Pope recalled that when Hobbes engaged in argument "he would leave the Company in a passion, saying his business was to Teach, not to Dispute." Hobbes himself said that "to be taught, I think not very laudable, though

[89] Glanvill, *Scire/i tuum nihil est*, p. 29; idem, *Scepsis scientifica*, "To the Royal Society," sig bl.

[90] Hobbes, "Behemoth," p. 190; idem, "Leviathan," pp. 164, 195-196; and see Sacksteder, "Hobbes: The Art of the Geometricians."

to teach, provided it be rightly done, and without hire, is honourable."[91]

Every group of intellectuals aimed to win the allegiance of citizens through the establishment of some disciplinary ingenuity. Hobbes's attack applied to all such groups: at the Restoration it was necessary to ground authority firmly in the control of civil power, not in these dangerous confederacies. Compare Hobbes's criticisms of Boyle and those he made of the lawyers. In 1666 he composed a *Dialogue between a Philosopher and a Student of the Common Laws of England*. Aubrey said he inspired this dialogue by giving Hobbes a copy of Bacon's *Maxims of the Law*, and pointing out that legal deduction was vulnerable since it was built on "old fashioned maximes (some right some wrong)."[92] The Civil War witnessed a prolonged struggle for legal reform: lawyers were the subject of radical attack just like churchmen and universities. Hobbes answered this challenge in the same way as he examined the priests: he attacked what he saw as the divisive theory of special legal skill developed earlier in the century by Edward Coke. He cited Coke's view that legal skill was "an artificial perfection of reason, gotten by long study, observation and experience, and not of every man's natural reason." In the *Dialogus physicus*, Hobbes had his interlocutor argue similarly that experimenters did not rely on "one hundred thousand of those everyday phenomena" but on "critical works of nature" produced by "artifice." In the 1666 *Dialogue*, Hobbes's philosopher denied all such claims: the "life of the law" was natural reason, not some "artificial" competence. No "infinite number of grave and learned men" made certainty: "It is not wisdom, but authority, that makes a law." The civil power gave sway over citizens in any and all of the sciences.[93]

[91] Hobbes, "De corpore politico," pp. 211-212; cf. idem, "Some Principles and Problems in Geometry," in Mandey, *Mellificium mensionis* (1682), pp. 172-173; Pope, *The Life of Seth*, pp. 125-126.

[92] For the composition of the *Dialogue*, see Hexter, "Hobbes and the Law"; Grover, "Legal Origins of Hobbes's Doctrine of Contract" (for St. Germain's *Doctor and Student*); B. Shapiro, "Law and Science" (for natural philosophical inductivism). For Aubrey and Hobbes, see Aubrey to Anthony Wood, 3/13 February 1673, in Hunter, *Aubrey and the Realm of Learning*, p. 52; for Bacon's maxims, see Kocher, "Bacon on the Science of Jurisprudence."

[93] For radical attacks on the law, see C. Hill, *Change and Continuity*, chap. 6; Veall, *The Popular Movement for Law Reform*. For Coke and Hobbes, see W. J. Jones, *Politics and the Bench*, pp. 32-52; C. Hill, *Intellectual Origins of the English Revolution*, chap. 5; Tanner, ed., *Constitutional Documents of James I*, p. 187; for law at the Restoration, see Carter, "Law, Courts and Constitution"; Havighurst, "Judiciary and Politics"; and Hobbes, "Dialogue between a Philosopher and a Student of the Common Laws," pp. 4-5, 44; idem, "Dialogus physicus," p. 241.

This argument directly challenged the way lawyers, clerics, and experimenters used the power of their special community, notably through the collaboration of witnesses. Hobbes wrote at the end of *Leviathan* that the "matters in question are not of *fact*, but of *right*, wherein there is no place for *witnesses*." Witnesses gave no authority; they were still private and fallible. This stood in contrast to the practices that experimenters and their allies used to make authority in the 1660s.[94] In chapter 2 we analyzed Boyle's use of witnesses and his technology of making virtual experiences in the "standing records" he presented. "How neer the nature of *Axioms* must all those *Propositions* be which are examin'd before so many *Witnesses*," Hooke wrote of his microscopical reports. Wilkins, More, and Stillingfleet all presented arguments that applied the same criteria of testimony to Scriptural accounts. Sprat and Boyle appealed to "the practice of our courts of justice here in England" to sustain the moral certainty of their conclusions and to support the argument that the multiplication of witnesses allowed "a concurrence of such probabilities." Boyle used the provision of Clarendon's 1661 Treason Act, in which, he said, *two* witnesses were necessary to convict. So the legal and priestly models of authority through witnessing were fundamental resources for the experimenters. Reliable witnesses were *ipso facto* the members of a trustworthy community: Papists, atheists, and sectaries found their stories challenged, the social status of a witness sustained his credibility, and the concurring voices of many witnesses put the extremists to flight. Hobbes challenged the basis of this practice: once again, he displayed the form of life that sustained witnessing as an ineffective and subversive enterprise.[95]

Hobbes's threat was a threat to the social space in which experimenters, priests, and lawyers could work. Those who replied to him defended their own disciplinary space. The Lord Chief Justice, Matthew Hale, wrote a significant answer: we have noted Hale's involvement in the air-pump trials, and his trial of pneumatic experiments. His histories of law fortified the privileges of legal specialists and an Ancient Constitution which demanded skilled interpreters. Hale told Hobbes bluntly that "the Production of long and Iterated Experience" uniquely qualified skilled professionals (and disqualified Hobbesian philosophers). Lawyers were "fitter Judges and Interpret[rs] of the Lawes of this Kingdome than any

94 Hobbes, "Leviathan," p. 712; B. Shapiro, *Probability and Certainty*, pp. 173-193.

95 Hooke, *Micrographia*, "The Preface," sig dl[r]; More, *Modest Enquiry*, pp. 483-489; Stillingfleet, *Origines sacrae*, pp. 171-176; Sprat, *History*, p. 100; Boyle, "Some Considerations about Reason and Religion," p. 182; idem, "New Experiments," p. 34.

other whose Studyes and Education have intirely or Principally applyed to the Study of Philosophy or Mathematiques." This training guaranteed the conciliation of tradition with sovereign power in dangerous disputes. Hobbes denied this automatic harmony: peaceful assent could not be assumed, but must be *made*.[96] Divines had the same complaint about Hobbesian means of gaining authority. Hobbes was seen as identifying natural law with civil power, and breaking its power by denying it could reach men's souls. This denied the competence of lawyers and of clerics. In his *True Intellectual System of the Universe*, the Cambridge Platonist Ralph Cudworth complained that truth was not made by power: it must be of God. "Truth is not factitious; it is a thing which cannot be arbitrarily made, but is." Cudworth wrote that it was wrong to claim that power could make anything "indifferently to be true or false." Priestcraft as the cure of souls, legal practice as the skilled interpretation of juridical principles, and natural philosophy as the organization of communal experiment all depended upon the construction of separate realms of power. Experimentalists exploited these principles; Hobbes undermined them. Cudworth wrote that "the Civil Sovereign is no Leviathan, but a God." Hobbes told the King in 1662 that "religion is not philosophy, but law."[97]

This was a contest about power and assent. Geometry was normative for social relations because it was consistent with the Hobbesian model of assent. No special skill—priestly, legal, or experimental—was necessary. Declaration of the right rules of action and their potent reinforcement were the necessary and sufficient conditions of control over subjects. Geometry had no sects: "Euclid taught geometry, but I never heard of a sect of philosophers, called Euclidians, or Alexandrians, ranged with any other of the sects." For Boyle, on the other hand, geometers were just one example of a specially competent group. They were not entitled to command or inspire any other community: "It will not much qualify our sense of the burning heat of a fever . . . to know, that the three angles of a triangle are equal to two right ones."[98] Hobbes treated geometry

[96] For Hale, see Pocock, *The Ancient Constitution*, pp. 162-181; Yale, "Hobbes and Hale"; for pneumatics, see Hale, *Difficiles nugae* (and our account in chapter 5); for Hale on Hobbes, see Hale, "Reflections by the Lrd. Cheife Justice Hale on Mr. Hobbes his Dialogue of the Lawe," pp. 500-502, 505.

[97] Cudworth, *True Intellectual System* (1678), pp. 718, 896-899; Hobbes, "Seven Philosophical Problems," pp. 5-6. For natural law and civic authority, see Tuck, "*Power* and *Authority*"; Hanson, *From Kingdom to Commonwealth*, chap. 5; Oakley, "Jacobean Political Theology"; Shapin, "Of Gods and Kings."

[98] Boyle, "Excellency of Theology," pp. 30-31; Hobbes, "Six Lessons," p. 346; idem, "Concerning Body," pp. 309-312.

as the form of knowledge which offered the answer to civil peace. Its virtue did not lie in any specific cognitive quality but in its relation to the activity of the social body. Not even skilled geometers were privileged. The laws of geometry compelled in the same sense as the laws of civil society. Both geometry and the commonwealth were artifactual. They were equally compelling, and equally vulnerable. In *Leviathan*, Hobbes made this point in the case of geometry:

> [M]en care not, in that subject, what be truth, as a thing that crosses no man's ambition, profit or lust. For I doubt not, but if it had been a thing contrary to any man's right of dominion, or to the interest of men that have dominion, *that the three angles of a triangle, should be equal to two angles of a square*; that doctrine should have been, if not disputed, yet by the burning of all books of geometry, suppressed, as far as he whom it concerned was able.[99]

The decisive criterion in the development of experimental practice was the allegiance to the social conventions of a specific form of life. Those whose practice followed these conventions were counted as members of the experimental group. But toleration was conditional on stringent rules of acceptable behaviour. For example, Hobbesian plenism was an integral part of his assault on priestcraft and experimental pneumatics. Yet plenism *alone* would not have differentiated Hobbes from members of the experimental community. The context of use was decisive. Hobbes did not use plenism within the experimental form of life. Plenists such as Power or Linus played a substantial role in the experimental programme. Contrast Hobbes with Henry Power. The extreme heterodoxy of Power's natural philosophy has been demonstrated by Charles Webster. Power was a committed plenist and spiritualist, influenced both by Thomas Browne and by his introduction to Cartesianism at Cambridge. He used Helmontian and Paracelsian experiences to evince active principles; he argued for the Cartesian dualist view of animals as "nothing else but engines or matter sett into a Continued & orderly motion."[100] His contacts with the experimenters in London were brief. But when he came in touch with Boyle and

[99] Hobbes, "Leviathan," p. 91; compare Glanvill, *Scepsis scientifica*, p. 98 (citing Hobbes).

[100] Webster, "Henry Power's Experimental Philosophy," p. 157; Cowles, "Henry Power"; Power to Browne, 10/20 February 1647, in Halliwell, *Collection of Letters*, p. 92; Power to Reuben Robinson, 25 September/5 October 1661, British Library Sloane MSS 1326 ff 20-21.

his colleagues, in 1661-1663, he obeyed the conventions which that community developed.

Croune told Power in September 1661 that the London experimental group "believe that to make any Hypothesis, & publickly owne it, must be after the triall of so many Expmts as cannot be made but in a long tract of time." Power learnt of the boundary which split individual beliefs from public matters of fact which the Royal Society would credit. In summer 1661 he attentively read Boyle's *Certain Physiological Essays*: he told his friends to read it. He cited a lengthy extract in his own *Experimental Philosophy*, and even quoted the essays back at Boyle in a letter to him of November 1662: "I beseech you to looke upon us as Countrey-Drudges of *much greater Industry than Reason*, fitt onely to Collect Expmts for other Heads."[101] Power's contacts with these others did affect his way of going on. Power said that Boyle's pneumatics "rubbed up all my old dormant notions." This change is apparent in his public treatment of plenism. Power did not waver in his professed endorsement of plenism. The Torricellian space was "fill'd up with the dilated particles of Ayr" and "a thin Aetherial Substance intermingled with them." But Power accepted the matter of fact of the air's permanent spring, citing Boyle's reports as proof. He used Boyle's work on fluidity in his own experiments on elastic fluids. He sent Boyle a refutation of Linus's funicular hypothesis, and Power and his collaborators were major actors in the work that led to the publication of Boyle's Law. Power saw *Hobbesian* plenism as an "exorbitant conceit."[102] Power discussed the publication of his book with Hooke and Wilkins in London, and the Society sent him instructions on numerous experiments he was to conduct in Yorkshire. He annotated his own copy of *Experimental Philosophy* with notes on the matters of fact that "Boyle has incomparably proved in the Mechanicall experiments of his Engine." In chapter 6 we noted that Power took part in the replication of anomalous suspension with the Society's pump in July 1663. Finally, Power drew

[101] Croune to Power, 14/24 September 1661, British Library Sloane MSS 1326 f 25; Power to Robinson, 25 September/5 October 1661, ibid., f 20^{v}; Power, *Experimental Philosophy*, "Preface," sig c3^{v}, citing Boyle, "Proëmial Essay," pp. 303-304; Power to Boyle, 10/20 November 1662, ibid., f 33^{v}, citing Boyle, "Proëmial Essay," p. 307. For comments on Power's change to experimentalism, see Webster, "Henry Power's Experimental Philosophy," p. 166; idem, "Discovery of Boyle's Law," p. 472; Hunter, *Science and Society*, p. 47.

[102] Power, *Experimental Philosophy*, sig b4^{r}, pp. 95, 121-123, 132 (against the "exorbitant conceit" of Hobbes), 133-142 (against Linus); Webster, "Discovery of Boyle's Law," pp. 472-479.

the line between metaphysics and physiology, which Boyle described in his own work on pneumatics. Power expressed his hope that some genuine experiments would evince the effects of the aether that he reckoned filled the "seeming vacuity" of the Torricellian space: "Perhaps some happy Experimenter hereafter may come to give us a better than this Speculative and Metaphysical Evidence of it."[103]

Power's work belonged within the space of experimental philosophy. The London experimenters sanctioned his work and credited the matters of fact he produced. Hobbes did not belong. Plenism was a resource he used against subversive politics and "ghostly men." This is a crucial distinction. Rival forms of life were at stake in a political context sensitive to the implications of subjects' assent. The gaining of assent to matters of fact was dependent on the structure of the community. Hobbes rejected this community's conventions. He asserted that its boundaries were as porous to political interests as the air-pump was to pure air. Neither Boyle's engine nor his institution made stable items to which assent could be won. The air-pump was always full, but this fact could never be revealed through experimental action: "That we come to know that to be a body, which we call air, it is by reasoning." Hobbes spelt out the difference between common error and the truths of philosophy: "It is not therefore a thing so very ridiculous for ordinary people to think all that space empty, in which we say is air; it being the work of reason to make us conceive that the air is anything." Hobbes alleged that if the experimental form of life were adopted, this difference would be lost, and the result would be political disaster. The "multitude" would "pass . . . for skilful in all parts of natural philosophy." In the 1660s it was necessary for civil order that reason should make subjects conceive correctly: the source of rebellion was "the fear of things invisible" and no porous experimental space could ever stifle that fear.[104]

[103] British Library Sloane MSS 1326 ff 36-38, 46-48; Webster, "Discovery of Boyle's Law," p. 472n; Power, *Experimental Philosophy*, p. 102; M. B. Hall, ed., *Henry Power's Experimental Philosophy*, p. 206. For another plenist, see Glanvill, *Plus ultra*, p. 61.

[104] Hobbes, "Concerning Body," pp. 523-525; "Leviathan," p. 98; and "Considerations on the Reputation of Hobbes," pp. 436-437. Hobbes was the "twin of fear": idem, "Vita, carmine expressa," p. lxxxvi: "meque metumque simul."

· VIII ·

The Polity of Science: Conclusions

Lords and Commons of England, consider what Nation it is whereof ye are, and whereof ye are the Governours.
MILTON, Areopagitica

SOLUTIONS to the problem of knowledge are solutions to the problem of social order. That is why the materials in this book are contributions to political history as well as to the history of science and philosophy. Hobbes and Boyle proposed radically different solutions to the question of what was to count as knowledge: which propositions were to be accounted meaningful and which absurd, which problems were soluble and which not, how various grades of certainty were to be distributed among intellectual items, where the boundaries of authentic knowledge were to be drawn. In so doing, Hobbes and Boyle delineated the nature of the philosophical life, the ways in which it was permissible or obligatory for philosophers to deal with each other, what they were to question and what to take for granted, how their activities were to relate to proceedings in the wider society. In the course of offering solutions to the question of what proper philosophical knowledge was and how it was to be achieved, Hobbes and Boyle specified the rules and conventions of differing philosophical forms of life. We conclude this book by developing some ideas about the relationships between knowledge and political organization.

There are three senses in which we want to say that the history of science occupies the same terrain as the history of politics. First, scientific practitioners have created, selected, and maintained a polity within which they operate and make their intellectual product; second, the intellectual product made within that polity has become an element in political activity in the state; third, there is a conditional relationship between the nature of the polity occupied by scientific intellectuals and the nature of the wider polity. We can elaborate each of these points by refining a notion we have used informally throughout this book: that of an intellectual *space*.[1]

[1] We are not aware of any specific debts for this usage. However, topographic

Our previous usages of terminology such as "experimental space" or "philosophical space" have been twofold: we have referred to space in an abstract sense, as a cultural domain. This is the sense customarily intended when one speaks of the boundaries of disciplines or the overlap between areas of culture. The cartographic metaphor is a good one: it reminds us that there are, indeed, abstract cultural boundaries that exist in social space. Sanctions can be enforced by community members if the boundaries are transgressed. But we have also, at times, used the notion of space in a physically more concrete sense. The receiver of the air-pump circumscribed such a space, and we have shown the importance attached by Boyle to defending the integrity of that space. Yet we want to elaborate some notions concerning a rather larger-scale physical space. If someone were to be asked in 1660, "Where can I find a natural philosopher at work?", to what place would he be directed? For Hobbes there was to be no special space in which one did natural philosophy. Clearly, there were spaces that were deemed grossly inappropriate. Since philosophy was a noble activity, it was not to be done in the apothecary shop, in the garden, or in the tool room. He told his adversaries that philosophers were not "apothecaries," "gardeners," or any other sort of "workmen." Neither was philosophy to be withdrawn into the Inns of Court, the physicians' colleges, the clerics' convocations, or the universities. Philosophy was not the exclusive domain of the professional man. Any such withdrawal into special professional spaces threatened the public status of philosophy. Recall Hobbes's indictment of the Royal Society as yet another restricted professional space. He asked, "Cannot anyone who wishes come?" and gave the answer, "The place where they meet is not public."[2] We have seen that the experimentalists also insisted upon the public nature of their activity, but Boyle's "public" and Hobbes's "public" were different usages. Hobbes's philosophy had to be public in the sense that it must not become the preserve of interested professionals. The special interests of professional groups had acted historically to corrupt knowledge. Geometry had escaped this appropriation only because, as a contingent historical matter, its theorems and findings had not been seen to have a bearing on such interests: "Because men care not, in that subject, what be truth, as a thing that crosses no man's

---

sensibilities in the study of culture characterize a number of modern French sociologists and historians; see, for example, Foucault, "Questions on Geography"; idem, "Médecins, juges et sorciers au 17e siècle."

[2] Hobbes, "Dialogus physicus," p. 240.

ambition, profit or lust."[3] Hobbes's philosophy also had to be public because its purpose was the establishment of public peace and because it commenced with social acts of agreement: settling the meanings and proper uses of words. Its public was not a witnessing and believing public, but an assenting and professing public: not a public of eyes and hands, but one of minds and tongues.

In Boyle's programme there was to be a special space in which experimental natural philosophy was done, in which experiments were performed and witnessed. This was the nascent *laboratory*. What kind of physical and social space was this laboratory? Consider the German experimental scene in figure 22. This picture comes from Caspar Schott's *Mechanica hydraulico-pneumatica* of 1657, and it shows experimental knowledge being constituted. This was the book that prompted Boyle's decision to begin the construction of an air-pump allegedly superior to Guericke's device shown here.[4] Guericke himself is shown in the left foreground. He holds a baton (possibly of his office as *Bürgermeister* in Magdeburg) in his right hand, and with his left he points another stick at his machine; he is not shown actually touching the pump with his hand. He is not dressed in any special way, such as might be necessitated by actual manipulations with this rather messy machine; nor is he dressed differently from the witnesses to the experiment, assembled separately from Guericke in the right foreground. The architectual space in which the scene is set is a courtyard or forum. We do not know whether it is meant that these experiments were specially brought to this public place to be tried, or whether the artist or engraver was merely using artistic conventions familiar to him to situate the objects and actions he was told to depict. (The paving-stones of public places were, of course, routinely used by Renaissance and post-Renaissance artists to lay down a perspective grid; see also the Royal Society emblematic scene in figure 2. Certainly, we do know that in the 1660s the cumbersome and fragile Royal Society pump was continually trundled about between Gresham College and Arundel House.) This picture shows the natural philosopher as presiding officer, and it shows the experimental witnesses, but it does not show any human being actually doing an

[3] Hobbes, "Leviathan," p. 91. Hobbes made no claim of the sort that geometry is essentially neutral.

[4] Unfortunately, we have not been able to locate any picture of a seventeenth-century English experimental scene in pneumatics. (But cf. Boyle, "Continuation of New Experiments," p. 206 and Plate 5, figure 1.) Other diagrams, e.g., our figure 21, attempt to show the technical construction of Guericke's machine without depicting the experimental scene in which knowledge was constituted.

FIGURE 22

Otto von Guericke's first pump demonstrated before witnesses. From Schott's Mechanica hydraulico-pneumatica *(Würzburg, 1657), p. 445. (Courtesy of Cambridge University Library.)*

experiment. The machines are worked by *putti* (cherubs). This was a standard convention of baroque illustrations. Here and elsewhere, it was implied that the resulting knowledge was divine.

What little we do know about English experimental spaces in the middle part of the seventeenth century indicates that their status as private or public was intensely debated. We briefly noted in chapter 2 that the word "laboratory" arrived in English usage in the seventeenth century, carrying with it apparently hermetical overtones: the space so designated was private, inhabited by "secretists." During the 1650s and 1660s new open laboratories were developed, alongside Boyle's rhetorical efforts to lure the alchemists into public space and his assaults on the legitimacy of private practice. The public space insisted upon by experimental philosophers was a space for collective witnessing. We have shown the importance of witnessing for the constitution of the matter of fact. Witnessing was regarded as effective if two general conditions could be satisfied: first, the witnessing experience had to be made accessible; second, witnesses had to be reliable and their testimony had to be creditable. The first condition worked to open up experimental space, while the second acted to restrict entry. What in fact resulted was, so to speak, a public space with restricted access. (Arguably, this is an adequate characterization of the scientific laboratory of the late twentieth century: many laboratories have no legal sanction against public entry, but they are, as a practical matter, open only to "authorized personnel.") Restriction of access, we have indicated, was one of the positive recommendations of this new experimental space in Restoration culture. Either by decision or by tacit processes, the space was restricted to those who gave their assent to the legitimacy of the game being played within its confines.

In chapter 5 we described differences in the engagements Boyle conducted with two sorts of adversaries: those who disputed moves within the experimental game and those who disputed the game. The latter could be permitted entry to the experimental community only at the price of putting that community's life at risk. Public stipulations about the accessibility of the experimental laboratory were tempered by the practical necessity of disciplining the experimental collective. This tension meant that Hobbes's identification of the Royal Society as a restricted place was potentially damaging, just as it is damaging in modern liberal societies to remark upon the sequestration of science. Democratic ideals and the exigencies of professional expertise form an unstable compound.[5] Hobbes's identification of restrictions on the experimental public shows why virtual witnessing was so vitally important, and why troubles in the experimental programme of physical replication were so energetically dealt with. Virtual witnessing acted to ensure that witnesses to matters of fact could effectively be mobilized in abstract space, while securing adequate policing of the physical space occupied by local experimental communities.

[5] This has often been noted by historians dealing with widely differing settings; see, for example, Daniels, "The Pure-Science Ideal and Democratic Culture"; Ezrahi, "Science and the Problem of Authority in Democracy"; Fries, "The Ideology of Science during the Nixon Years"; Gillispie, "The *Encyclopédie* and the Jacobin Philosophy of Science."

For Hobbes, the activity of the philosopher was not bounded: there was no cultural space where knowledge could be had where the philosopher should not go.[6] The methods of the natural philosopher were, in crucial respects, identical to those of the civic philosopher, just as the purpose of each was the same: the achievement and protection of public peace. Hobbes's own career was a token of the philosophical enterprise so conceived. For Boyle and his colleagues, the topography of culture looked different. Their cultural terrain was vividly marked out with boundary-stones and warning notices. Most importantly, the experimental study of nature was to be visibly withdrawn from "humane affairs." The experimentalists were not to "meddle with" affairs of "church and state." The study of nature occupied a quite different space from the study of men and their affairs: objects and subjects would not and could not be treated as part of the same philosophical enterprise. By erecting such boundaries, the experimentalists thought to create a quiet and a moral space for the natural philosopher: "civil war" within their ranks would be avoided by observing these boundaries and the conventions of discourse within them. They would not speak of that which could not be mobilized into a matter of fact by the conventionally agreed patterns of community activity—thus the importance of legislation against speech about entities that would not be made sensible: either those that indisputably *did* exist (e.g., God and immaterial spirits) or those that probably did not (e.g., the aether). As a practical matter, Hobbes could hardly deny that the experimentalists had established a community with some politically important characteristics: a community whose members endeavoured to avoid metaphysical talk and causal inquiry, and which displayed many of the attributes of internal peace. But this community was not a society of *philosophers*. In abandoning the philosophical quest, such a group was contributing to civil disorder. It was the philosopher's task to secure public peace; this he could only do by rejecting the boundaries the experimentalists proposed between the study of nature and the study of men and their affairs.

The politics that regulated transactions between the philosophical community and the state was important, for it acted to characterize and to protect the knowledge the philosopher produced.

[6] According to Hobbes, men "cannot have any idea of [God] in their mind, answerable to his nature" ("Leviathan," p. 92), and, for that reason, theology was explicitly excluded from the philosophical enterprise ("Concerning Body," p. 10).

The politics that regulated transactions within the philosophical community was equally important, for it laid down the rules by which authentic knowledge was to be produced. We remarked in chapter 4 that Hobbes assumed philosophical places to have "masters": Father Mersenne had been such a master in Paris, and Hobbes spoke of Boyle and some few of his friends "as masters of the rest" in the Royal Society. It was fitting that philosophical places should have masters who determined right philosophy, just as it was right and necessary that the commonwealth should have such a master. Indeed, Leviathan could legitimately act as a philosophical master. Hobbes found it no argument against the King's right to determine religious principles that "priests were better instructed," and he also rejected the argument "that the authority of teaching *geometry* must not depend upon kings, except they themselves were geometricians."[7] Insofar as a philosophical master was not Leviathan, he was someone else who had found out fundamental matters: the correct principles upon which a unified philosophical enterprise could proceed. He was a master by virtue of his exercise of pure mind, not by his craft-skills or ingenuity. In the body politic of the Hobbesian philosophical place, the mind was the undisputed master of the eyes and the hands.

In the body politic of the experimental community, mastery was *constitutionally restricted.* We have seen how Hooke described the experimental body in terms of the relationships that ought to subsist between intellectual faculties: "The *Understanding* is to *order* all the inferior services of the lower Faculties; but yet it is to do this only as a *lawful Master*, and not as a *Tyrant*." The experimental polity was an organic community in which each element crucially depended upon all others, a community that rejected absolute hierarchical control by a master. Hooke continued:

> So many are the *links*, upon which the true Philosophy depends, of which, if any one be *loose*, or *weak*, the whole *chain* is in danger of being dissolv'd; it is to *begin* with the Hands and Eyes, and to *proceed* on through the Memory, to be *continued* by the Reason; nor is it to stop there, but to *come about* to the Hands and Eyes again, and so, by a *continual passage round* from one Faculty to another, it is to be maintained in life and strength, as much as the body of man is.[8]

[7] Hobbes, "Philosophical Rudiments," p. 247.

[8] Hooke, *Micrographia*, "The Preface," sig b2[r].

The experimental polity was said to be composed of free men, freely acting, faithfully delivering what they witnessed and sincerely believed to be the case. It was a community whose freedom was responsibly used and which publicly displayed its capacity for self-discipline. Such freedom was safe. Even disputes within the community could be pointed to as models for innocuous and managed conflict. Moreover, such free action was said to be requisite for the production and protection of objective knowledge. Interfere with this form of life and you will interfere with the capacity of knowledge to mirror reality. Mastery, authority, and the exercise of arbitrary power all acted to distort legitimate philosophical knowledge. By contrast, Hobbes proposed that philosophers should have masters who enforced peace among them and who laid down the principles of their activity. Such mastery did not corrode philosophical authenticity. The Hobbesian form of life was not, after all, predicated upon a model of men as free-acting, witnessing, and believing individuals. Hobbesian man differed from Boylean man precisely in the latter's possession of free will and in the role of that will in constituting knowledge. Hobbesian philosophy did not seek the foundations of knowledge in witnessed and testified matters of fact: one did not ground philosophy in "dreams." We see that both games proposed for natural philosophers assumed a causal connection between the political structure of the philosophical community and the genuineness of the knowledge produced. Hobbes's philosophical truth was to be generated and sustained by absolutism. Boyle and his colleagues lacked a precise vocabulary for the polity they were attempting to erect. Almost all of the terms they used were highly contested in the early Restoration: "civil society," a "balance of powers," a "commonwealth." The experimental community was to be neither tyranny nor democracy. The "middle wayes" were to be taken.[9]

Scientific activity, the scientist's role, and the scientific community have always been dependent: they exist, are valued, and supported insofar as the state or its various agencies see point in them. What sustained the experimental space that was created in the mid-seventeenth century? The nascent laboratory of the Royal Society and other experimental spaces were producing things that were widely wanted in Restoration society. These wants did not simply preexist, waiting to be met; they were actively cultivated by the experimen-

[9] The phrase is Hooke's: ibid., sig b1^{v}, similar locutions typify much Royal Society publicity.

talists. The experimentalists' task was to show others that their problems could be solved if they came to the experimental philosopher and to the space he occupied in Restoration culture.[10] If the experimentalists could effectively cultivate and satisfy these wants, the legitimacy of experimental activity and the integrity of laboratory and scientific role would be ensured. The wants addressed by the experimental community spread across Restoration economic, political, religious, and cultural activity. Did gunners want their artillery pieces to fire more accurately? Then they should bring their practical problems to the physicists of the Royal Society. Did brewers want a more reliable ale? Then they should come to the chemists. Did physicians want a theoretical framework for the explanation and treatment of fever? Then they should inspect the wares of the mechanical philosopher. The experimental laboratory was advertised as a place where practically useful knowledge was produced.[11] But the laboratory could also supply solutions to less tangible problems. Did theologians desire facts and schemata that could be deployed to convince otherwise obdurate men of the existence and attributes of the Deity? They, too, should come to the laboratory where their wants would be satisfied. Through the eighteenth century one of the most important justifications for the natural philosopher's role was the spectacular display of God's power in nature.[12] Theologians could come to the place where the Leyden jar operated if they wanted to show cynics the reality of God's majesty; natural theologians could come to the astronomer's observatory if they wanted evidence of God's wise and regular arrangements for the order of nature; moralists could come to the natural historian if they wanted socially usable patterns of natural hierarchy, order, and the due submission of ranks. The scientific role could be institutionalized and the scientific community could be legitimized insofar as the experimental space became a place where this multiplicity of interests was addressed, acquitted, and drawn together. One of the more remarkable features of the early

[10] For this section we are deeply indebted to recent work by Bruno Latour, especially his "Give Me a Laboratory" and *Les microbes: guerre et paix.*

[11] From the best modern historical research it now appears that none of the utilitarian promissory notes could be, or were, cashed in the seventeenth century; see Westfall, "Hooke, Mechanical Technology, and Scientific Investigation"; A. R. Hall, "Gunnery, Science, and the Royal Society." If science did not deliver technological utility, it becomes even more important to ask about its other perceived values, including social, political, and religious uses.

[12] See particularly Schaffer, "Natural Philosophy"; idem, "Natural Philosophy and Public Spectacle."

experimental programme was the intensity with which its proponents worked to publicize experimental spaces as useful: to identify problems in Restoration society to which the work of the experimental philosopher could provide the solutions.

There was another desideratum the experimental community sought to mobilize and satisfy in Restoration society. The experimental philosopher could be made to provide a model of the moral citizen, and the experimental community could be constituted as a model of the ideal polity. Publicists of the early Royal Society stressed that theirs was a community in which free discourse did not breed dispute, scandal, or civil war; a community that aimed at peace and had found out the methods for effectively generating and maintaining consensus; a community without arbitrary authority that had learnt to order itself. The experimental philosophers aimed to show those who looked at their community an idealized reflection of the Restoration settlement. Here was a functioning example of how to organize and sustain a peaceable society between the extremes of tyranny and radical individualism. Did civic philosophers and political actors wish to construct such a society? Then they should come to the laboratory to see how it worked.

This book has been concerned with the identification of alternative philosophical forms of life, with the display of their conventional bases, and with the analysis of what hinged upon the choice between them. We have not taken as one of our questions, "Why did Boyle win?" Obviously, many aspects of the programme he recommended continue to characterize modern scientific activity and philosophies of scientific method. Yet, an unbroken continuum between Boyle's interventions and twentieth-century science is highly unlikely. For example, the relationship between Boyle's experimental programme and Newton's "mathematical way" is yet to be fully explored. Nevertheless, modern historians who find in Boyle the "founder" of truly modern science can point to similar sentiments among late seventeeth-century and eighteenth-century commentators. Despite these qualifications the general form of an answer to the question of Boyle's "success" begins to emerge, and it takes a satisfyingly historical form. This experimental form of life achieved local success to the extent that the Restoration settlement was secured. Indeed, it was one of the important elements in that security.

Insofar as we have displayed the political status of solutions to problems of knowledge, we have not referred to politics as something that happens solely outside of science and which can, so to speak, press in upon it. The experimental community vigorously developed and deployed such boundary-speech, and we have sought to situate this speech historically and to explain why these conventionalized ways of talking developed. What we cannot do if we want to be serious about the historical nature of our inquiry is to use such actors' speech unthinkingly as an explanatory resource. The language that transports politics outside of science is precisely what we need to understand and explain. We find ourselves standing against much current sentiment in the history of science that holds that we should have less talk of the "insides" and "outsides" of science, that we have transcended such outmoded categories. Far from it; we have not yet begun to understand the issues involved. We still need to understand how such boundary-conventions developed: how, as a matter of historical record, scientific actors allocated items with respect to *their* boundaries (not ours), and how, as a matter of record, they behaved with respect to the items thus allocated. Nor should we take any one system of boundaries as belonging self-evidently to the thing that is called "science."

We have had three things to connect: (1) the polity of the intellectual community; (2) the solution to the practical problem of making and justifying knowledge; and (3) the polity of the wider society. We have made three connections: we have attempted to show (1) that the solution to the problem of knowledge is political; it is predicated upon laying down rules and conventions of relations between men in the intellectual polity; (2) that the knowledge thus produced and authenticated becomes an element in political action in the wider polity; it is impossible that we should come to understand the nature of political action in the state without referring to the products of the intellectual polity; (3) that the contest among alternative forms of life and their characteristic forms of intellectual product depends upon the political success of the various candidates in insinuating themselves into the activities of other institutions and other interest groups. He who has the most, and the most powerful, allies wins.

We have sought to establish that what the Restoration polity and experimental science had in common was a form of life. The practices involved in the generation and justification of proper knowledge were part of the settlement and protection of a certain kind of social order. Other intellectual practices were condemned and rejected because they were judged inappropriate or dangerous to the polity that emerged in the Restoration. It is, of course, far from original to notice an intimate and an important relationship between the form of life of experimental natural science and the political forms of liberal and pluralistic societies. During the Second World War, when liberal society in the West was undergoing its most virulent challenge, that perception was formed into part of the problematic of the academic study of science. What sort of society is able to sustain legitimate and authentic science? And what contribution does scientific knowledge make to the maintenance of liberal society?[13] The answer then given was unambiguous: an open and liberal society was the natural habitat of science, taken as the quest for objective knowledge. Such knowledge, in turn, constituted one of the sureties for the continuance of open and liberal society. Interfere with the one, and you will erode the other.

Now we live in a less certain age. We are no longer so sure that traditional characterizations of how science proceeds adequately describe its reality, just as we have come increasingly to doubt whether liberal rhetoric corresponds to the real nature of the society in which we now live. Our present-day problems of defining our knowledge, our society, and the relationships between them centre on the same dichotomies between the public and the private, between authority and expertise, that structured the disputes we have examined in this book. We regard our scientific knowledge as open and accessible in principle, but the public does not understand it. Scientific journals are in our public libraries, but they are written in a language alien to the citizenry. We say that our laboratories constitute some of our most open professional spaces, yet the public does not enter them. Our society is said to be democratic, but the public cannot call to account what they cannot comprehend. A form of knowledge that is the most open in principle has become the most closed in practice. To entertain these doubts about our science is to question the constitution of our society. It is no wonder that scientific knowledge is so difficult to hold up to scrutiny.

In this book we have examined the origins of a relationship between our knowledge and our polity that has, in its fundamentals, lasted for three centuries. The past offers resources for understanding the present, but not, we think, for foretelling the future. Nevertheless, we can venture one prediction as highly probable.

[13] Merton, *The Sociology of Science*, chaps. 12-13; Needham, *The Grand Titration*; Zilsel, *Die sozialen Ursprünge der neuzeitlichen Wissenschaft*.

The form of life in which we make our scientific knowledge will stand or fall with the way we order our affairs in the state.

We have written about a period in which the nature of knowledge, the nature of the polity, and the nature of the relationships between them were matters for wide-ranging and practical debate. A new social order emerged together with the rejection of an old intellectual order. In the late twentieth century that settlement is, in turn, being called into serious question. Neither our scientific knowledge, nor the constitution of our society, nor traditional statements about the connections between our society and our knowledge are taken for granted any longer. As we come to recognize the conventional and artifactual status of our forms of knowing, we put ourselves in a position to realize that it is ourselves and not reality that is responsible for what we know. Knowledge, as much as the state, is the product of human actions. Hobbes was right.

• APPENDIX •

Hobbes's *Physical Dialogue* (1661)

TRANSLATED BY SIMON SCHAFFER

THIS is a virtually complete translation of Thomas Hobbes's response to Boyle's *New Experiments Physico-Mechanical* of 1660. To our knowledge this is the first translation from the original Latin. Two editions of the *Dialogus physicus* appeared in Hobbes's lifetime. The first was published in London in August 1661 by Andrew Crooke; the other was included as the sixth part (separately paginated) of the 1668 Amsterdam edition of Hobbes's *Opera philosophica*, published by Johan Blaeu. In Molesworth's *Latin Works* of 1839-1845, the *Dialogus physicus* appears as pp. 233-296 of volume IV. Molesworth pages are indicated in the margins; page breaks are signalled by a stroke in the text. The accompanying item, *De duplicatione cubi*, is not translated.

Differences between 1661 and 1668 editions were slight. Most differences were grammatical and were reconciled by Molesworth, whose transcription is quite accurate. A few substantive differences between earlier and later editions, especially in the dedication to Sorbière, are indicated in the translation.

The annotations to this translation point out those passages that correspond or relate to passages in *De corpore* (1655, 1656), *Problemata physica* (1662), and *Decameron physiologicum* (1678). These are especially frequent in the later sections of the *Dialogus* on heat and hydrostatics.

Certain Latin terms are indicated where the original has particular significance. *Conatus* is consistently rendered as "endeavour," following standard usage in Hobbes's own versions of his work. *Aer purus* and *antitupia* (pure air and resistance or spring) are technical terms whose significance is discussed in chapters 4 and 5. *Pondus* (weight) is always distinguished from *gravitas* (gravity), since Hobbes makes a distinction between the two in *Dialogus physicus* (although this is by no means consistent elsewhere). *Experientia* is rendered "experience" and distinguished from *experimentum* (experiment), though this distinction was not common in the period. Hobbes makes liberal use of the verb *supponere* (to suppose): this is mainly used as an axiom of demonstration that may have no veracity, contrasted to *credere* (to believe). For useful remarks on Hobbes's Latin, see H. W. Jones's introduction to his translation of Hobbes's *White's De Mundo Examined*, pp. 9-19.

A Physical Dialogue of the Nature of the Air: A Conjecture taken up from Experiments recently made in London at Gresham College[1]

235 / TO THE MOST FAMOUS AND MOST BELOVED SAMUEL SORBIÈRE[2]

Among the various spectacles of an amusing nature, most learned Sorbière, which a man well-known in breeding and ingenuity recently displayed in a concave glass sphere at the London Academy, the first things worthy of your attention are those that pertain to the nature of the air; and chiefly to your art in which you excel, of preserving human life as much as nature allows. In the following dialogue I have described this sphere, together with the whole machine and its use, as far as I could without a picture. Among its other marvels worthy of inquiry, however, I recommend that this one be considered apart from the rest: that an animal inside it is killed very quickly because of some change made in the air in which it is enclosed. And most say that the cause of death is that the air within the sphere, in which live all animals with lungs, is sucked out. But I am not of the opinion that the air can be sucked out, nor that even if it were sucked out the animal would die so quickly. Indeed, the action which that death follows may be either some suction, and because of this suction, the shutting off of the air which kills the animal by shutting off respiration, or else a compulsion of the air from each part towards the centre of the sphere in
236 which / the animal is enclosed. And so it may die by being stifled by the tenacity of the compressed air, as if drowned by water, as it were, having imbibed air that was more tenacious than usual into the interior of the lungs and there stopping the course of the blood between the pulmonary

[1] In 1661 the title was *A Physical Dialogue, or a Conjecture about the Nature of the Air taken up from Experiments recently made in London at Gresham College*. In 1668 the title was as shown.

[2] In 1661 the dedication was "To the most famous Samuel Sorbière, most excellent Doctor of Medicine." In 1668 the dedication was as shown.

artery and vein. But why should I anticipate what you are about to read? I do not want to prejudice you. Yet I thought to add this on the manner of killing in writing to you of a matter which is to be determined from the structure of the human body.[3] Besides the experiments on the nature of the air, which were many, and which you might have said were like offerings made intentionally by nature to confirm my physics, they also have others leading to other parts of physics; so that it is not to be doubted that there may be some great consequence for the advancement of the sciences from their meeting, that is, when they have either discovered the true science of motion for themselves or else they have accepted mine. For they may meet and confer in study and make as many experiments as they like, yet unless they use my principles they will advance nothing. Indeed, Aristotle judged rightly that *to be ignorant of motion is to be ignorant of nature*. If ingenuity were sufficient for the sciences, for a long time now no science would have been lacking to us. For this new Academy abounds with most excellent ingenious men. But ingenuity is one thing and method [*ars*] is another. Here method is needed. The causes of those things done by motion are to be investigated through a knowledge of motion, the knowledge of which, the noblest part of geometry, is hitherto untouched; unless I have led the way a little along the path of those who try not for victory but for truth. But as yet it seems I live in vain. For those living for ingenuity vie with each other.[4]

(*Added in 1668 edition*:) Many politicans and the clergy vie with me about the royal right. Mathematicians of a new kind, to whom it is proper to reckon *unity* indiscriminately either in lines or squares, dispute with me about geometry. Those
237 Fellows of Gresham who are most believed, / and are as mas-

[3] For Hobbes's later comments to Sorbière on the *Dialogus physicus*, see letter of March 1662, in Tönnies, *Studien*, p. 73

[4] Hobbes alluded to Lucretius, *De rerum natura*, book II, lines 7-11: "Nil dulcius est bene quam munita tenere / edita doctrina sapientum templa serena, / despicere unde queas alios passimque videre / errare atque viam palantis quaerere vitae, / certare ingenio, contendere nobilitate." The term *ingeniosi* was used by Hartlib for members of the Royal Society in 1662; see G. H. Turnbull, *Hartlib, Dury and Comenius*, p. 33n. For a similar statement on motion and geometry, see Hobbes, "Seven Philosophical Problems," pp. 3-4. For John Wallis's comments on this preamble, see *Hobbius heauton-timorumenos*, p. 5: "We find him now (with a *frustra dum vivo*) adjourning his hopes (of being Dictator) at lest till he be Dead."

ters of the rest, dispute with me about physics. They display new machines, to show their vacuum and trifling wonders, in the way that they behave who deal in exotic animals, which are not to be seen without payment. All of them are my enemies. One part of the clergy compelled me to flee from England to France; and another part of the clergy compelled me to flee back from France to England. The algebraists revile me. But you ask how the Greshamites have harmed me. You know. I thought to have found some method of interposing two mean proportional lines between any two straight lines; and having worked on it in the country I wrote down the method. I sent it to a friend in London, so that he should give our geometers access to it. The next day it happened I noticed it was wrong, and I wrote a recantation of it. It was one of them who, seeing the same fault in the meantime, which was easily done, refuted it. They reproduced this refutation in the archives of the Society, while knowing it to be condemned by the author himself. What a noble and generous deed! Thus, it is true those living for ingenuity vie with each other very fiercely, no less by guile than by strength.[5]

(*The 1661 edition concludes*): You will die, therefore, you will say, for the public good. I think so, but not to such an extent that for that reason I must desire death one minute sooner. We live as long and as well as we can; and let us love each
238 other.[6] Farewell. /

TO THE READER

Whoever you are, who searches for physics, that is, the science of natural causes, not within yourself but in the books of the masters, you are to be warned lest you understand too little or you do not rightly reckon what you understand.[7] Nature does all things by the conflict of bodies pressing each other mutually with their motions. So, in the conflict of two

[5] Hobbes referred briefly to his dispute with Wallis and the events of 1660-1662 (reported in Hobbes, "Mathematicae hodiernae" [1660], and Wallis, *Hobbius heautontimorumenos* [1662]) in *Dialogus physicus*, p. 287. In March 1662 Hobbes tried to mobilize John Pell in defence against Brouncker and Wallis; see Halliwell, *Collection of Letters*, pp. 96-97, and Scott, *Mathematical Work of Wallis*, chap. 10. The passage added in the 1668 edition beginning "Many politicians . . ." is on sig $a2^{v}$-$a3^{r}$.

[6] Hobbes alluded to Catullus, *Odes*, V, line 1: "vivamus . . . atque amemus."

[7] Hobbes, "Concerning Body," "To the Reader," p. xiii: philosophy, "the child of the world and your mind, is within yourself."

bodies, whether fluid or hard, if you understand how much motion performs in each body, that is by what path and quantity, as a not unsuitable reader you will come to physics and you will find the very probable causes of motion rightly calculated. If you are content with the worthless statements of others, you will seem to yourself to understand what cannot be understood, so that you will err the more, however rightly you reason. In physics books, many things present themselves which cannot be grasped, such as those things said of rarefaction and condensation, of immaterial substances, of essences and many other things: which if you try to explain with their words, it is useless, and if with your own, you will say nothing. Having been warned of these
239 things, read, judge and forgive. Farewell. /

A Physical Dialogue of the Nature of the Air

A. I see you as I wished.

B. And I am glad to hear you; for indeed I see nothing, since the brightness of very clear days blinds me.

A. Sit down by me, therefore, until that excessive motion of the organ of vision settles down.

B. You advise well. Truly, I am of the opinion that lassitude of this kind due to solar heat has the habit of increasing mental cloudiness a little. But I do not see enough of the way in which either light or heat produces such effects. Since the time you first demonstrated it to us, I have no longer doubted that not only all feeling but also all change is some motion in the feeling body and in the moving body, and that this motion is generated by some external mover. For previously almost everyone denied it; for whether standing, sitting, or lying down, they nevertheless understood well enough that they were feeling.[8]

A. From the same cause they could have doubted whether their own blood moved; for no one feels the motion of their blood unless it pours forth.

[8] Hobbes, "Six Lessons," pp. 339-340: "I do glory, not complain, that whereas all the Universities of Europe hold sensation to proceed from species, I hold it to be a perception of motion in the organ." For a translation of a similar passage, see Hobbes, *White's De Mundo Examined*, p. 323, and "Human Nature," pp. 4-7.

B. Indeed, everyone doubted it before Harvey. Now, however, the same people both confess that Harvey's opinion
240 is true / and they are also beginning to accept your beliefs about the motion by which vision is produced. For in our Society there are few who feel otherwise.[9]

A. What is this Society of yours?

B. About fifty men of philosophy, most conspicuous in learning and ingenuity, have decided among themselves to meet each week at Gresham College for the promotion of natural philosophy. When one of them has experiences or methods or instruments for this matter, then he contributes them. With these things new phenomena are revealed and the causes of natural things are found more easily.

A. Why do you speak of fifty men? Cannot anyone who wishes come, since, as I suppose, they meet in a public place, and give his opinion on the experiments which are seen, as well as they?

B. Not at all.

A. By what law would they prevent it? Is this Society not constituted by public privilege?

B. I do not have an opinion. But the place where they meet is not public.

A. So if it pleased the master of the place, they could make one hundred men from the fifty.

B. Perhaps, but surely the glory and thanks will be due to these first, noblest, and most useful of the institute.

A. Indeed, if what they discovered were markedly useful for the defence or the ornament of the country or the human race; otherwise both they and, because of them, their philosophy, would be condemned.

B. Certainly, such a thing is to be hoped of from these considerations, or else further natural science is to be despaired of. For the rest, the will to endeavour, even if in
241 vain, is praiseworthy. /

[9] For Hobbes's assertion that Harvey was the only man to see his doctrine established in his lifetime, see "Concerning Body," p. viii.

A. You say rightly, that this will is to be directed only to the sciences themselves, not to the glory of the ingenious. But I ask further, what method will you follow in investigating the causes of things?

B. First, experiments are produced, and then on another day, whatever the cause of the phenomenon is suspected to be, someone orally explains it, if he can. For we do not have enough trust in written natural histories, since even if they were very certain, such as could serve our institute, they would be deprived of the circumstances which are necessary to the discovery of natural causes.

A. Indeed, it is right not to believe in histories blindly. But are not those phenomena, which can be seen daily by each of you, suspect, unless all of you see them simultaneously? Those experiments you see in the meetings, which experiments indeed are well known to be few, you will believe to be sufficient; but are there not enough, do you not think, shown by the high heavens and the seas and the broad Earth?

B. There are some critical works of nature, not known to us without method and diligence; in which one part of nature, as I will say, by artifice, that is, produces its way of working more manifestly than in one hundred thousand of these everyday phenomena. Moreover, such are our experiments, in which one discovered cause can be fitted to an infinite number of common phenomena.

A. What are they? But first I wish to hear who are those learned men who make up your Academy. For in France and Italy they call societies of such a kind Academies. They say that the assembly now held in Paris at the house of M. Montmor is such a one. And when I was in Paris, we
242 held / a meeting which was not very different at the convent of the Minims, although we did not meet with a fixed number nor on fixed days, at the house of that excellent man, and notable for the promotion of good methods, P. Marin Mersenne, who published our discoveries in a book called *Cogitata physico-mathematica*. For whoever might have demonstrated a problem, would produce it for him to be examined by him and by others. I think you also do the same.

B. Not at all, but as I have said, orally. Since you ask who they are, I will name a few of those in that number you know by sight or from writings; it is not necessary for the others. There is C.

A. I knew the man. He is honest, subtle and ingenious.

B. And D.

A. So you will not lack natural histories, if you like believing in them.[10]

B. There are also the indefatigable E, F, G.

A. They are ciphers.

B. And H, I, K.

A. Algebraists are not pleasing among physicists. Now tell me about these critical experiments of yours.

B. The first is on the vacuum and the nature of the air, with a machine of such a kind that I am worried whether I will be able to describe to you clearly enough in words, for I do not have a picture.[11] It is a kind of spherical concave glass vessel of a size that can take about fifty pints of water, which they call the *Receiver.* At the bottom of this is placed a hollow straight tube, sticking out of the receiver, with a tap by which the transit of air is prevented or allowed at will. A hollow cylindrical brass vessel is connected below the receiver, fourteen inches long, the diameter of whose cavity is
243 three inches. At the top of the cylinder / is an oblique hole inserted at the side so that when necessary it can be opened and shut. They call the perforated part the *valve.* In the cavity of this cylindrical vessel is inserted in one part a tube that sticks out of the receiver and in the other part is forced a solid wooden cylinder whose surface this touches exactly to prevent the air entering and which matches the cavity so that it cannot be pushed in or pulled out without great effort. This solid cylinder is called the *Sucker,* by which in fact the air is prevented from escaping from the cylinder. Have you understood?

A. Yes. From the two concave vessels, one glass and spherical, the other brass and cylindrical, one concave vessel

[10] Wallis satirized these lines in *Hobbius heauton-timorumenos,* pp. 149-151. We have not attempted an identification. The phrase "if you like believing in them" was omitted from the 1668 edition.

[11] For the picture which B lacks, see figure 1.

is made; in which connection the transit of the air is allowed or prevented at will; and of course it is by the valve that the air from the cylindrical vessel can be let out into the open air when necessary.

B. You have it. Now in the cylinder they push in and pull back the sucker, for which strength is required, with a sort of small machine made of iron with teeth, such as we use for drawing back crossbows. Besides, there is a sufficiently wide orifice at the top of the receiver, with a lid and a tap, which can be opened or shut to admit the ambient air. Now imagine that the transit between the receiver and the brass cylinder is not impeded, and the sucker is connected to the top of the cylinder; then the transit of the air is shut off by turning the tap and the sucker pulled back a little. What do you think then follows from this? Would not the space left by the sucker be a vacuum? For whence is it refilled if not from the receiver, since the sucker, exactly filling the concave cylinder, prevents the transit of the ambient air?

A. I think it cannot be known whence it can be refilled
244 nor what might follow, unless the nature of the air is / known first. And so I fear lest they conclude from some supposed properties of the air that the space left by the retraction of the sucker is a vacuum, and thence conversely that, given that this space is a vacuum, they might wish to prove that the nature of the air is such as they may suppose; that is, lest they demonstrate without a principle of demonstration.

B. But what do you imagine to be the nature of that air which, when supposed, could fill that space?

A. I? I suppose the air is fluid, that is, easily divisible into parts that are always still fluid and still air, such that all divisible quantities are there in any quantity. Nor do I suppose as much, but I also believe that we only understand an air purified from all effluvia of earth and water, such as may be considered an aether. Nor is there anyone who has yet advanced a reason why this should not be so. On the contrary, in fact, if a part of the air, whose quantity is less than any water-drop you have seen, is fluid, how is it to be proved to you by anyone that a part half the size of its parts, or, if you wish, one hundred thousand thousandth, might not be of the same nature, still fluid and still air, I will say pure air [*aer purus*] ?

B. But most of us distinguish the nature of fluids from nonfluids by the size of the parts of which any body consists, and, as it were, is composed. So we do not only look on air, water and all liquids as fluids, but also ashes and dust. And we do not deny that fluids can be made of nonfluids. For we do not stomach that infinite divisibility.

A. Infinite division cannot be conceived, but infinite di-
245 visibility can easily be. On the contrary, I / do not accept the distinction between fluids and nonfluids, which you take from the size of the parts; for if I accepted this, the ruins or rubbish lying in Paul's Church might be called fluids by me. But if you were to deny them to be fluid because of the large size of the stones, then define for me the size that the parts of the ruined wall must have to be called fluid. Truly, you who cannot accept infinite divisibility, tell me what appears to you to be the reason why I should think it more difficult for almighty God to create a fluid body less than any given atom whose parts might actually flow, than to create the ocean. Therefore, you make me despair of fruit from your meeting by saying that they think that air, water, and other fluids consist of nonfluids: as if they were to call fluid a wall whose ruined stones fell around the place. If such is to be said, then there is nothing that is not fluid. For even marble can be divided into parts smaller than any Epicurean atom.[12]

B. If I concede this to you, then what follows?

A. It follows from this that it is not necessary for the place that is left by the pulling back of the sucker to be empty. For when the sucker is drawn back, by however much larger is the space left, by so much less is the space left to the external air, which being pushed back by the motion of the sucker towards the outside, similarly moves the air next to it, and this the next, and so continuously: so that of necessity the air is forced into the place left by the sucker and enters between the convex surface of the sucker and the concave

[12] For the doctrine of fluidity and divisibility, see Boyle, "New Experiments," p. 15, and Hobbes, "Concerning Body," pp. 100 (on division) and 426 (on the definition of fluidity).

surface of the cylinder. For supposing the parts of the air are infinitely subtle, it is impossible but that they insinuate themselves by this path left by the sucker. For firstly, the contact of these surfaces cannot be perfect at all points, since
246 the surfaces themselves cannot be made infinitely / smooth. Then, that force which is applied to draw back the sucker distends the cavity of the cylinder a little bit. Finally, if any hard atoms get in between the edges of the two surfaces, pure air gets in that way, with however weak an endeavour. I could also have counted that air, which from the same cause might have insinuated itself through the valve of the cylinder. Thus, you see that the consequence of the retraction of the sucker for the existence of empty space is removed. It also follows that the air that is pushed into the place left by the sucker, since it is pushed with a great force, is moved with a very swift circular motion between the top and the bottom of the cylinder, since there is nothing yet that could weaken its motion. For you know that there can be nothing that can impart motion to itself or diminish it.[13]

B. It leaves that place full of pure air, as you say, that is, as I understand it, with an aetherial body. What do you think would happen now, if, when the tap is turned, air were drawn to cross from the receiver into the cylinder beneath?

A. I think that the disordered air would travel round everywhere in both vessels with the same motion, with some speed, but somewhat weaker than before by however much of the motion is communicated to the air.

B. But when the tap was turned, we observed that a sound was made as if air were breaking into the cylinder.[14]

A. This is not astonishing, because of the collision of the air in the cylinder with the air in the receiver. But how do you explain this?

B. In two ways. First, and by preference, thus. We suppose that there is an elastic force in the air in which we live,
247 that is, air consists of, or at least abounds / with, parts endowed with this nature, so that atmospheres compressed by a weight impinging on them endeavour as much as they can

[13] For definitions of these terms, see Hobbes, "Concerning Body," pp. 206-211, and the comments on "*conatus*" in Hobbes, *White's De Mundo Examined*, pp. 148-150.

[14] Boyle, "New Experiments," p. 11.

in moving away to free themselves from compression, and it appears as if those corpuscles are removed or else give way for whatever cause. However, what we are saying is understood better if you imagine the air here near the Earth to be like a heap of corpuscles, which, lying on top of each other, resemble wool, whose thin and flexible hairs can be easily bent or twisted like so many strings, so that they perpetually endeavour to extend and restore themselves. Just as if someone were to compress wool in the hand, each of its threads is endowed with a power or principle of dilatation, by whose strength, when the hand is relaxed, the wool distends and restores itself with a spontaneous motion. Thus the explanation of most of the phenomena of the vacuum and the nature of the air is not difficult with this elastic force of the parts of the air.[15] The other way is—

A. The other way is differentiated very little from this one. Meanwhile, I ask you, is this not the rule for all hypotheses, that all things that are supposed must all be of a possible, that is, conceivable, nature?[16]

B. Absolutely. And the force that is supposed here, by which things when pressed restore themselves, since it is easily seen in many things, can very easily be conceived to be in air.

A. This is indeed true. For we see steel plates tensed in a crossbow return with a very speedy motion to their accustomed straightness by the force or principle of restitution when the impediment is removed. But I cannot believe it to have been a philosopher who first exhibited the experiment of crossbows or longbows or whatever elastic machines. It is for a philosopher to find the true or at least very probable
248 causes of such things.[17] How / could compressed wool or steel plates or atoms of air give your experimental philos-

[15] Ibid., pp. 11-12.

[16] Hobbes, "Concerning Body," p. 425: "Every supposition, except such as be absurd, must of necessity consist of some supposed possible motion." Compare Hobbes to Newcastle, 29 July/8 August 1636, cited and discussed in Gargani, *Hobbes e la scienza*, pp. 209-218.

[17] Hobbes, "Concerning Body," pp. 478-479; idem, "Seven Philosophical Problems," pp. 33-34, cf. p. 37: "If nature have betrayed herself in any thing, I think it is in this, and in that other experience of the crossbow; which strongly and evidently demonstrates the internal reciprocation of the motion, which you suppose to be in the internal parts of every hard body."

ophers the cause of restitution? Or do you offer a likely cause why in a crossbow the steel plate regains its usual straightness so swiftly?

B. I cannot give a very certain cause for this thing. I certainly know that the removal of the impediment may not be involved in the cause, since all cause of motion consists of some action in the moving body. Again, I do not believe that by the removal of the impediment the plate rebounds when impelled by the ambient air, nor by any weight of the atmosphere, since the contiguous air may not be compressed by the action of the crossbow, and if it could, yet the heavy plate would be open to the same thing. Besides, it is impossible that the plate be moved by itself, that is, that it be its own source of motion, and indeed this is not conceded by our colleagues. So what remains, unless the endeavour to straightness be itself true local motion, but in imperceptible spaces, though very swift, such that it produces a very swift motion?

A. You speak rightly, and you have perfectly demonstrated the theorem by a wonderfully easy method, as befits a philosopher. Now I ask what might be the motion of the parts of a body that endeavours to its own restitution?

B. That motion cannot be straight, since, if it were straight, the whole body (so to speak) would be carried away by the motion of the crossbow itself, in the way that a missile is usually carried off. Therefore it is necessary that the endeavour be circular, such that every point in a body restoring itself may perform a circle.

A. Truly that is not necessary, but it is necessary that it be such a motion that whatever is moved returns to the place
249 whence it began to be moved. But, / then, what is the reason
why a woollen hair extends itself after compression?

B. Whatever true cause I told you, you would not then acquiesce to its truth, but would ask me further what was then the cause of this cause, whence it would go on to infinity.

A. That is by no means true. For when you will have come to some external cause, there I will leave off asking you. So say what cause can bring about that motion of the particles which make up the nature of steel or wool or air?

B. I answer you, that particles even smaller than those particles of air which I compared to woollen hairs, effect that motion of restitution, returning into themselves, with their own natural motion of which there is no beginning.

A. Thus, the parts of each aerial corpuscle were moved apart by returning into themselves with that motion, before which that corpuscle would have been made up of those smaller ones.

B. It cannot be made otherwise.

A. Do not your Fellows also think so?

B. Perhaps one or another, but not the rest.

A. I believe you. For this motion of restitution comes from Hobbes, and is first and solely explained by him in the book *De corpore*, chap. 21, art. 1.[18] Without which hypothesis, however much work, method or cost be expended on finding the invisible causes of natural things, it would be in vain. You now see that this spring of the air that they suppose is either impossible or they must have recourse to the Hobbesian hypothesis, which because perhaps they have not understood, they have rejected.

B. I do not know what is to be answered to this. But if with this hypothesis of yours you set out as clearly the other
250 phenomena of the machine, which they / have done by supposing the gravity of the atmosphere, I will judge yours to be true. But they also have another hypothesis, which they think can save the same phenomena, the Cartesian. The Cartesian view is that air is nothing else but a mass of flexible corpuscles endowed with various sizes and shapes, elevated from the Earth and water by heat, especially that of the Sun, and swimming in the aetherial matter that flows in every direction round the Earth: and that those corpuscles are thus moved and turned in curves by the motion of that aetherial matter, so that, extended and moving circularly, they repel all the rest from themselves: while the same turning motion lets slip cooling bodies and restores languid ones.

A. Indeed, I remember that Descartes here spoke of the nature of water, whose parts he compared with eels. But if

[18] Hobbes, "Concerning Body", pp. 317-319; idem, "De corpore," pp. 258-260. Compare idem, "Decameron physiologicum," pp. 108, 135.

I remember well he says that the nature of the air is like tree branches.[19] But it is of little importance who was the author of that supposition. For that hypothesis itself, in which is supposed a motion of subtle matter, very swift yet without a cause, and moreover with various innumerable circulations of corpuscles produced by the sole motion of that matter, is scarcely that of a sane man.[20] But let us return to the former hypothesis, in which air is assigned gravity. First, it would have been proper to explain what *gravity* might be. Everyone knows that gravity is an endeavour from all places to the centre of the Earth. Furthermore, the endeavour is a motion, even if imperceptible. With what machines are you investigating the efficient cause of this endeavour or imperceptible motion? For that was to be found out first; next, how the phenomena of your machine might be saved by the gravity of the atmosphere.[21] I am easily persuaded that in the atmosphere are many particles both of earth and water, min-
251 gled with the body of the aether. / But it is inconceivable that, moving up, down, and every way in the middle of the aether, and resting each upon the other, they should then gravitate. Since every body is heavy, wood and other bodies lighter than water nevertheless add something to the total weight. And in aetherial substance, whatever is not heavy cannot gravitate unless while sinking. So while they are not sinking, if gravity be a downwards endeavour, how can they be said to gravitate, or to compress air even if it be woollen?

B. These things need greater meditation than that I should agree immediately. Yet may we go on to our experiments, so that we may see whether their causes may be rendered by your suppositions? And first—

A. First, I must propound to you the suppositions themselves, and explain them so as you understand. You know

[19] Hobbes referred to Descartes, "Météores," in *Oeuvres*, vol. VI, pp. 233-235; compare Descartes, "Principia philosophiae," ibid., vol. VIII-1, pp. 222-224 (part 4, arts. 36-38).

[20] Compare Descartes, *Oeuvres*, vol. III, pp. 287-292, 300-326, 341-348, and Hervey, "Hobbes and Descartes"; Brandt, *Hobbes' Mechanical Conception*, pp. 160ff.; Gargani, *Hobbes e la scienza*, pp. 233-237; Hobbes, *Latin Works*, vol. V, pp. 277-307.

[21] Hobbes, "Seven Philosophical Problems," pp. 11-12: "In natural causes all you are to expect, is but probability; which is better yet, than making gravity the cause, when the cause of gravity is that which you desire to know." For a comparable treatment of "*gravitas*" and "*pondus*," see Hobbes, *White's De Mundo Examined*, p. 74; idem, "Rosetum geometricum," p. 56.

the hypothesis of the annual motion of the Earth round the Sun, such that its axis is always carried in parallels to it, was first introduced by Copernicus. Indeed, what he said of the axis is also true of all other straight lines considered in the body of the Earth.[22]

B. I know this, and that the hypothesis is taken as true today by almost all the learned.

A. Hobbes calls this motion *simple circular motion*, with which all points on the Earth, when the whole makes its circle, also describe their circle (as is demonstrated by him in the book *De corpore*, chap. 2, art. 1). In the same chapter, art. 10, he shows a motion that is also *simple circular* is generated from simple circular motion. So since the cause of the annual motion is thought by these learned to be the Sun, he also ascribes such a motion to the Sun. And, indeed, he
252 uses these hypotheses to save not these but other / phenomena.[23] But in speaking of the vacuum and the nature of the air, he assumes another hypothesis, this one, that the Earth has its own motion, due to its own nature or creation, which is also *simple circular*. And from this supposition he demonstrates clearly many things about natural causes; and, indeed, what that may be you will understand thus. Take up a basin in your hand, in whose bottom is a little water, however little, yet visible. Could you not move the water by moving the basin, so that it ran in a circle, raising itself a little around the concave surface of the basin?

B. I can, and very easily. For I shall agitate the basin, grasping it on both sides with my hands, but so that it makes very small circles lest the water spill out. When I do this, the water that was in the bottom will without doubt rise and flow round the concave surface of the bowl.

A. But where will be the centre of that circular motion, since you say you move the basin in a circle on both sides by hand?

B. The centre? You say this unexpectedly. But I answer that there are many centres, not one; as many points as can

[22] Hobbes, "Concerning Body," pp. 426-432.

[23] Ibid., pp. 317-319, 329-330; idem, "De corpore," pp. 258-260, 268-269. The text here should read "chap. 21." See also Mintz, "Galileo, Hobbes, and the Circle of Perfection."

be considered to be in the body of the basin, I believe, and (which follows from this) just as many circles, themselves equal among themselves.

A. You have, therefore, described that motion that Hobbes calls *simple circular*; except he understands by *circular* any motion whatever which returns unto itself.[24]

B. I also understand it thus. Nothing is conceived more easily. Therefore, as you told me, such is supposed to be the
253 *simple circular* motion of the Earth, congenital to its nature. /

A. Would you not believe the space left retained the same motion if this were annihilated by divine omnipotence or if half this Earth were removed to some other distant place beyond the fixed stars?

B. I believe so, and (since I see where you aim) I say moreover that if one of its atoms were left here, then even that would be moved with the same *simple circular* motion.

A. Therefore those particles of earth and water, which interspersed in our air make your atmosphere, have that same congenital *simple circular* motion.

B. It necessarily follows.

A. But if the Sun, whether principally or on its own, raises those particles from the Earth, as your colleagues believe, and I with them, it does not seem incredible to me that, by however much the air is closer to the Earth, by so much is it fuller of those earthy parts.

B. It cannot be doubted.

A. So you have understood my hypotheses: first, that many earthy particles are interspersed in the air, to whose nature simple circular motion is congenital; second, the quantity of these particles is greater in the air near the Earth than in the air further from the Earth.

B. The hypotheses are by no means absurd. It remains that you show their use in saving the phenomena of which

[24] Hobbes, "Seven Philosophical Problems," p. 8: "It is the same motion which country people use to purge their corn." Wallis commented (*Hobbius heauton-timorumenos*, p. 154) that "he hath one great Engine, which he calls his Simple Circular Motion," and compared it with "that of the Good-Womans Hand that turns the Wheel when she Spins," labelling this motion the *Vertigo Hobbiana* (ibid., p. 157).

I am now about to speak. First, since when the receiver has been nearly exhausted, as we say, or, as you prefer, the suction has been frequently repeated, I have seen the handle fall back from the hands of whoever happens to pull back the sucker, and to be carried back towards the top of the
254 cylinder: / explain, if you can, therefore, why this should be necessary.[25]

A. Since pure air was thrust in by the retraction of the sucker, but the earthy parts were not thrust in, after the retraction there was a greater ratio of earthy particles that were near the sucker outside the cylinder to the pure air in which they exercised their motion than before. Having less space to exercise their natural motion, those particles so moved, therefore struck and pushed each other. So it was necessary that the particles that were next to the surface of the sucker should have pushed on the sucker. Which is the phenomenon itself. However, this is to be noted, that by the rising of the sucker the air that was inside the cylinder would be expelled by the same path by which it entered.

B. I see this could easily come about; nor do I see anything wonderful in this except the hypothesis itself. Yet I admit this is considerably less wonderful than is our hypothesis of the elastic force of the air.

A. It is not foreign to reason that the causes of the wonderful works of nature should also be wonderful; nor do I judge it to be for a man of philosophy to suppose the sizes of bodies such as the Sun and the stars to be wonders, yet indeed not to allow wonder at little things, since it is of the same infinite worth to create both, as much the greatest as the least, and it would be impossible to render the causes of wonderful effects without wonderful hypotheses. They make a legitimate hypothesis from two things: of which the first is, that it be conceivable, that is, not absurd; the other, that, by conceding it, the necessity of the phenomenon may be inferred.[26] Your

[25] Boyle, "New Experiments," pp. 17, 71-73.

[26] Hobbes, "Decameron physiologicum," p. 133: "[A]s he made some bodies wondrous great, so he made others wondrous little. For all his works are wondrous." See Hobbes to Newcastle, 29 July/8 August 1636, in Gargani, *Hobbes e la scienza*, p. 212: "The most that can be atteyned unto is to have such opinions, as no certayne experience can confute, and from which can be deduced by lawfull argumentation, no absurdity."

255 hypothesis lacks the first of these; unless / perhaps we concede what is not to be conceded, that something can be moved by itself. For you suppose that the air particle, which certainly stays still when pressed, is moved to its own restitution, assigning no cause for such a motion, except that particle itself.

B. You know that experiment of the Torricellian vacuum. They invert and immerse a hollow glass cylinder exactly closed in one place, and open at another, filled with mercury, in an open vessel in which is contained as much mercury as is necessary to cover the mouth of the cylinder. So will not the space that is left by the mercury in the cylinder remain a vacuum?

A. It is not necessary. If a bladder full of air is pushed down to the bottom of the sea, and, being broken, were to give out air, do you not think that the air, now free, remaining at the bottom of the sea, would ascend very vigorously to the surface of the water?

B. Certainly, it will ascend, boiling manifestly.

A. But why? You do not want to answer me thus, that it surely happens because air is less heavy than water: yet show me by what motor it is carried in penetrating the body of the water that is less mobile than is that of air itself.

B. Water endeavours downward much more than air. So it is necessary, as it seems to me at least, that water presses the air by the endeavour it has towards the centre of the Earth, greater than air has, and the air, being pressed, presses the bottom, and the bottom, being pressed, pushes back on the air, with such an endeavour that, making its way through the water, it necessarily emerges.

A. What would happen if the water, being underneath
256 the air above it in a closed vessel, were supposed / to rise with the same endeavour with which it naturally tends downwards?

B. The air, penetrating the water again, would fall to a lower place in the cylinder.

A. Why should not the same happen if we were to put mercury instead of water in the cylinder?

B. The same would happen.

A. Now consider that in the Torricellian experiment, the mercury descends into the vessel underneath, which also contains mercury; but the mercury that is in the vessel ascends with the same endeavour with which it descends in the cylinder, and in ascending it presses the air lying above, which air (the whole world being supposed to be full) could not escape the pressure of the rising mercury more than if the surrounding bodies were enclosed in one and the same cylinder. From which it is necessary that the air penetrates the body of the mercury itself, or else crosses between the convex surface of the mercury and the concave surface of the cylinder. So you see the reason of this phenomenon can be rendered without the supposition of a vacuum, either by the elasticity or the simple circular motion of the atoms.

B. But if the air were really heavy, or if there were such a motion of the earthy particles as you suppose, would they not contribute anything to the ascent or the descent of the mercury in the cylinder?

A. Indeed, they would contribute, not surprisingly, so that the mercury would descend a little less than if the external air were pure and weightless.

B. We know that at the bottom of very high mountains the mercury that is in the cylinder falls more [sic] than it
257 does at the top of the mountain. /

A. But those particles interspersed in the air are moved as we suppose such that they are more crowded at the bottom of the mountain than at the top. For we suppose this, too.

B. We poured water into an open vessel; we placed a long straight narrow tube in the water. And we observed that the water did ascend from the vessel beneath into the upright tube.

A. No wonder. For the particles interspersed in the air near the water struck the water with their motion, so that the water could not but ascend, and indeed to do so sensibly into a pipe so very slender.[27]

[27] For the Torricellian phenomenon and the siphon, see Hobbes, "Concerning Body," pp. 420-425; idem, "Seven Philosophical Problems," pp. 23-24; idem, "Decameron physiologicum," pp. 92-93.

B. I return to the phenomena of our machine. If, after the frequently repeated pushing in and pulling out of the sucker, someone endeavours to remove the cover of the upper orifice from the receiver, he will find it gravitates very much, as if a weight of many pounds hung from it.[28] Whence does this happen?

A. From the very strong circular endeavour of the air that is in the receiver, made by the violent entry of the air in between the convex surface of the sucker and the concave surface of the cylinder, generated by that repeated pushing in and pulling out of the sucker, which you incorrectly call the exsuction of the air. For because of the plenitude of nature, the cover cannot be removed, since the air, which is next to the cover in the receiver, must be removed as well. But if this air were at rest, the cover would be removed very easily. Yet since it circulates very swiftly, it follows with some difficulty, that is, it seems to be very heavy.

B. It is very likely. For when new air is gradually admitted into the receiver, it also gradually loses that apparent gravity.
258 After some strokes of the receiver, we also saw / that water, when put down into the receiver, bubbled as if boiling over a fire.[29]

A. This also happens because of the speed of the air, as was said, circulating in the receiver; unless perhaps you find that the water, while it boils, is also hot. For if we were certain that it were growing hotter, it would be fit to think of another cause of the phenomenon.

B. On the contrary, we are certain that it does not sensibly get any hotter.

A. So what do you think the greater or lesser gravity of the atmosphere could contribute to such a motion of the water?

B. Indeed, I do not suppose they attribute that motion to the atmosphere.

A. From this experiment, it is manifest that by this exsuction of the air, as you call it, from the receiver, a vacuum is not produced. For the water could not be moved unless it

[28] Boyle, "New Experiments," p. 15.

[29] Ibid., p. 115.

were moved by something moving and contiguous. So the demonstration of this phenomenon by my supposition seems not to contain anything unsound. Besides, tell me, could you get to see the water boiling?

B. Why not?

A. Do not your colleagues grant that vision is produced by a continuous action from the object to the eye? Do they not also consider all action to be motion, and all motion to be of a body? So how could the motion be derived from the object, the water, indeed, to your eyes, through a vacuum, that is, through a nonbody?

B. Our colleagues do not affirm the receiver to be so empty that no air at all is left there.

A. It does not matter at all whether the whole receiver be empty, or its larger part. For whichever were supposed, the derivation of the motion from the object to the eye would
259 be interrupted.[30] /

B. It seems to be so; I have nothing with which I might answer. So I go on with the experiments. With the same operation of moving the sucker backwards and forwards, furthermore, animals will die in two or three minutes if they are enclosed in the receiver, just as if the air were sucked out; which, granted a vacuum, is not surprising; if denied, then I do not know how it could happen.[31]

A. Do you believe that those animals were killed so quickly due to the fact that they lacked air? Then how do divers live underwater, when some, being used to it from childhood, go without air for a whole hour?[32] The very vi-

[30] Hobbes, "Concerning Body," pp. 523-525. Continuous transmission of light through air was used by Hobbes as a fundamental argument from reason for the presence of air: "That we come to know that to be a body, which we call air, it is by reasoning; but it is from one reason alone, namely, because it is impossible for remote bodies to work upon our organs of sense but by the help of bodies intermediate without which we could have no sense of them till they come to be contiguous. . . . It is not therefore a thing so very ridiculous for ordinary people to think all that space empty, in which we say is air; it being the work of reason to make us conceive that the air is anything." And recall Franciscus Linus's argument in chapter 5.

[31] Boyle, "New Experiments," pp. 97-99.

[32] Hobbes, "Concerning Body," p. 515.

olent motion by which the enclosed bladders are distended and broken kills those animals contained in the receiver.

B. I leave aside the machine again, and I wish to know the causes of all these experiments. If someone endeavours to remove the air contained in a vial empty of all bodies save air, gripping the open mouth tightly on both sides with his lips, at first he feels that his lips are removed from there with difficulty, and next, if he dips the mouth of the inverted vial into however much water, he will see the water ascend higher in the vial than is the surface of the water beneath. I ask why the water ascends against its nature, apart from the fact that in rising it fills the empty space that was made inside the vial?

A. The avoidance of a vacuum [*fuga vacui*] cannot be the cause of the thing. If he had sucked out the air, then either more air would have entered when the vial was transferred from the lips to the water; or after it was transferred the water would not have entered. For water rises more easily than air [*sic*]. So what brought it about that the water should rise? The endeavour of the air to leave the vial. Which you will understand thus. Whoever sucks the vial, attracts nothing into the lungs, unlike those who are breathing, nor swal-
260 lows anything in the belly, unlike the infant / who sucks the mother's breast. So which place is it which might receive the air sucked out? None. So it is not sucked out. So does the suction do nothing, you will ask. Indeed, it does a lot. For from that it comes to pass that the lips, sucking the neck of the vial, stick to it strongly, so that they cannot easily be detached by beginning the disjunction from the outside edge of contact. Second, it produces the ascent, so that the air that is inside the vial endeavours to leave by that part where the suction was begun, that is, through the mouth of the vial. So when the mouth of the vial is immersed in however much water, if the endeavour of the air that it has from the suction were greater than the force with which water gravitates, then it is necessary that the air should leave by penetrating the water, and the water rise in its place, until the strength of the suction decreasing, the endeavour of the air to leave, and that of water to subside become equal.[33]

[33] Boyle, "New Experiments," pp. 73-75, on the cause of suction; Hobbes, "Seven Philosophical Problems," p. 24; idem, "Decameron physiologicum," pp. 89-90.

B. Nothing is more probable. Now tell me the cause of that wonderful force by which lead balls or arrows are let loose from the barrels of those things they call wind-guns [*sclopetus ventaneus*], whose construction almost all those who are used to associate with philosophers know.

A. The wind-gun, like your machine, has two chambers and a sucker. Your machine has an opening with a tap connected to the chamber; but this gun has connected to the chamber an opening with a valve, which air struck by the sucker easily opens and the air enters into the other chamber, and because of the great force with which it was struck it circulates, tending towards the exit, until given exit by the little flaps with its utmost force, it breaks out through the bottom of the other chamber, with such a force as many strong strokes of the sucker confer. So it is not surprising if it throws out a ball placed in the exit through quite long
261 distances.[34] /

B. But how can these strokes be sufficiently strong in striking the air, since the sucker must be such that it equally fills the first chamber?

A. In the sucker itself, as you know, there is an opening with a valve, which valve is easily opened by the external air when the sucker is pulled back. And when it is pushed in, that valve which is connected to both chambers is opened.

B. I remember that to be so. Nor do I doubt that you have rendered the true and unique cause of that effect, which is also the same as that by which the motion of the air excited those wonders in our machine. I do not know, however, what cause our colleagues render, or are about to render, for that effect. I again return to the experiments of our machine. They hang an inflated bladder from one of the arms of the scales of a balance; from the other, enough lead to produce equilibrium; and they put it into the receiver so that it hangs from the cover. Then when the air is sucked out, we see the bladder weighing more. So air is being weighed in empty space, and consequently they conclude that its gravity is something because of the bladder weighing

[34] Compare the account of the wind-gun in Hobbes, *White's De Mundo Examined*, p. 49, with that in idem, "Concerning Body," pp. 519-520.

more. Indeed, they understand to a certain extent how much it is.[35]

A. They can be certain that the scale in which the bladder is, is more depressed than the other, their eyes bearing witness. But they cannot be certain that this happens because of the natural gravity of the air, especially if they do not know what is the efficient cause of gravity. But what cause of gravity do they assign?

B. As yet, none, but they seek for it with this experiment itself. Yet since the bladder nevertheless hangs down, even if it does not weigh more, show by what cause it hangs down.

A. I do not wish to deny that the bladder, whether it be inflated by bellows or by blowing from the mouth, may be
262 heavier than when the same bladder is not / inflated, because of the greater quantity of atoms from the bellows or sooty corpuscles from the breath being blown in. However, they gather nothing sufficiently certain from the experiment made with the inflated bladder. They should have put two vessels of equal weight on the scales, of which one should have been exactly closed and the other open. For in this way, the air would have been weighed, not blown in, but only enclosed. So when you see air weighed like this, we will then think about what might be said of the phenomenon you bring back.[36] That which concerns the cause of gravity does not, indeed, seem at all likely to me, since that cause, which could bring together homogeneous bodies and separate heterogeneous ones from the beginning, could not bring together the same homogeneous substance when separated by force and tear apart heterogeneous substance brought together by force. Yet the motion that can do this cannot be anything but that simple circular motion that Hobbes defined in the book *De corpore*, chap. 15, and calls somewhere *fermentation*, and demonstrates this about that property: that homogeneous substances come together and heterogeneous substances separate. Moreover, he supposes its beginning to be in the Sun.[37]

[35] Boyle, "New Experiments," p. 13; Hobbes, "Concerning Body," p. 519.
[36] For Boyle on balancing in fluids, see "New Experiments," p. 78.
[37] Hobbes, "Concerning Body," pp. 203-217.

B. Your hypothesis pleases me more than that of the elastic force of the air. For I see that the truth of the vacuum or of the plenum depends upon the former's truth, whereas from the truth of the latter, nothing follows for either part of the question. It is said that the structure of the air is similar to compressed wool. It is well. Wool is made from hairs. Right. But of what shape? If parallelepipeds, then there could be no compression of the parts: if not parallelepipeds, then there would be some spaces left in between those hairs: which if empty, then they assume a vacuum in order to prove
263 that a vacuum is possible, and if full, / then what they think is a vacuum they call a plenum. Now I go on to other experiments, and first I will refer to those things that happen to a flame enclosed in the receiver. We saw a burning candle placed in the receiver, and hanging in the middle of it, extinguished in the space of half a minute after its mouth was closed and the sucker was begun to move backwards and forwards.[38]

A. When a burning candle has been hung down in a pit from which they dig coal, even though the hole was neither closed nor dark, but such that the sky might be seen in a small amount of water at its bottom as if in a mirror, I may say that even without the working of any sucker, I have seen the candle extinguished in the space of half a minute before it reached half the depth of the pit.[39]

B. We have seen wood coals burning steadily, put into the receiver, as we said, immediately grow faint from the beginning of the suction, and after the space of three minutes the flame could not be seen any more.

A. I have seen steadily burning earth coals, put down in the same pit, as I said just now, first growing faint; then in the space of three or four minutes the flame could not be seen any more; and yet in the same amount of time, when taken out of the pit, catch fire again.

B. The same thing also happens to our coals when air is admitted. It would be wonderful if our machine did not act like that pit.

[38] Boyle, "New Experiments," pp. 26-30.

[39] Hobbes, "Concerning Body," p. 524.

A. Doubtless it does act like it, except that those pits do not display the experiment every time. For in both cases the extinction of the fire has the same cause, which you will understand thus: what path do you judge is taken from there by the air which the force of the sucker pulled back in the
264 brass cylinder compels to enter between / the convex surface of the sucker and the concave surface of the cylinder?

B. First, it has a path following those straight lines that make up the concave surface of the cylinder; next, along the lines that make up the surface of the bottom of the cylinder itself. So the parts of the air entering along diametrically opposite straight lines will be moved in every direction by contrary motions. Therefore, pressing upon each other, they will necessarily endeavour along inward lines; and, because of the pressure being equal on every side, they will have some sensible motion towards neither side, each part, moreover, running against each other with a forceful endeavour.

A. Therefore, it will be necessary that while these endeavours last, all that air has a greater consistency than if its parts were held together by contact alone.

B. But neither the candle nor the coals were placed in the cylinder, but in the receiver.

A. I know. But when the tap is opened, the lines by which the motions of the entering air are defined mutually intersect at their common entrance, and, consequently, while in the reverse direction, the endeavour of that air goes on in the same way in the receiver and also in the cylinder; and the consistency of the air will be the same on both sides, some mean between the consistency of pure air and that of water. Consider, therefore, that there might be in their nature a force equally consistent with the extinction of candles or flames or the life of animals, at least those that owe their lives to lungs; and that it might be necessary that that circular endeavour makes a forceful motion, even if invisible, in every place in the receiver. It is not foreign to reason to deal in the same way with the force by which candles are extinguished and that by which burning coals are extinguished
265 in pits / (even though that phenomenon be not constant); and to say that sometimes air, simultaneously blown from all parts of the walls of the pit, fills the pit with a very quick

and contrary motion. For all the same things will follow as in the receiver. So it is not surprising if the effects in both cases are the same.

B. Why do coals, when extinguished, ever revive? And how do animals, having become inactive, ever regain life?

A. What do your colleagues feel about this matter?

B. There were some of them that said that there remained in those coals, even though they seemed extinguished, some fiery particles, which being blown by the admitted air set on fire the rest of the mass again.

A. Indeed, from what they said they seem not to have thought but to have chosen at random. Do you believe that in a burning coal there is some part that is not coal but fire, or that in red-hot iron there is a part that is not iron but fire?[40] A fire can be produced in a great city from one spark. And if the body of the fire be different from the thing burning, there cannot be more parts of fire in the whole fire than in that one spark. We see that bodies of different kinds can be set on fire by the light of the Sun, by the refraction or the reflection made by burning mirrors; and yet I do not believe that anyone thinks that fiery particles ejected from the Sun can pass through the substance of a crystal globe. In the air between there is no fire. Yet if the motion in the smallest parts of combustible bodies were such that it dispersed and scattered those minute particles, so that it moved air fairly strongly to the eye, then fire would do the same,
266 in no other way than by hitting or strongly rubbing the eye, / and a phantasm of light would usually arise. But Hobbes has sufficiently explained the nature and cause of fire in the book *De corpore*, chap. 27.[41] The nature of fire is derived from such a motion by which the force of percussion makes a phantasm of light arise in the eye; although the force of that motion may be diminished in the evacuated receiver, as you call it, being pushed down by the consistency of the air moving around inside. But that motion is not extinguished; and so that pressure being lifted, it will have enough force

[40] Hobbes, "Human Nature," p. 8: "Our heat is pleasure or pain, according as it is great or moderate; but in the coal there is no such thing."

[41] Hobbes, "Concerning Body," pp. 445-465, esp. 448-453; idem, "Decameron physiologicum," p. 119.

to excite the phantasm of light, although weaker. The same is to be supposed of the life of animals, which do indeed seem to be dead inside the receiver or the pit, yet the internal motion of the calorific parts, peculiar to life, is not yet extinguished, and thence soon after they regain life.

B. But when is it that we can truly say of a man that he is dead, or, which is the same thing, that he has breathed out his soul? For it is known that some men taken for dead have revived the next day when exhumed.[42]

A. It is difficult to determine the point in time at which the soul is separated from the body. So go on to other experiments.

B. If a moderately inflated bladder is put inside the receiver, it is distended further by the pumping of the sucker, and, in the end, if the operation is continued, it is broken open.[43] How?

A. Because every skin is made up of small threads, which because of their shapes cannot touch accurately in all points. The bladder, being a skin, must therefore be pervious not only to the air but to water, such as sweat. Therefore, there
267 is the same compression of the air / compressed inside the bladder by force as there is outside, whose endeavour, its motion following paths that intersect everywhere, tends in every direction towards the concave surface of the bladder. Whence it is necessary that it swells in every direction and, the strength of the endeavour increasing, it is at last torn open.

B. If a magnetized needle hangs freely inside the receiver, it will nevertheless follow the motion of iron which is moved about outside the receiver. Likewise, objects placed inside will be seen by those outside, and sounds made inside will be heard outside. All these things happen in the same way after as well as before the exsuction of air, except that the sounds are a little weaker after than before.[44]

A. These are very clear signs that the receiver is always full, and that the air cannot be sucked out of it. That the

[42] Compare John Bramhall's comment, in Hobbes, "Answer to Bramhall," p. 350: "God only knows what becomes of man's spirit when he expireth."

[43] Boyle, "New Experiments," pp. 18-20.

[44] Ibid., pp. 32-33, 62-63.

sounds are felt to be weaker from there is a sign of the consistency of the air. For the consistency of the air is due to its motion along diametrically opposite lines.

B. And if one of two equal and similar pendulums, at equal heights, is suspended freely in air, the other in the evacuated receiver, and they are simultaneously drawn back from the perpendicular position, then their beats and returns are completed simultaneously; at least no obvious difference appears.[45]

A. I believe it. For the receiver was not, as you thought, emptier after the suction than before.

B. If two hard bodies of well-levelled smooth marble are supposed to be put together along their smooth surfaces, as you know, they stick together so that when suspended in air the lower marble cannot be separated from the upper without a great weight or some other large force. Our colleagues attribute this to the weight of some atmospheric column, whose pressure as a result is terminated in the lower surface
268 of the lower marble / that consequently also sustains it. But lest the weight of the same column press directly on the surface of the same marble, the adjacent marble above prevents it. So if marbles thus cohering were transferred into the receiver and suspended therein, the air being sucked out, were the lower marble to cease sticking to the upper, it would not be possible to doubt that the assigned cause was true. They were moved into the receiver, but without the success expected. For by no further means would they cease to cohere, unless it happened that they were not joined together well enough.[46]

A. Indeed, since there was nothing in this which should be done by the weight of the atmosphere. No stronger or more evident argument could be devised against those who assert the vacuum than this experiment. For if either of two cohering bodies were pushed along the line in which their contiguous surfaces lie, then they would be easily separated, the air always successively flowing into the vacated space. But it would be impossible to tear them apart simultaneously,

[45] Ibid., p. 61.

[46] Hobbes, "Concerning Body," pp. 415-419; idem, "Decameron physiologicum," pp. 90-91; Boyle, "New Experiments," pp. 69-70.

so that they give up all contact at once, in a full world. It would then be necessary either for a motion to be produced from one end to the other in an instant, or else for two bodies to be in the same place at the same time; to say either of which is absurd. Now, see with how many and how great difficulties the cause they assign is burdened. And first, what things follow from that atmospheric column which they wish to rest on the surface of the upper marble. For it is fully acknowledged by them, as by everyone else, that all weight is an endeavour along straight lines from all places to the centre of the Earth; and thence it is not made through a cylinder or column but through a pyramid, whose vertex is in the centre of the Earth; the base, a part of the surface of
269 the atmosphere. Thus, if that pyramid were cut by / the coherent marbles, the shape of that pyramid would be such as the edge of the intersecting marbles would define. So the endeavour of all the weighing points will be propagated to the surface of the upper marble, before it could be propagated further, suppose, towards the Earth. Now, after the endeavour would have been propagated to the Earth, the air once again as a result would endeavour according to lines coming back to the lower surface of the lower marble. For incident perpendicular lines are reflected perpendicularly. So whenever the upper marble is suspended thus, such that it may not endeavour downwards, all the endeavour of the pyramid resting on it is put upon the upper marble. So it will not be propagated to the Earth. So it will not produce an effect on the lower marble. So nothing arises as a result of atmospheric endeavour sustained by the lower marble to prevent its separation from contact with the upper one.

B. This is very certain. But those who rendered such a cause for it were perhaps not geometers. Yet I wonder that this paralogism was not seen by professors of geometry, who ought not to be ignorant of the paths of reflection. But cannot that elastic force which they say is in the air contribute anything to sustaining the marble?

A. By no means. For the endeavour of the air is no greater towards the centre of the Earth than to any other point in the universe. Since all heavy things tend from the edge of the atmosphere to the centre of the Earth, and thence again to the edge of the atmosphere by the same lines

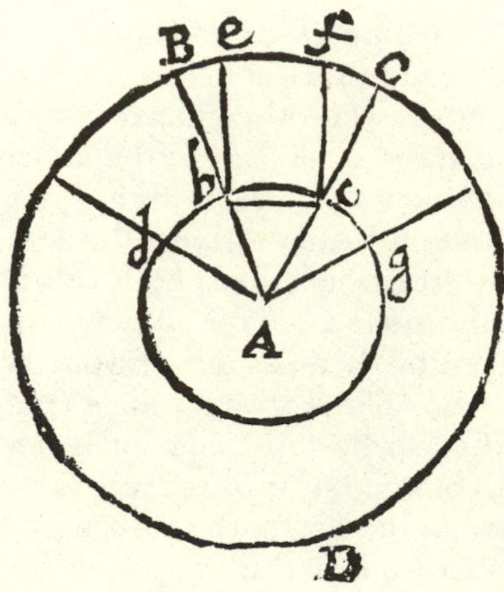

of reflection, the endeavour upwards would be equal to the endeavour downwards, and thence by mutually annihilating each other they would endeavour neither way. It is not to be doubted that water is heavy; and if / a large enough column of water were placed on some man lying on dry [?] land that neither his body nor anything else could support, I far from doubt that the man would be crushed by such a weight. But the weight of the same column pressing on him will not be felt by another similar man lying at the bottom of the sea. For all places in a spherical shape of matter, however subtle, are arched or curved by the endeavour (if such it has) from the circumference to the centre, such that it may not collapse, nor, consequently, lie in ruins.[47] Which I will demonstrate clearly to you, if you think it to be worth it, by copying this figure.

B. I think it worth it. Therefore demonstrate it.

A. The circle BCDB is described, centre of the Earth A, radius AB, which is the semidiameter of the atmosphere. And let some part of the whole atmosphere be supposed somewhere *bc*; on which rests the atmospheric column *efbc*. I say that the column *efbc* does not press by weight on its section *bc*. For were it to press, because of gravity it would press on the atmospheric matter. Thus, the endeavour of

[47] For another use of the arching mechanism, see Hobbes, "Seven Philosophical Problems," pp. 13, 17-18; *White's De Mundo Examined*, p. 226; and his discussion of heavy bodies falling in "Concerning Body," pp. 511-512.

the column *efbc* would press the part *bc* towards the centre A, that is, along the straight lines *b*A, *c*A, and the others in between. For all endeavour of a heavy body is everywhere from the circumference to the Earth's centre. For which reason *bc*, which is too large to be able to descend into the triangle *bc*A, cannot descend unless it makes the straight lines *b*A, *c*A, diverge, so that there be a place for descent. But this cannot happen, since the endeavour of the atmosphere in
271 *dh* and *gc* / makes *b*A and *c*A converge just as much as the weight in *bc* makes them diverge. Therefore, the part *bc* (because of its size) cannot descend, however heavy it is; therefore, it can neither press nor gravitate. Which was to be demonstrated.

B. All this, which I had only fancied myself to have come to learn of philosophy in our Academy, you have taken away from me, now I have been awakened by your demonstration.

A. Both these fantasies, the gravity of the air as well as the elastic force or spring [*antitupia*] of the air, were dreams.[48] For if spring were allowed by them to be something in the threads of the air, and they were to search for something by which, when somewhat curved yet at rest, the threads would be moved again to straightness: if they wish to be taken for physicists, they would have to assign some possible cause for it. They would behave somewhat the same as those who venture to answer the question, *How many rings?*, even though they have not heard the first stroke. Besides, if I should deny it to be possible by human art to make the surfaces of two hard bodies touch so accurately that not the least creatable particle could be let through, then I do not see how their hypothesis could be rightly sustained, nor how our negation could be rightly argued to be unproven.

B. I will lay before you no more than one further experiment from our machine (though the most wonderful of all). We have seen the sucker pulled back by manual force when it had been pushed all the way to the top of the cylinder, after all the entering air had been shut out so that the space in the cylinder was a great enough vacuum. Then a weight of more than one hundred pounds was suspended

[48] For *antitypy*, see Hobbes, "Decameron physiologicum," p. 108, and the comments in Bernstein, "*Conatus*, Hobbes and the Young Leibniz."

from the sucker. As soon as it had gained liberty, we have
272 seen the sucker / together with the suspended weight ascend by itself all the way to the top of the cylinder. Now if the place left by the retraction of the sucker were empty, how could that vacuum, that is, that nothing, pull this weight at all? Unless that space were full of air, could the sucker draw up with such cords and grip with such hooks? How large was that elastic force of the external air, such that a weight of more than one hundred pounds was compelled to ascend again into the brass cylinder, and with the sucker touching it in all points? Our colleagues come to a standstill here.[49] How will you clear this up?

A. I have already cleared it up. For the air pushed back by the retraction of the sucker, not finding a place in the world (which we suppose full) that might receive it, unless it were to make one for itself by pushing adjacent bodies from their places, is at length forced into the cylinder by a continuous push with such speed between the convex surface of the sucker and the concave surface of the cylinder as could answer to those great forces that you have experienced as necessary to the pulling back of the sucker. For that air which speedily enters retains the same on entering, and then distends in every direction the sides of the brass cylinder endowed with an elastic force. So the air strongly moving in the cylinder endeavours against all parts of the concave surface of the cylinder: yet in vain, until the sucker is pulled back; but as soon as the sucker, having slipped from the hand, ceases to push the air, that air, which was earlier driven in, because of its endeavour against every point of the inner surface of the cylinder and of the elastic force of air, will insinuate itself between the same surfaces with the same speed as that with which it has been impelled, that is, with that which answers to the forces of impulsion. So if as large a force of weight be hung from the sucker as the size of the manual force by which it was impelled, then the speed with
273 which the air leaves the cylinder, / having no place in a full world that might receive it, will impel the sucker back to the top of the cylinder, because of the same cause that made the sucker impel the air a little earlier.

B. This is likely. I shall omit the rest of the experiments

[49] Boyle, "New Experiments," pp. 71-73.

of the machine, which seem to be reducible to the same hypotheses of yours without difficulty.

A. So you admit there to be nothing yet from your colleagues for the advancement of the science of natural causes, except that one of them has found a machine that can excite the motion of the air so much that parts of the sphere simultaneously tend from everywhere towards the centre, and so that the hypotheses of Hobbes, which indeed were probable enough beforehand, may by this be rendered more probable.

B. Nor is it shameful to admit it; for it is something to advance so far, if nothing further is allowed.

A. Why *so far*? Why such apparatus and the expense of machines of difficult manufacture, just so as you could get as far as Hobbes had already progressed? Why did you not rather begin from where he left off? Why did you not use the principles he established? Since Aristotle had rightly said that *to be ignorant of motion is to be ignorant of nature*, how did you dare take such a burden upon yourselves, and to arouse in very learned men, not only of our country but also abroad, the expectation of advancing physics, when you have not yet established the doctrine of universal and abstract motion (which was easy and mathematical)? Moreover, other causes are added to those because of which you could not nor will you be able to advance even a little: such as the hatred of Hobbes [*odium Hobbii*], since he had very freely written the truth about Academies (for since that time angry mathematicians and physicists have publicly declared that they
274 would not accept any truth that came from him: / "Whatever be Hobbes's doctrine we will not accept it," said Owen, vice-chancellor of Oxford): and since there are very few of those who profess the sciences who are not pained by the discovery of difficult truths by others rather than themselves.[50] But passing over these things, let us go on to the phenomena of physics whose causes you deduce not from that machine but from elsewhere. Imagine a hollow glass sphere from which sticks up a neck, also hollow. A brass pipe is put in through the neck, which, crossing through the middle, almost reaches

[50] John Wallis dedicated his *Elenchus geometriae Hobbianae* (1655) to John Owen: see Owen, *Correspondence*, pp. 86-88. For Henry Stubbe's attempt to mediate between Hobbes and Owen, see J. Jacob, *Stubbe*, p. 22.

the bottom. Furthermore, the gap between the neck and the inserted pipe may be closed, so that there may not be a passage for air. A tap is driven through the pipe and the neck together, by which the transit of air or water may be freely allowed or prevented. You can see a diagram of the whole instrument at the end of chap. 26 in Hobbes's book *De corpore*.[51] If water is injected with great force through the pipe into the glass sphere (as happens in syringes), and the operation is repeated (indeed in this way the water can fill almost three-quarters of the sphere), it will be seen that, if by turning the tap the exit be opened, all the water gradually rising will be thrown out of it. Hobbes assigns this cause to the phenomenon. The air, with which the spherical glass was full in the beginning, being moved by those earthy particles in the simple circular motion we described a little earlier, being compressed by the force of injection, that of it which is pure leaves by penetrating the injected water for the outer air, leaving a place for the water. So it follows that those earthy particles are left less space in which to exercise their natural motion. Thus impinging on one another, they force
275 the water to leave: and, in leaving, the external air / (since the universe is supposed to be full) penetrates it, and successively takes up the place of the air that leaves, until the same quantity of air being replaced, the particles regain the liberty natural to their motion.[52] But which of your hypotheses explains the same phenomenon?

B. I do not know. But why cannot the water, which, when it was injected, compressed the air particles, be again pushed out by the same particles uncoiling themselves?

A. Because the uncoiled need no more space than the compressed. Just as in a vessel full of water, in which are a multitude of eels, the eels always take up the same space whether wound up or uncoiled. So they cannot propel the water with an elastic force, which is nothing but the motion of bodies uncoiling themselves.

B. That comparison of air with water full of eels will not, I believe, displease our colleagues. Others, no less authoritative among us, are of the opinion that it would not be very

[51] Hobbes, "De corpore," pp. 342-346, and "Tabulae," chap. 26, fig. 2; see also idem, "Concerning Body," pp. 420-425.

[52] For Boyle on siphons, see "New Experiments," pp. 80-81.

repugnant if by vacuum were understood a place empty of all corporeal substances. For supposing air to be made up of particles that cannot be put together without interstices, they see it to be necessary that these interstices be full of corporeal substances, or (as I will say more openly) of bodies. But they do not believe what the *plenists* understood of such a vacuum, especially recently.

A. Why do they not believe it?

B. Because the plenists arguing against the vacuum take their arguments from the fact that liquid rises in a tube
276 sucked by mouth: and from the fact that water / does not descend in a gardener's pot closed at the top and perforated with numerous holes underneath. For they say that such arguments tend towards this, that there is no place left in these lower regions that is not either full of a visible body or of air.[53]

A. None of those whom you call *plenists* understands the vacuum as anything but a place in which there is no corporeal substance at all. If someone speaking negligently were to say, "in which there is no visible body or air," then he would be saying that he understands by air all that body which fills all the space left by the Earth and the stars. I decide not to consider the beliefs of those others who for their own aims deny this. Whoever sucks water into his mouth through a tube first sucks the air in between, which removes the distended air outside; which, being removed, cannot have a place (in a plenum) except by moving that next to it: and so by continual pressure the water is driven into the tube, and replaces the air that is sucked out. But in perforated pots the water remains stationary, because that which leaves through such a small hole is so little that it cannot spread out along the length, so that in descending it might make an entrance for the air around the edges of the holes; nor can air pushed by the water that leaves have any other place (in a full world) except that the water might leave.[54] So you see the natural cause of the ascent of water in the tube by suction and also of the nondescent through the holes of the pot. Furthermore, you see how foolish it would be for the

[53] For an example of such arguments, see Seth Ward, *In Thomae Hobbii philosophiam exercitatio epistolica* (1656), pp. 119-125.

[54] Hobbes, "Concerning Body," pp. 414-415.

explanation of such efforts to summon metaphorical terms such as *fuga vacui, horror naturae*, etc., which the schools used once to use to defend their reputation.

B. Indeed, I believe the causes assigned by you to the phenomena of the vacuum to be right. However, I might not have easily conceded what you say, that there is no one
277 who understands by the vacuum / something empty of visible body and of air. Both Democritus and Epicurus seem to me to understand it thus.

A. If what they understood by vacuum is to be judged from the doctrine of Lucretius, then they understood the same as what I said above, a place empty of all visible and invisible bodies. But indeed they were not *plenists* but *vacuists*.[55] So much for the nature of the air. Let us pass on to water. If you put some water in a basin, and in the water an oblong piece of woollen rag, of which one part may be immersed in the water, and the other hangs outside the basin, the water gradually ascending through the rag soaks it up to the edge of the basin; and if the part that is outside the basin hangs down below the surface of the water that is in the basin, then the water will run down it. What cause of this effect is rendered by your colleagues?

B. I have not yet heard anything about this, unless this phenomenon and that other one of the curved or two-legged siphon have the same cause.

A. Indeed, that is impossible. For the water will not ascend from the basin into the siphon unless both legs are filled with water. The cause of the ascent into the rag is the motion of the earthy atoms that are near the water, a simple circular motion, let me say, communicated to the air in which they are moved: which atoms striking the water drive it up into the woollen material, and being struck in this way they moisten it more and more until it is completely wet. And indeed when completely wet, then if the part of the rag outside the basin be below the surface of the water that is in the basin, the water flows through the rag because of the excess gravity of the water in the part of the rag outside over that which is in the part inside the basin. For the quantity
278 of gravity / in each kind of body does not follow the mass of

[55] Ibid., pp. 415-419.

the gravitating body but the height: though the weight is to be judged otherwise.

B. Furthermore, this experiment of the gardener's pot confirms your hypothesis of earthy particles in the air going round the Earth with a simple circular motion: and yet more that other one of the machine you just described, in which water injected by force is thrown out again. In fact, I knew the designer of the machine.

A. Is that not one of your colleagues?

B. Not at all. For he is a mechanic, not a philosopher.

A. If indeed philosophy were (as it is) the science of causes, in what way did they have more philosophy, who discovered machines useful for experiments, not knowing the causes of the experiments, than this man who, not knowing the causes, designed machines? For there is no difference, except that the one who does not know acknowledges that he does not know, and the others do not so acknowledge.[56]

B. After examining the nature of air and of water, let us go on (if you wish) to the nature of fire. And first, what is fire? Body or accident?

A. Is it not your business to say something about what kind of thing fire might be, and its effects, which you see and feel and name every day?

B. I have never seen fire without burning matter. If wood, coal, iron, or in short any matter whatever, glows and grows hot, then I call it fire: and so do you. So fire seems to me to be a body, or rather many fiery particles simultaneously in the wood or other burning bodies.

A. But are not those particles burning?

279 *B*. Not burning, but pure fire. /

[56] Contrast John Wilkins, *Mathematical Magick* (1648), p. 8: "According to ordinary signification, the word [mechanic] is used in opposition to the liberal arts: whereas in propriety of speech those employments alone may be styled *illiberal*, which require onely some bodily exercise, as manufactures, trades, etc. And on the contrary, that discipline which discovers the generall causes effects and properties of things, may truly be esteemed as a *species* of Philosophy." Hobbes used this *against* the experimenters.

A. But you have just said that fire is burning bodies. Thus, by fire you understand bodies that glow and grow hot in wood or other matter that glows and grows hot: so, that it may not be fire unless it be a body and in another fire, which may again be in a third fire, and so on infinitely. And therefore, the fire to be separate from the burning body, which is an absurd statement. So when we ask what fire might be, we ask for nothing else than the cause by which wood or other matter shines and produces heat: that is, we ask for the causes of light and heat, or rather our sensation that we perceive light and heat.

B. Assuredly this cannot be denied. But then what are the true causes of light and heat?

A. Those themselves which Hobbes did not obscurely derive in the book *De corpore*, chap. 27, from his hypotheses by the demonstrative method; from which (since the book is extant) it is not necessary to deduce this.[57] What I have just taught you is sufficient, that fire is not distinct from the burning body. Yet what have you supposed about cold and ice? Do you indeed now suppose cold and ice to be cold and icy corpuscles in cold and icy matter?

B. We have not yet found anything certain about the causes of cold and ice. However, we have deduced from experiments that when water is frozen it rarefies.[58]

A. Rarefies? I do not understand. If you are saying that the same amount of water fills more space when frozen than when not frozen, then you are saying something that you will never be able to conceive in the mind. For the same amount of a body always has the same quantity, not surprisingly, the space that it fills in whatever shape always being equal.[59] But if you say that particles of air enter the water
280 at the same time as the water during freezing, to / occupy a

[57] Hobbes, "Concerning Body," p. 449; see also his "Decameron physiologicum," p. 119.

[58] On the form of fire and combustion, see Hobbes, "Concerning Body," p. 449. For Boyle on freezing, see his "New Experiments," pp. 95-96.

[59] Hobbes, "Concerning Body," p. 509; idem, "Leviathan," p. 679: "As if a body were made without any quantity at all, and that afterwards more or less were put into it, according as it is intended the body should be more or less dense." This issue was crucial in the critique of Ward and Wallis in the 1650s: see Hobbes's coinage of the terms "wallifaction" and "wardensation" in "Six Lessons," pp. 224-225.

larger space than the water alone, you are saying nothing wonderful.

B. I say that the same amount of water ascends higher in a glass cylinder when frozen than when not frozen.

A. I understand. There was something in the cause by which frozen water is lighter than that not frozen. For everyone knows that a floating piece of ice projects somewhat from the water in which it floats. Archimedes demonstrated that all bodies of equal mass heavier than water, sink; and those that are lighter, project above it; while those that are of equal gravity float such that their top surface coincides with the surface of the water.

B. I see that it is necessary, therefore, for the explanation of the generation of ice, not only that it be known what makes frozen water lighter than it was before, but also what it might be that makes the same thing harder.[60]

A. Whatever lessens or impedes its endeavour towards the centre of the Earth makes it lighter; truly, that cannot be anything but the motion of some body endeavouring against that downwards, if not diametrically, at least obliquely; and that, too, whether the ice is made in a vessel from snow mixed with salt, or in open spaces, such as the northern or southern ocean. But some such opposed motion makes hardness. For we call bodies hard, of which when one part is moved the whole necessarily moves away: as stone is said to be hard, because if you press on one of its parts, the whole will move away, or else the part pressed upon will not move away, at least sensibly. Thus, one motion can produce both the lightness of ice, and its hardness, or the consistency
281 of its parts. Indeed, that motion / (which concerns frozen seas) is easily conceived to be a strong motion of the air, moving closely over the Earth and the ocean in every direction through meridian circles meeting at opposite poles.[61] For such a motion of the air fastens together the protruding upper particles of the water, that is, it makes the whole hard: and at the same time it removes each particle further from the centre of the Earth than the length of the Earth's radius. So it is necessary that the upper particles of water will be

60 Hobbes, "Seven Philosophical Problems," pp. 39-40.

61 Hobbes, "Concerning Body," pp. 472-474; idem, "Decameron physiologicum," pp. 122-125. Hobbes said that this wind "shaves the globe."

held up a little by such a motion of the air: whence all the compacted or frozen water is rendered lighter. Meanwhile, indeed, those earthy atoms, compacted by their simple circular motion, will simultaneously shake all the water particles, so that none of their parts may be moved without the rest, that is, they make all the water hard. It happens similarly in a vessel surrounded by snow and salt. The air that is in the snow, leaving as the snow liquifies, moves in every direction over the surface of the water contained in the vessel, and produces the same effect on the water we have just described in the freezing of the sea; in fact, it makes the water lighter and harder. Yet the transparency (for this is also worthy of note) is diminished a little, because of the mixing of the air with the upper water. The transparency whitens because of the confused position of all the parts.[62]

B. But why does the transparency whiten through the confused situation of the parts? Indeed, I know that glass broken into minute parts is no longer transparent, but white; and spray made by the conflict of water particles is white, and many similar things. Besides, I know all transparent bodies are polished and homogeneous; and thence to be fit to reflect the rays of light, so that the luminous parts may be, as it were, depicted in their arrangement, and that they make a distinct image of the whole; but I do not know the
282 cause.[63] /

A. Since each transparent body is like a lens, so also its parts, however small, are like so many lenses, unless they are flat and in the same plane, and they show just as many bright objects, but very minutely, whose crowded images display not one large bright colour but one nearly approaching that of light, which we call *white*. So the surfaces of bodies that are naturally white consist of innumerable surfaces that are not perceptible to sight because of their small size, but which are convex, and so reflect the light such that as many rays can reach the eye from every part as suffice to produce vision; which indeed cannot be produced by one simple surface.[64]

[62] For "diaphanous" and "transparent" bodies, see Hobbes, "Concerning Body," pp. 463, 480.

[63] Ibid., p. 480.

[64] On reflection and the meaning of "species," see Hobbes, *White's De Mundo*

B. Yet what surface must an object have to appear black?

A. As *white* to light, so *black* is similar to darkness. And so the surface of a black body must be such that no ray (or very few) of those that fall on it from the luminous object may be reflected to the eye wherever it is placed.

B. Then what kind is that?

A. That which is composed of very minute parts that escape the sight, but upright. If they were very minute and the same were upright, then all the rays falling on them from a luminous object, wherever situated, would be reflected at the subjected body, and therefore would not come to the eye: and so it does not seem *black*, unless it is distinguished by some circumstance of vision.

B. One of our colleagues, well known to you and a friend of yours, rendered this very cause frequently at the meeting,
283 but he did not persuade them. For the answer was / that if it were thus, then all hairy cloth ought to be black. And to many he seemed to have been answered correctly.

A. What, did they think that those hairs were bodies so small that they could not be seen? It is fitting to conjecture from this what good reasoners they are: and what may be expected from them in natural philosophy. Perhaps they did not wish to admit that cause of blackness to be true, since the same was first assigned by Hobbes in the book *De corpore*, chap. 27, last article.[65] Yet what cause of *hardness* have you heard assigned by them?

B. Three of them: first, the magnitude of the parts; second, that the parts of the surface mutually touch; third, the entangled position of the parts. Which suffice for the hardness of any body whatsoever.

A. Since the particles (such as are the atoms supposed by Lucretius as well as by Hobbes), already hard now, may be easily compressed by any of the stated causes, it is not to be doubted that a whole made of them would become hard.

Examined, p. 104, and A. Shapiro, "Kinematic Optics," p. 164; Bernhardt, "Hobbes et la mouvement de la lumière," p. 11. On "speculum" as lens or burning-glass, see Hobbes, *White's De Mundo Examined*, p. 85. On the cause of whiteness, see idem, "Concerning Body," pp. 463-464.

[65] Hobbes, "Concerning Body," pp. 464-465.

But those who wish to assign a cause of *hardness* must indicate a cause of those *primary hard bodies* [*duri primi*].

B. So it seems. Some of us, who wished to refute those who attribute the cohesion of parts to some kind of gluing, asked them (and indeed rightly, so it seemed to us) what that gluing might be if not glue.

A. Using the same right, I ask similarly about the first and last of the stated causes, what might be in hard particles that are completely hard that might produce hardness. For the argument you used against those who supposed that
284 there was glue militates equally against you yourselves. /

B. Primary hard particles were perhaps created thus, and indeed the others greater or less from the beginning.

A. Let it be so. Indeed, the first things are rightly referred to the first cause. But if they say hard bodies are made out of primary hard bodies, why do they not also think that fluid bodies are made out of primary fluid bodies? Could the greatest fluids, such as the aether, be created, and the least not be? He that first made a body hard or fluid could, if he had pleased, have made it greater or less than any other given body. If a fluid is made from nonfluid parts, as you have said, and such a hard body from hard parts, does it not follow that neither fluids nor hard bodies may be made from primary fluids?

B. It seems so. What then are the principles of fluidity and hardness?

A. What of fluidity, indeed, if not rest; and of hardness, some motion fit to produce that effect? By rest I understand the rest of two parts with respect to each other, when indeed they touch each other but neither presses upon the other. For whole fluids may be moved while retaining their fluidity, and hard bodies be at rest, although their parts may be moved.[66]

B. By what motion and in what way?

A. For the sake of an example, the air that is impelled by the pulling back of the sucker in the brass cylinder of

[66] Ibid., p. 470; idem, "Seven Philosophical Problems," p. 32.

your machine is not all moved, but remains in the same place: in truth since all its particles entered with a great motion, so also they are moved inside with the same great motion, and all of them strike against each other at the side of the cylinder opposite the entrance. And thence their motions are very fast through very small spaces, and circular, because of the mutual opposition. And hence it is manifest that there is a strong compression in the air moved and enclosed in this way, namely as large as that force by which it was driven
285 in could effect; / and also from such great compression will be made some degree of consistency, although less than the consistency of water. Now, if there were a simple circular motion in all the same particles of air, besides the motion by which each presses on the other, and that were strong enough: it would be almost impossible for one of them to be moved from its little circle, but, the other particles resisting, all would be moved together, that is, all would be hard. For all those things are hard of which no part moves away without the whole moving away. So you see hardness can be made in the very fluid air by this simple circular motion of the particles, which two contrarily directed motions gave them before. You have also seen that some degree of hardness can be given to something by compression alone: which is indeed confirmed in the generation of flesh between the muscles of the human body. For the matter of the flesh that is contained in the muscles is carried there either through the arteries or the nerves. Not by the arteries, in which nothing is carried except blood: for flesh is not made of blood, while healthy flesh can be washed clean by blood. Wherefore the material of flesh is carried to the muscles through the nerves. Yet the matter that is contained in the nerves is a very tenuous spirit: which while it makes flesh in the muscles, consists of innumerable little threads so minute and breakable, that they nevertheless escape the sight.[67] Whence indeed could this be made, unless the spirit from the brain, crossing through the long and very curved channels of the nerves, were condensed together by compression? And such indeed could be the efficient cause of *primary hardness*. However, the cause of *secondary hardness*, that is, of hard-

[67] Hobbes, "Concerning Body," pp. 407, 392. For Hobbes's switch in the 1640s between viewing the brain or the heart as the seat of sensation, see A. Shapiro, "Kinematic Optics," p. 148, and Gargani, *Hobbes e la scienza*, p. 219.

ness from the cohesion of primary hard bodies, could be that same simple circular motion together with their super-
286 ficial or else entangled contact. And if / we suppose with them, that the cause of hardness is the size or the thickness of parts: what reason could we give why frozen water might be harder or firmer than the same water before freezing?

B. They may have seen one. I am a narrator of other philosophies to you, not a defender.

A. Moreover, what reason can be offered for the transparency of those bodies through which all visible bodies appear no less distinctly than through the purest air? If glass or crystal consisted of hard particles, hooked, entangled, or with pores disconnected in whatever way, it would be impossible for rays of light to pass through the transparent sphere without various refractions, by which the arrangement of the parts would be disturbed, and vision would become confused, which daily experience shows to be false.

B. The experiments (as many as I remember) that I have now presented are those worthy of notice of the ones we have made or have received.

A. Yet what do they decide about questions of the nature of other things? And first, by what instrument, by what motion, does the magnet attract or repel iron? What directs it along the meridian and points it north? Which mover, by what motion, transfers the water of the seas and rivers to the clouds? Or from the roots to the tops of trees? Where are the lungs of the winds? Which mover, by what motion, produces the tides of the ocean and tidal differences? What variety of bodies produces the variety of smells and tastes? Why do liquors that look the same to the eye operate very differently on the rest of the organs? What is light? By what thing and what kind of motion is it generated, refracted, and reflected? By what motions do medi-
287 cines operate? By what Gorgon does wood / petrify, and other things do no less?

B. They have not yet decided on these things. Nor is it so long since we began to inquire into nature that you may expect such progress. Is it not enough that we have now almost uncovered the doctrine of the vacuum and of the nature and weight of the air, and soon. I hope, we will have perfected it, after our experiments, which we will make on that very high mountain on the island of Tenerife?[68]

A. It is well. We may await the offspring of the mountain. Meanwhile, content with Hobbesian physics, I will observe the nature and variety of motion. I will also use the same Hobbesian rules of politics and ethics for living.

B. Indeed, you are quite right about politics. For like our physics, that is experimental: for it is well confirmed by almost twenty years of superior experience.

[*Here follows a proof, which we omit, of the duplication of the cube.*]

295/6 *B.* Now, as it seems to me / I leave much that is more certain than before: and I agree with and approve of everything you have said, except I do not think you have correctly given a cause why the sucker rises speedily to the top of the cylinder when it is pulled back and then slips out of the hand. For it is incredible that there should be such a force in the motion of earthy particles (which you suppose) that could effect it. While you rely on the air that was struck being expelled from there, I, on the contrary, believe therefore that the sucker ascends suddenly because, when the retraction ceases, the air that was pushed in with a great force is expelled by the same force between the convex surface of the sucker and the concave surface of the cylinder, and thence (the plenitude of the world being supposed) the external air, restored to its place, simultaneously restores the sucker to the place from which it had been drawn back.

A. I judge the same. I have erred: and you have rightly corrected my error.

END

[68] In December 1660 the Royal Society prepared "some questions in order to the tryal of the quicksilver experiment upon Tenerife." Brouncker and Boyle presented a list of these questions, including the trial with the Torricellian tube and with bladders, in January 1661. Evelyn presented a report on Tenerife in March 1661, and this was subsequently published in Sprat, *History*. See Birch, *History*, vol. 1, pp. 5, 8-10, 18; Sprat, *History*, pp. 200-213.

· BIBLIOGRAPHY ·

Full details of manuscript sources, entries in seventeenth-century periodicals, and references to state and parliamentary papers have been provided in the notes, and these sources have not been included in the Bibliography. Abbreviations of modern journal titles follow the conventions used in the *American National Standard for the Abbreviation of Titles of Periodicals* and in the *ISIS Critical Bibliography*.

Works of Robert Boyle

All citations of Boyle's published writings are from *The Works of the Honourable Robert Boyle*, ed. Thomas Birch, 2d ed., 6 vols. London: J. & F. Rivington, 1772. Individual essays cited are listed below alphabetically (by first main word in title), with location in the Birch edition, date of original publication, and date of composition (if known, relevant, and not given in the text). Our practice is to cite individual essay titles rather than the overall title of collected essays (e.g., *Certain Physiological Essays*). For publication details, see John F. Fulton, *A Bibliography of the Honourable Robert Boyle*, 2d ed. Oxford: Clarendon Press, 1961.

"An Account of Philaretus [i.e., Mr. R. Boyle] during his Minority," I, xii-xxvi (composed circa 1647-1648).

"Animadversions upon Mr. Hobbes's Problemata de Vacuo," IV, 104-128 (1674).

"Of the Cause of Attraction by Suction, a Paradox," IV, 128-144 (1674).

"The Christian Virtuoso," V, 508-540 (1690); "Appendix to the First Part, and the Second Part," VI, 673-796 (1744).

"Continuation of the Experiments concerning Respiration," III, 371-391 (1670).

"A Continuation of New Experiments Physico-Mechanical touching the Spring and Weight of the Air, and their Effects," III, 175-276 (1669); ". . . The Second Part," IV, 505-593 (1680 in Latin; English translation in 1682).

"A Defence of the Doctrine touching the Spring and Weight of the Air . . . against the Objections of Franciscus Linus," I, 118-185 (1662).

"A Discourse of Things above Reason," IV, 406-469 (1681).

"A Discovery of the Admirable Rarefaction of Air," III, 496-500 (1671).

"A Disquisition on the Final Causes of Natural Things," V, 392-444 (1688).

"An Essay of the Intestine Motions of the Particles of Quiescent Solids," I, 444-457 (1661).

"Essays of the Strange Subtilty, Great Efficacy, Determinate Nature of

Effluviums . . . together with a Discovery of the Perviousness of Glass," III, 659-730 (1673).

"An Examen of Mr. T. Hobbes his Dialogus Physicus de Natura Aëris," I, 186-242 (1662).

"An Examen of Mr. Hobbes's Doctrine about Cold," II, 687-698 (1665).

"The Excellency of Theology, compared with Natural Philosophy," IV, 1-66 (1674; written 1665).

"An Experimental Discourse of Quicksilver growing Hot with Gold," IV, 219-230 (1676).

"The Experimental History of Colours," I, 662-778 (1663).

"Experimental Notes of the Mechanical Origin or Production of Fixedness," IV, 306-313 (1675).

"Experiments and Considerations about the Porosity of Bodies, in Two Essays," IV, 759-793 (1684).

"Experiments and Notes about the Producibleness of Chymical Principles; being Parts of an Appendix, designed to be Added to the *Sceptical Chymist*," I, 587-661 (1679).

"A Free Inquiry into the Vulgarly Received Notion of Nature," V, 158-254 (1686; written 1665-1666).

"The General History of the Air," V, 609-743 (1692).

"An Historical Account of a Degradation of Gold, made by an Anti-Elixir: A Strange Chemical Narrative," IV, 371-379 (1678).

"The History of Fluidity and Firmness," I, 377-442 (1661).

"An Hydrostatical Discourse, occasioned by the Objections of the Learned Dr. Henry More," III, 596-628 (1672).

"Hydrostatical Paradoxes, made out by New Experiments," II, pp. 738-797 (1666).

"A Letter concerning Ambergris," III, 731-732 (1673).

"New Experiments about the Differing Pressure of Heavy Solids and Fluids," III, 643-651 (1672).

"New Experiments about Explosions," III, 592-595 (1672).

"New Experiments Physico-Mechanical, touching the Spring of the Air," I, 1-117 (1660).

"New Experiments of the Positive or Relative Levity of Bodies under Water," III, 635-639 (1672).

"New Experiments about the Pressure of the Air's Spring on Bodies under Water," III, 639-642 (1672).

"New Experiments about the Relation betwixt Air and the Flamma Vitalis of Animals," III, 584-589 (1672).

"New Experiments about the Weakened Spring, and Some Unobserved Effects of the Air," IV, 213-219 (1675).

"New Experiments touching . . . Flame and Air," III, 563-584 (1672).

"New Pneumatical Experiments about Respiration," III, 355-391 (1670).

"The Origin of Forms and Qualities, according to the Corpuscular Philosophy," III, 1-137 (1666).

"A Physico-Chymical Essay, containing an Experiment, with Some Considerations touching the Different Parts and Redintegration of Salt-Petre," I, 359-376 (1661).

"A Proëmial Essay . . . with Some Considerations touching Experimental Essays in General," I, 299-318 (1661).

"The Sceptical Chymist," I, 458-586 (1661).

"Some Considerations about the Reconcileableness of Reason and Religion," IV, 151-191 (1675).

"Some Considerations touching the Usefulness of Experimental Natural Philosophy," II, 1-201 (1663; written circa 1650); ". . . The Second Tome," III, 392-457 (1671).

"Some Specimens of an Attempt to Make Chymical Experiments Useful to Illustrate the Notions of the Corpuscular Philosophy. The Preface," I, 354-359 (1661).

"Two Essays, concerning the Unsuccessfulness of Experiments," I, 318-353 (1661).

Works of Thomas Hobbes

The great majority of our citations are from the nineteenth-century Molesworth editions of Hobbes's English and Latin works. (These will eventually be superseded for the philosophical works by Howard Warrender's new edition, only one volume of which (*De cive*) has appeared at time of writing.)

The English Works of Thomas Hobbes of Malmesbury, ed. Sir William Molesworth, 11 vols. London: John Bohn, 1839-1845.

Thomae Hobbes Malmesburiensis opera philosophica quae Latine scripsit omnia . . . , ed. Sir William Molesworth, 5 vols. London: John Bohn, 1839-1845.

These are cited in notes as *English Works* and *Latin Works*. Individual works referred to are listed alphabetically below, followed by other Hobbes sources used. We provide locations in the Molesworth editions and dates of original publication. For publication details, see Hugh MacDonald and Mary Hargreaves, *Thomas Hobbes: A Bibliography*. London: The Bibliographical Society, 1952.

ENGLISH WORKS

"An Answer to a Book Published by Dr. Bramhall," IV, 279-384 (1682; written circa 1668).

"The Art of Rhetoric," VI, 419-536 (1637, 1681; this is an abridgement of Aristotle's *Rhetorica*).

"Behemoth: The History of the Causes of the Civil Wars of England," VI, 161-418 (1679; written 1668).

"Considerations upon the Reputation, Loyalty, Manners, and Religion of Thomas Hobbes," IV, 409-440 (1662).

"Decameron physiologicum; or Ten Dialogues of Natural Philosophy," VII, 69-177 (1678).

"De corpore politico: or the Elements of Law, Moral and Politic . . . ," IV, 77-228 (1650; written 1640).

"A Dialogue between a Philosopher and a Student of the Common Laws of England," VI, 1-160 (1681; printed together with "The Art of Rhetoric").

"Elements of Philosophy. The First Section, Concerning Body," I (1656; translation of 1655 *De corpore*).

"An Historical Narration concerning Heresy and the Punishment Thereof," IV, 385-408 (1680; written 1666-1668).

"Human Nature: or the Fundamental Elements of Policy," IV, 1-76 (1650; written 1640; this is part of the *Elements of Law*; the Latin version, *Elementorum philosophiae sectio secunda de homine*, was not published until 1658).

"Leviathan: or, The Matter, Form, and Power of a Commonwealth," III (1651).

"Philosophical Rudiments concerning Government and Society," II, 1-319 (1651; this is an English translation of *De cive* of 1642).

"Seven Philosophical Problems and Two Propositions of Geometry," VII, 1-68 (1682; this is an English translation of *Problemata physica* of 1662).

"Six Lessons to the Professors of the Mathematics, One of Geometry, the Other of Astronomy . . . in the University of Oxford," VII, 181-356 (1656).

"*Stigmai* . . . , or Marks of the Absurd Geometry, Rural Language, Scottish Church Politics, and Barbarisms of John Wallis . . . ," VII, 357-400 (1657).

"Three Papers Presented to the Royal Society against Dr. Wallis," VII, 429-448 (1671).

LATIN WORKS

"De principiis et ratiocinatione geometricarum," IV, 385-484 (1666).

"Dialogus physicus de natura aeris, conjectura sumpta ab experimentis nuper Londini habitis in Collegio Greshamensi. Item de duplicatione cubi," IV, 233-296 (1661; see Appendix for English translation by Schaffer).

"Elementorum philosophiae sectio prima de corpore," I, 1-431 (1655).

"Examinatio et emendatio mathematicae hodiernae," IV, 1-232 (1660).

"Lux mathematica excussa collisionibus Johannes Wallisii . . . ," V, 89-150 (1672).

"Objectiones ad Cartesii Meditationes de Prima Philosophia," V, 249-274 (1641; English translation 1680).

"Principia et problemata aliquot geometrica . . . ," V, 151-214 (1674).

"Problemata physica . . . ," IV, 297-359 (1662).
"Rosetum geometricum sive propositiones aliquot frustra antehac tentatae. Cum censura brevi doctrinae Wallisianae de motu," V, 1-88 (1671).
"Thomae Hobbes Malmesburiensis vita," I, xiii-xxi (1681).
"Thomae Hobbes Malmesburiensis vita, carmine expressa, authore seipso," I, lxxxi-xcix (1681; written 1672).
"Tractatus opticus," V, 215-248 (1644; published by Mersenne in his *Cogitata physico-mathematica*).

LATIN WORKS: AMSTERDAM EDITION

Thomae Hobbes Malmesburiensis opera philosophica, quae Latine scripsit, omnia. Amsterdam: Johan Blaeu, 1668. (Eight separately paginated items, variously ordered, including a slightly revised version of *Dialogus physicus, Problemata physica, Mathematicae hodiernae,* and the Appendix to *Leviathan.*)

OTHER HOBBES SOURCES

Brown, Harcourt. "The Mersenne Correspondence: A Lost Letter by Thomas Hobbes," *Isis* 34 (1943), 311-12.
de Beer, G. R. "Some Letters of Thomas Hobbes," *Notes Rec. Roy. Soc. Lond.* 7 (1950), 195-210.
Hobbes, Thomas. *Critique du De Mondo de Thomas White,* ed. Jean Jacquot and Harold Whitmore Jones. Paris: Vrin, 1973. (Written 1642-1643, but not published in Hobbes's lifetime.)
———. *Thomas White's De Mundo Examined,* ed. and trans. Harold Whitmore Jones. London: Bradford University Press, in association with Crosby Lockwood Staples, 1976. (As above.)
———. "Little Treatise," in Hobbes, *Elements of Law,* ed. Ferdinand Tönnies, pp. 193-210 (written 1640). London: Simpkin, Marshall, 1889.
———. "Some Principles and Problems in Geometry," in Venturus Mandey, *Mellificium mensionis: or, the Marrow of Measuring,* separately paginated. London, 1682. (A translation of Hobbes, "Principia et problemata . . ." of 1674.)
———. "Thomas Hobbes: *Tractatus opticus,*" ed. F. Alessio, *Riv. Crit. Stor. Fil.* 18 (1963), 147-228. (This is Hobbes's second optical treatise of the 1640s.)
Mintz, Samuel I. "Hobbes on the Law of Heresy: A New Manuscript," *J. Hist. Ideas* 29 (1968), 409-414.
Nicastro, Onofrio. *Lettere di Henry Stubbe a Thomas Hobbes (8 Luglio 1656-6 Maggio 1657).* Siena: Università degli Studi Facoltà di Lettere e Filosofia, 1973.
Tönnies, Ferdinand. *Studien zur Philosophie und Gesellschaftslehre im 17. Jahrhundert,* ed. E. G. Jacoby. Stuttgart: Frommann-Holzboog, 1975. (A major source of Hobbes correspondence.)

Seventeenth-Century and Related Primary Sources

Ailesbury, Thomas Bruce, Earl of. *The Memoirs of Thomas Bruce, Earl of Ailesbury*, 2 vols. London: Roxburghe Club, 1890.

Anon. *An Excerpt of a Book shewing that Fluids Rise not in the Pump, in the Syphon and in the Barometer by the Pressure of the Air but propter fugam vacui: at the Occasion of a Dispute in a Coffee-House with a Doctor of Physick.* London, 1662.

Aubrey, John. "The Life of Thomas Hobbes," in *'Brief Lives,' Chiefly of Contemporaries, Set Down by John Aubrey, between the Years of 1669 & 1696*, 2 vols., ed. Andrew Clark, vol. 1, pp. 321-403. Oxford: Clarendon Press, 1898.

Barlow, Thomas. "The Case of Toleration in Matters of Religion," in *Several Miscellaneous and Weighty Cases of Conscience*, ed. Sir Peter Pett. London, 1692.

Barrow, Isaac. *The Usefulness of Mathematical Learning Explained and Demonstrated*, trans. John Kirkby. London, 1734; orig. publ. 1664-1666.

Barry, Frederick, ed. *The Physical Treatises of Pascal* [being a modern translation of Blaise Pascal, *Traités de l'équilibre des liqueurs et de la pesanteur de la masse de l'air*, ed. Florin Périer (Paris, 1663)]. New York: Columbia University Press, 1937.

Baxter, Richard. *A Sermon of Repentance.* London, 1660.

Birch, Thomas. *The History of the Royal Society of London for the Improving of Natural Knowledge, from its First Rise*, 4 vols. London, 1756-1757.

———. "The Life of the Honourable Robert Boyle," in Boyle, *Works* (see above), vol. 1, pp. vi-clxxi.

Burnet, Gilbert. *History of His Own Time*, 6 vols. Oxford: Clarendon Press, 1823.

Cavendish, Margaret, Duchess of Newcastle. *The Cavalier in Exile: Being the Lives of the First Duke & Dutchess of Newcastle.* London: Newnes, 1903; orig. publ. 1667.

———. *Observations upon Experimental Philosophy.* London, 1663; 2d ed., London, 1668.

Charleton, Walter. *Physiologia Epicuro-Gassendo-Charletoniana: or, a Fabrick of Science Natural, upon the Hypothesis of Atoms. . . .* London, 1654.

Chillingworth, William. *The Religion of Protestants a Safe Way to Salvation.* Oxford, 1638.

Clarendon, Edward Hyde, Earl of. *The History of the Rebellion and the Civil Wars in England . . .*, new ed. Oxford: Oxford University Press, 1843.

Clarkson, Laurence. *The Lost Sheep Found. . . .* London, 1660.

Clerke, Gilbert. *De plenitudine mundi brevis & philosophica dissertatio. . . .* London, 1660.

———. *Tractatus de restitutione corporum, in quo experimenta Torricelliana & Boyliana explicantur & rarefactio Cartesiana defenditur. . . .* London, 1662.

Coke, Roger. *Justice Vindicated from the False Fucus Put by T. White, Gent., Mr. T. Hobbs and Hugo Grotius*. London, 1660.

Conway, Anne. *Conway Letters: The Correspondence of Anne, Viscountess Conway, Henry More, and Their Friends, 1642-1684*, ed. Marjorie Hope Nicolson. New Haven: Yale University Press, 1930.

Cowley, Abraham. *A Proposition for the Advancement of Experimental Philosophy*. London, 1661.

Cudworth, Ralph. *The True Intellectual System of the Universe*. London, 1678.

Culverwell, Nathaniel. *An Elegant and Learned Discourse of the Light of Nature*. London, 1652.

Descartes, René. *Oeuvres de Descartes*, ed. Charles Adam and Paul Tannery, new ed., 11 vols. Paris: Vrin, 1973-1976.

Du Chesne, Joseph [= J. Quercetanus]. *The Practise of Chymicall and Hermeticall Physicke for the Preservation of Health . . . translated . . . by T. Timme*. London, 1605.

Du Moulin, Peter. *The Devill of Mascon: or a True Relation of the Chiefe Things which an Uncleane Spirit Did, and Said at Mascon in Burgundy in the House of F. Perreaud*. Oxford, 1658.

———. *A Vindication of the Sincerity of the Protestant Religion*. London, 1664.

Eachard, John. *The Grounds and Occasions of the Contempt of the Clergy and Religion Enquired Into*. London, 1670.

———. *Mr. Hobb's State of Nature Considered, in a Dialogue between Philautus and Timothy*. London, 1672.

Edwards, John. *A Compleat History of All the Dispensations and Methods of Religion*. London, 1699.

Edwards, Thomas. *Gangraena; or, a Fresh and Further Discovery of the Errors . . . of the Sectaries of this Time*. London, 1646.

Evelyn, John. *The Diary of John Evelyn*, ed. E. S. de Beer. London: Oxford University Press, 1959.

———. *The State of France as it Stood in the IXth Year of this Present Monarch Lewis XIII*. London, 1652.

Falkland, Lucius Cary, Viscount. *A Discourse of Infallibility*, 2d ed. London, 1660; orig. publ. 1645.

Galilei, Galileo. *Dialogues concerning Two New Sciences*, trans. Henry Crew and Alfonso de Salvio. New York: Macmillan, 1914; orig. publ. 1638.

Glanvill, Joseph. "Against Modern Sadducism in the Matter of Witches and Apparitions," in idem, *Essays on Several Important Subjects in Philosophy and Religion*, separately paginated. London, 1676.

———. *A Blow at Modern Sadducism*. London, 1668.

———. *Philosophia pia; or, A Discourse of the Religious Temper, and Tendencies of the Experimental Philosophy* . . . London, 1671.

———. *Plus ultra; or the Progress and Advancement of Knowledge since the Days of Aristotle*. London, 1668.

———. *A Praefatory Answer to Mr. Henry Stubbe*. London, 1671.

Glanvill, Joseph. *Scepsis scientifica, or Confest Ignorance the Way to Science.* London, 1665.

———. *Scire/i tuum nihil est: or, the Author's Defence of The Vanity of Dogmatizing.* London, 1665.

———. *The Vanity of Dogmatizing: or Confidence in Opinions Manifested in a Discourse of the Shortness and Uncertainty of Our Knowledge.* London, 1661.

Grew, Nehemiah. *Musaeum Societatis Regalis, or a Catalogue & Description of the Natural and Artificial Rarities belonging to the Royal Society.* London, 1681.

Guericke, Otto von. *Neue (sogenannte) Magdeburger Versuche über den leeren Raum,* ed. and trans. Hans Schimank. Düsseldorf: VDI-Verlag, 1968.

Hale, Sir Matthew. *Difficiles nugae, or Observations touching the Torricellian Experiment and the Various Solutions of the Same.* London, 1674; 2d ed., London, 1675.

———. *An Essay touching the Gravitation or Non-Gravitation of Fluid Bodies and the Reasons Thereof.* London, 1673.

———. "Reflections by the Lrd. Cheife Justice Hale on Mr. Hobbes his Dialogue of the Lawe," in Sir William S. Holdsworth, *History of English Law,* 17 vols., vol. v (1924), pp. 499-513. London: Methuen, 1903-1972.

Hales, John. *A Tract concerning Schism and Schismatiques.* London, 1642.

Hall, Marie Boas, ed. *Henry Power's Experimental Philosophy.* New York: Johnson Reprint Corp., 1966.

Hall, Thomas. *Histrio-mastix. A Whip for Webster. . . .* London, 1654.

Halliwell, James Orchard, ed. *A Collection of Letters Illustrative of the Progress of Science in England from the Reign of Queen Elizabeth to that of Charles the Second.* London: Historical Society of Science, 1841.

Harrington, James. *A System of Politicks.* London, 1658.

Hartlib, Samuel, comp. *Chymical, Medicinal and Chyrurgical Addresses made to Samuel Hartlib, Esquire.* London, 1655; composed 1642-1643.

Hooke, Robert. *An Attempt for the Explication of the Phaenomena, Observable in an Experiment Published by the Honourable Robert Boyle.* London, 1661.

———. *The Diary of Robert Hooke, M.A., M.D., F.R.S., 1672-1680,* ed. Henry W. Robinson and Walter Adams. London: Taylor & Francis, 1935.

———. *Lectures De potentia restitutiva, or of Spring, Explaining the Power of Springing Bodies.* London, 1678.

———. *Micrographia: or Some Physiological Descriptions of Minute Bodies made by Magnifying Glasses.* London, 1665; reprinted as vol. XIII of R. T. Gunther, *Early Science in Oxford* (see below).

———. *Philosophical Experiments and Observations,* ed. William Derham. London, 1726.

———. *The Posthumous Works of Robert Hooke, M.D. S.R.S. Geom. Prof. Gresh., &c.,* ed. Richard Waller. London, 1705.

Huet, Pierre Daniel. *Lettre touchant les expériences de l'eau purgée.* Paris, 1673.

Huygens, Christiaan. *Oeuvres complètes de Christiaan Huygens*, 22 vols. The Hague: Nijhoff, 1888-1950.

Kendall, George. *Sancti sanciti. Or, The Common Doctrine of the Perseverance of the Saints*. London, 1654.

Kenyon, J. P., ed. *The Stuart Constitution, 1603-1688: Documents and Commentary*. Cambridge: Cambridge University Press, 1966.

Leibniz, Gottfried Wilhelm. *Philosophical Papers and Letters*, trans. and ed. Leroy E. Loemker, 2d ed. Dordrecht: Reidel, 1969.

L'Estrange, Roger. *Considerations and Proposals in Order to the Regulation of the Press*. London, 1663.

Linus, Franciscus. *Tractatus de corporum inseparabilitate; in quo experimenta de vacuo, tam Torricelliana, quàm Magdeburgica, & Boyliana, examinantur*. . . . London, 1661.

Lucretius. *On the Nature of the Universe*, trans. James H. Mantinband. New York: Frederick Ungar, 1965.

Lucy, William. *Observations, Censures, and Confutations of Notorious Errours in Mr. Hobbes His Leviathan*. London, 1663.

Mayow, John. *Tractatus quinque medico-physici*. Oxford, 1674; trans. by A. Crum Brown and Leonard Dobbin as *Medico-physical Works*. Edinburgh: Alembic Club, 1907.

Mersenne, Marin. *Correspondance du P. Marin Mersenne religieux minime*, ed. Cornélis de Waard et al., 15 vols. Paris: Beauchesne; Presses Universitaires de France; Centre National de la Recherche Scientifique, 1932-1983.

———. *La verité des sciences, contre les s[c]eptiques ou pyrrhoniens*. . . . Paris, 1625.

Milton, John. *Complete Prose Works of John Milton*, ed. Don M. Wolfe et al., 8 vols. New Haven: Yale University Press, 1953-1982.

Monconys, Balthasar de, *Journal des voyages*, 2 vols. Lyons, 1665-1666.

More, Henry. *An Antidote against Atheisme, or an Appeal to the Naturall Faculties of the Minde of Man, whether there be not a God*. London, 1653; 3d ed. in More, *Collection* (1662) (see below).

———. *A Collection of Several Philosophical Writings of Dr. Henry More*, 2d ed. London, 1662.

———. *Divine Dialogues, containing Sundry Disquisitions and Instructions concerning the Attributes and Providence of God*, 2 vols. London, 1668.

———. *Enchiridion metaphysicum: sive, de rebus incorporeis succincta & luculenta dissertatio*. London, 1671.

———. *An Explanation of the Grand Mystery of Godliness*. London, 1660.

———. *The Immortality of the Soul*. London, 1659.

———. *A Modest Enquiry into the Mystery of Iniquity*. London, 1664.

———. *Philosophicall Poems*. Cambridge, 1647.

———. *Remarks upon Two Late Ingenious Discourses*. London, 1676.

Newton, Isaac. *The Correspondence of Isaac Newton*, ed. H. W. Turnbull, J. D. Scott, A. Rupert Hall, and Laura Tilling, 7 vols. Cambridge: Cambridge University Press, 1959-1977.

Newton, Isaac. *Opticks*. New York: Dover, 1952; based on 4th ed., London, 1730.

———. *Unpublished Scientific Papers of Isaac Newton*, ed. A. Rupert Hall and Marie Boas Hall. Cambridge: Cambridge University Press, 1962.

Noël, Etienne. *Le plein du vide*. Paris, 1648.

North, Roger. *The Lives of the Right Hon. Francis North, Baron Guilford, The Hon. Sir Dudley North, and The Hon. and Rev. Dr. John North*, 3 vols. London: Colburn, 1826.

Oldenburg, Henry. *The Correspondence of Henry Oldenburg*, ed. A. Rupert Hall and Marie Boas Hall, 11 vols. Madison: University of Wisconsin Press/London: Mansell, 1965-1977.

Owen, John. *Correspondence of John Owen*, ed. P. Toon. Cambridge: James Clarke, 1970.

Papin, Denis. *Nouvelles expériences du vuide*. Paris, 1674.

Pascal, Blaise. *Oeuvres complètes*, ed. Louis Lafuma. Paris: Editions du Seuil, 1963.

[Patrick, Simon]. *A Brief Account of the New Sect of Latitude-Men together with Some Reflections upon the New Philosophy*. London, 1662.

Pepys, Samuel. *The Diary of Samuel Pepys, M.A., F.R.S.*, ed. Robert Latham and William Matthews, 11 vols. London: G. Bell, 1970-1983.

[Pet]T., [Pete]R. *A Discourse concerning Liberty of Conscience*. London, 1661.

Pett, Peter, ed. *The Genuine Remains of Dr. Thomas Barlow*. London, 1693.

Petty, William. *The Advice of W. P. to Mr. S. Hartlib for the Advancement of Some Particular Parts of Learning*. London, 1648.

Plattes, Gabriel. *Caveat for Alchymists*. London, 1655; included in Hartlib (above).

Pope, Walter. *The Life of the Right Reverend Father in God Seth, Lord Bishop of Salisbury*. London, 1697.

Power, Henry. *Experimental Philosophy*. London, 1664.

Renaudot, Théophraste. *Conference concerning the Philosopher's Stone*. London, 1655; included in Hartlib (above).

Rigaud, Stephen Jordan, ed. *Correspondence of Scientific Men of the Seventeenth Century* . . . , 2 vols. Oxford: Oxford University Press, 1841.

Rogers, John. *A Christian Concertation with Mr. Prin, Mr. Baxter, Mr. Harrington, for the True Cause of the Commonwealth*. London, 1659.

Sanderson, Robert. *Several Cases of Conscience discussed in Ten Lectures in the Divinity School at Oxford*. London, 1660.

Schott, Caspar. *Mechanica hydraulico-pneumatica*. . . . Würzburg, 1657.

———. *Technica curiosa sive mirabilia artis*. Würzburg, 1664.

Scriba, Christoph J., ed. "The Autobiography of John Wallis, F.R.S.," *Notes Rec. Roy. Soc. Lond.* 25 (1970), 17-46.

Shadwell, Thomas. "The Virtuoso," in *The Complete Works of Thomas Shadwell*, ed. Montague Summers, 5 vols., vol. III, pp. 95-182 (orig. publ. 1676). New York: Benjamin Blom, 1968.

Sorbière, Samuel de. *A Voyage to England, containing Many Things Relating to the State of Learning, Religion, and Other Curiosities of that Kingdom. . . .* London, 1709; orig. publ. as *Relation d'un voyage en Angleterre*, Paris, 1664.

Sprat, Thomas. *The History of the Royal-Society of London, for the Improving of Natural Knowledge.* London, 1667.

———. *Observations on Mons. de Sorbiere's Voyage into England. Written to Dr. Wren. . . .* London, 1708; orig. publ. 1665.

Stillingfleet, Edward. *Origines sacrae.* London, 1662.

Strong, Sandford Arthur. *A Catalogue of Letters and Other Historical Documents, Exhibited in the Library at Welbeck.* London: John Murray, 1903.

Stubbe, Henry. *Censure upon Certaine Passages Contained in a History of the Royal Society as being Destructive to the Established Religion and Church of England.* Oxford, 1670.

———. *Lord Bacons Relation of the Sweating-Sickness Examined.* London, 1671.

———. *Malice Rebuked, or a Character of Mr. Richard Baxters Abilities and a Vindication of the Hon. Sir Henry Vane from His Aspersions.* London, 1659.

Sylvester, Matthew. *Reliquiae Baxterianae, or Mr. Richard Baxter's Narrative of the Most Memorable Passages of His Life and Times.* London, 1696.

Tanner, Joseph R., ed. *Constitutional Documents of the Reign of James I.* Cambridge: Cambridge University Press, 1930.

Thirsk, Joan, ed. *The Restoration.* London: Longman, 1976.

Wagstaffe, John. *The Question of Witchcraft Debated*, 2d ed. London, 1671.

Wallis, John. *Elenchus geometriae Hobbianae.* Oxford, 1655.

———. *Hobbius heauton-timorumenos. Or a Consideration of Mr Hobbes his Dialogues. In an Epistolary Discourse, Addressed, to the Honourable Robert Boyle, Esq.* Oxford, 1662.

Walton, Izaak. *The Lives of Dr. John Donne, Sir Henry Wotton, Mr. Richard Hooker, Mr. George Herbert and Dr. Robert Sanderson*, 2 vols. London: J. M. Dent, 1898; orig. publ. 1670.

Ward, Richard. *The Life of the Learned and Pious Dr Henry More.* London, 1710.

Ward, Seth. *In Thomae Hobbii philosophiam exercitatio epistolica.* Oxford, 1656.

———. *Vindiciae academiarum, containing Some Briefe Animadversions upon Mr Websters Book. . . .* Oxford, 1654.

White, Thomas. *An Exclusion of Scepticks from All Title to Dispute.* London, 1665.

———. *The Grounds of Obedience and Government.* London, 1655.

Wilkins, John. *Mathematical Magick. Or, The Wonders that may be Performed by Mechanicall Geometry.* London, 1648.

Wood, Anthony à. *The Life and Times of Anthony Wood, Antiquary, at Oxford, 1632-95, as Described by Himself*, ed. Andrew Clark, 5 vols. Oxford: Oxford Historical Society, 1891-1900.

Worthington, John. *The Diary and Correspondence of Dr. John Worthington, Master of Jesus College, Cambridge*, ed. James Crossley and Richard C. Christie, 3 vols., Chetham Society series, vols. 13, 36, 114. Manchester: Chetham Society, 1847-1886.

Secondary Sources

Aaron, R. I. "A Possible Draft of *De Corpore*," *Mind* 54 (1945), 342-356.

Aarsleff, Hans. *From Locke to Saussure: Essays on the Study of Language and Intellectual History*. London: Athlone Press, 1982.

Abbott, Wilbur C. "English Conspiracy and Dissent, 1660-1674," *Amer. Hist. Rev.* 14 (1909), 503-528, 696-722.

Abernathy, George R., Jr. "Clarendon and the Declaration of Indulgence," *J. Eccles. Hist.* 11 (1960), 55-73.

———. *The English Presbyterians and the Stuart Restoration, 1648-1663*. Transactions of the American Philosophical Society, vol. 55, part 2. Philadelphia: American Philosophical Society, 1965.

Agassi, Joseph. "Who Discovered Boyle's Law?" *Stud. Hist. Phil. Sci.* 8 (1977), 189-250.

Aiton, E. J. "Newton's Aether-Stream Hypothesis and the Inverse Square Law of Gravitation," *Ann. Sci.* 25 (1969), 255-260.

Albury, William R. "Halley and the *Traité de la lumière* of Huygens: New Light on Halley's Relationship with Newton," *Isis* 62 (1971), 445-468.

Alpers, Svetlana. *The Art of Describing: Dutch Art in the Seventeenth Century*. London: John Murray, 1983.

Anderson, Paul Russell. *Science in Defense of Liberal Religion: A Study of Henry More's Attempt to Link Seventeenth-Century Religion with Science*. New York: Putnam's Sons, 1933.

Anderson, R.G.W. *The Playfair Collection and the Teaching of Chemistry at the University of Edinburgh, 1713-1858*. Edinburgh: Royal Scottish Museum, 1978.

Anon. "Hobbes," *Encyclopaedia Britannica*, 3d ed., vol. VIII, pp. 601-603. Edinburgh, 1797.

Applebaum, Wilbur. "Boyle and Hobbes: A Reconsideration," *J. Hist. Ideas* 25 (1964), 117-119.

Ashley, Maurice. *John Wildman, Plotter and Postmaster: A Study of the English Republican Movement in the Seventeenth Century*. London: Jonathan Cape, 1947.

Auger, Léon. *Un savant méconnu: Gilles Personne de Roberval, 1602-1675*. Paris: A. Blanchard, 1962.

Axtell, James L. "The Mechanics of Opposition: Restoration Cambridge *v.* Daniel Scargill," *Bull. Inst. Hist. Res.* 38 (1965), 102-111.

Aylmer, G. E. "Unbelief in Seventeenth-Century England," in *Puritans and Revolutionaries: Essays in Seventeenth-Century History Presented to Christopher Hill*, ed. Donald Pennington and Keith Thomas, pp. 22-46. Oxford: Clarendon Press, 1978.

Barnes, Barry, and Bloor, David. "Relativism, Rationalism and the Sociology of Knowledge," in *Rationality and Relativism*, ed. Martin Hollis and Steven Lukes, pp. 21-47. Oxford: Basil Blackwell, 1982.

Barnouw, Jeffrey. "Hobbes's Causal Account of Sensation," *J. Hist. Phil.* 18 (1980), 115-130.

Bate, Frank. *The Declaration of Indulgence, 1672: A Study in the Rise of Organised Dissent.* London: Constable, 1908.

Baxandall, Michael A. *The Limewood Sculptors of Renaissance Germany.* New Haven: Yale University Press, 1980.

———. *Painting and Experience in Fifteenth-Century Italy: A Primer in the Social History of Pictorial Style.* London: Oxford University Press, 1974.

Beaujot, Jean-Pierre, and Mortureux, Marie-Françoise. "Genèse et fonctionnement du discours: *Les pensées diverses sur la comète* de Bayle et les *Entretiens sur la pluralité des mondes* de Fontenelle," *Lang. Fr.* 15 (1972), 56-78.

Bechler, Zev. "Newton's 1672 Optical Controversies: A Study in the Grammar of Scientific Dissent," in *The Interaction between Science and Philosophy,* ed. Y. Elkana, pp. 115-142. Atlantic Highlands, N.J.: Humanities Press, 1974.

Beddard, Robert V. "The Restoration Church," in *The Restored Monarchy, 1660-1688,* ed. J. R. Jones, pp. 155-175. London: Macmillan, 1979.

Bennett, J. A. "Robert Hooke as Mechanic and Natural Philosopher," *Notes Rec. Roy. Soc. Lond.* 35 (1980), 33-48.

Bernhardt, Jean. "Hobbes et le mouvement de la lumière," *Rev. Hist. Sci.* 30 (1977), 3-24.

Bernstein, Howard R. "*Conatus,* Hobbes, and the Young Leibniz," *Stud. Hist. Phil. Sci.* 11 (1980), 25-37.

Bloor, David. "Durkheim and Mauss Revisited: Classification and the Sociology of Knowledge," *Stud. Hist. Phil. Sci.* 13 (1982), 267-297.

———. *Knowledge and Social Imagery.* London: Routledge and Kegan Paul, 1976.

———. *Wittgenstein: A Social Theory of Knowledge.* London: Macmillan, 1983.

Bosher, Robert S. *The Making of the Restoration Settlement: The Influence of the Laudians, 1649-1662.* London: Dacre Press, 1951.

Bourne, Henry Richard Fox. *English Newspapers: Chapters in the History of Journalism,* 2 vols. London: Chatto & Windus, 1887.

Bowle, John. *Hobbes and His Critics: A Study in Seventeenth-Century Constitutionalism.* London: Jonathan Cape, 1951.

Boylan, Michael. "Henry More's Space and the Spirit of Nature," *J. Hist. Phil.* 18 (1980), 395-405.

Brandt, Frithiof. *Thomas Hobbes' Mechanical Conception of Nature.* Copenhagen: Levin & Munksgaard/London: Librairie Hachette, 1928.

Brevold, Louis I. "Dryden, Hobbes, and the Royal Society," *Mod. Philol.* 25 (1928), 417-438.

Breidert, Wolfgang. "Les mathématiques et la méthode mathématique chez Hobbes," *Rev. Int. Phil.* 129 (1979), 415-432.

Brockdorff, Cay von. *Des Sir Charles Cavendish Bericht für Joachim Jungius über den Grundzügen der Hobbes'schen Naturphilosophie.* Kiel: Hobbes-Gesellschaft, 1934.

Brown, Harcourt. *Scientific Organizations in Seventeenth Century France (1620-1680).* Baltimore: Williams & Wilkins, 1934.

Brown, Keith. "Hobbes's Grounds for Belief in a Deity," *Philosophy* 37 (1962), 336-344.

Brown, Louise F. "The Religious Factors in the Convention Parliament," *Engl. Hist. Rev.* 22 (1907), 51-63.

Brugmans, Henri L. *Le séjour de Christian Huygens à Paris et ses relations avec les milieux scientifiques français.* Paris: Librairie E. Droz, 1935.

Brush, Stephen G. *Statistical Physics and the Atomic Theory of Matter, from Boyle and Newton to Landau and Onsager.* Princeton: Princeton University Press, 1983.

Buck, Peter. "Seventeenth-Century Political Arithmetic: Civil Strife and Vital Statistics," *Isis* 68 (1977), 67-84.

Bulmer, Ralph. "Why is the Cassowary not a Bird? A Problem of Zoological Taxonomy among the Karam of the New Guinea Highlands," *Man,* n.s. 2 (1967), 5-25.

Burnham, Frederic B. "The More-Vaughan Controversy: The Revolt against Philosophical Enthusiasm," *J. Hist. Ideas* 35 (1974), 33-49.

Burtt, Edwin Arthur. *The Metaphysical Foundations of Modern Physical Science,* rev. ed. Garden City, N.Y.: Anchor, 1954; orig. publ. 1924.

Cabanes, Charles. *Denys Papin, inventeur et philosophe cosmopolite.* Paris: Société Française d'Éditions Littéraires et Techniques, 1935.

Canny, Nicholas. *The Upstart Earl: A Study of the Social and Mental World of Richard Boyle, First Earl of Cork.* Cambridge: Cambridge University Press, 1982.

Cantor, G. N., and Hodge, M.J.S., eds. *Conceptions of Ether: Studies in the History of Ether Theories 1740-1900.* Cambridge: Cambridge University Press, 1981.

Capp, Bernard. *Astrology and the Popular Press: English Almanacs 1500-1800.* London: Faber and Faber, 1979.

———. *The Fifth Monarchy Men: A Study in Seventeenth-Century Millenarianism.* London: Faber and Faber, 1972.

Carter, Jennifer. "Law, Courts and Constitution," in *The Restored Monarchy, 1660-1688,* ed. J. R. Jones, pp. 71-93. London: Macmillan, 1979.

Christensen, Francis. "John Wilkins and the Royal Society's Reform of Prose Style," *Mod. Lang. Quart.* 7 (1946), 179-187, 279-290.

Christie, J.R.R., and Golinski, J. V. "The Spreading of the Word: New Directions in the Historiography of Chemistry 1600-1800," *Hist. Sci.* 20 (1982), 235-266.

Clark, G. N. *The Seventeenth Century,* 2d ed. Oxford: Clarendon Press, 1947.

Cohen, I. Bernard. "Hypotheses in Newton's Philosophy," *Physis* 8 (1966), 163-184.

Collins, H. M. *Changing Order: Replication and Induction in Scientific Practice.* London: Sage, 1985.

———. "The Seven Sexes: A Study in the Sociology of a Phenomenon, or the Replication of Experiments in Physics," *Sociology* 9 (1975), 205-224.

———. "Son of Seven Sexes: The Social Destruction of a Physical Phenomenon," *Soc. Stud. Sci.* 11 (1981), 33-62.

———. "The TEA Set: Tacit Knowledge and Scientific Networks," *Sci. Stud.* 4 (1974), 165-186.

———. "Understanding Science," *Fund. Sci.* 2 (1981), 367-380.

Collins, H. M., and Harrison, R. G. "Building a TEA Laser: The Caprices of Communication," *Soc. Stud. Sci.* 5 (1975), 441-450.

Collins, H. M., and Pinch, T. J. *Frames of Meaning: The Social Construction of Extraordinary Science.* London: Routledge and Kegan Paul, 1982.

Conant, James Bryant. *On Understanding Science: An Historical Approach.* Oxford: Oxford University Press, 1947.

———, ed. "Robert Boyle's Experiments in Pneumatics," in *Harvard Case Histories in Experimental Science,* 2 vols.; vol. I, pp. 1-63. Cambridge, Mass.: Harvard University Press, 1970; orig. publ. 1948.

Cope, Jackson I. *Joseph Glanvill: Anglican Apologist.* St. Louis: Washington University Press, 1956.

Cope, Jackson I., and Jones, Harold Whitmore. "Introduction" [to their edition of Thomas Sprat, *The History of the Royal-Society* . . .], pp. xii-xxxii. St. Louis: Washington University Press, 1959.

Cowles, Thomas. "Dr. Henry Power, Disciple of Sir Thomas Browne," *Isis* 20 (1933), 349-366.

Cristofolini, Paolo. *Cartesiani e sociniani: studio su Henry More.* Urbino: Argalia Editore, 1974.

Daly, James. *Cosmic Harmony and Political Thinking in Early Stuart England.* Transactions of the American Philosophical Society, vol. 69, part 7. Philadelphia: American Philosophical Society, 1979.

Damrosch, Leopold, Jr. "Hobbes as Reformation Theologian: Implications of the Free-Will Controversy," *J. Hist. Ideas* 40 (1979), 339-352.

Daniels, George H. "The Pure-Science Ideal and Democratic Culture," *Science* 156 (1967), 1699-1705.

Daston, Lorraine J. "The Reasonable Calculus: Classical Probability Theory, 1650-1840." Ph.D. thesis, Harvard University, 1979.

Daumas, Maurice. *Les instruments scientifiques aux XVIIe et XVIIIe siècles.* Paris: Presses Universitaires de France, 1953.

Davies, Godfrey. *The Restoration of Charles II, 1658-1660.* San Marino, Calif.: Huntington Library, 1955.

Dear, Peter. "*Totius in verba*: Rhetoric and Authority in the Early Royal Society," *Isis* 76 (1985), 145-161.

Debus, Allen G. *The Chemical Philosophy: Paracelsian Science and Medicine in the Sixteenth and Seventeenth Centuries*, 2 vols. New York: Science History Publications, 1977.

———. *The English Paracelsians*. London: Oldbourne, 1965.

DeKosky, Robert K. "William Crookes and the Quest for Absolute Vacuum in the 1870s," *Ann. Sci.* 40 (1983), 1-18.

Delorme, Suzanne. "Pierre Perrault, auteur d'un traité *De l'origine des fontaines* et d'une théorie de l'experimentation," *Arch. Int. Hist. Sci.* 3 (1948), 388-394.

de Waard, Cornélis. *L'expérience barométrique: ses antécédents et ses explications*. Thouars: J. Gamon, 1936.

Diamond, William Craig. "Natural Philosophy in Harrington's Political Thought," *J. Hist. Phil.* 16 (1978), 387-398.

Dobbs, Betty Jo Teeter. *Foundations of Newton's Alchemy, or "The Hunting of the Greene Lyon."* Cambridge: Cambridge University Press, 1975.

Douglas, Mary. "Self-Evidence," in idem, *Implicit Meanings: Essays in Anthropology*, pp. 276-318. London: Routledge and Kegan Paul, 1975.

Duffy, Eamon. "Primitive Christianity Revived: Religious Renewal in Augustan England," *Stud. Church Hist.* 14 (1977), 287-300.

Dugas, René. "Sur le cartésianisme de Huygens," *Rev. Hist. Sci.* 7 (1954), 22-33.

Duhem, Pierre. *The Aim and Structure of Physical Theory*, trans. Philip P. Wiener. New York: Atheneum, 1962; orig. publ. 1906.

Edgerton, Samuel Y., Jr. *The Renaissance Discovery of Linear Perspective*. New York: Harper & Row, 1976.

Eisenstein, Elizabeth L. *The Printing Press as an Agent of Change: Communications and Cultural Transformations in Early-Modern Europe*. Cambridge: Cambridge University Press, 1980.

Eklund, Jon. *The Incompleat Chymist: Being an Essay on the Eighteenth-Century Chemist in His Laboratory*. . . . Washington: Smithsonian Institution Press, 1975.

Ellis, H. F. *So This is Science!* London: Methuen, 1932.

Elzinga, Aant. "Christiaan Huygens' Theory of Research," *Janus* 67 (1980), 281-300.

———. *On a Research Program in Early Modern Physics, with Special Reference to the Work of Ch. Huygens*. Göteborg: Institution for the Theory of Science, University of Gothenburg 1971.

Ezrahi, Yaron. "Science and the Problem of Authority in Democracy," in *Science and Social Structure: A Festschrift for Robert K. Merton*, ed. Thomas F. Gieryn. Transactions of the New York Academy of Sciences, series II, vol. 39, pp. 43-60. New York: New York Academy of Sciences, 1980.

Fanton d'Andon, Jean-Pierre. *L'horreur du vide: expérience et raison dans la physique pascalienne*. Paris: Centre National de la Recherche Scientifique, 1978.

Farley, John, and Geison, Gerald L. "Science, Politics and Spontaneous

Generation in Nineteenth-Century France: The Pasteur-Pouchet Debate," *Bull. Hist. Med.* 48 (1974), 161-198.

Farrington, Benjamin. *The Philosophy of Francis Bacon.* Liverpool: Liverpool University Press, 1970.

Feiling, Keith. "Clarendon and the Act of Uniformity, 1662-3," *Engl. Hist. Rev.* 44 (1929), 289-291.

Feyerabend, Paul. *Against Method.* London: Verso, 1978.

Fisch, Harold. "The Scientist as Priest: A Note on Robert Boyle's Natural Theology," *Isis* 44 (1953), 252-265.

Fleck, Ludwik. *Genesis and Development of a Scientific Fact*, ed. Thaddeus J. Trenn and Robert K. Merton, trans. Fred Bradley and Trenn. Chicago: University of Chicago Press, 1979; orig. publ. 1935.

Foucault, Michel. *The Archaeology of Knowledge*, trans. A. M. Sheridan Smith. London: Tavistock, 1972.

———. "Médicins, juges et sorciers au 17e siècle," *Médecine de France* 200 (1969), 121-128.

———. "Questions on Geography," in idem, *Power-Knowledge: Selected Interviews and Other Writings, 1972-1977*, ed. Colin Gordon, pp. 63-77. Brighton: Harvester, 1980.

Frank, Robert G., Jr. *Harvey and the Oxford Physiologists: A Study of Scientific Ideas.* Berkeley: University of California Press, 1980.

———. "The John Ward Diaries: Mirror of 17th-Century Science and Medicine," *J. Hist. Med.* 29 (1974), 147-179.

Fraser, Peter. *The Intelligence of the Secretaries of State and Their Monopoly of Licensed News, 1660-1688.* Cambridge: Cambridge University Press, 1956.

Freudenthal, Gad. "Early Electricity between Chemistry and Physics: The Simultaneous Itineraries of Francis Hauksbee, Samuel Wall, and Pierre Polinière," *Hist. Stud. Phys. Sci.* 11 (1981), 203-229.

Freudenthal, Gideon. *Atom und Individuum im Zeitalter Newtons: zur Genese der mechanistischen Natur- und Sozialphilosophie.* Frankfurt am Main: Suhrkamp, 1982.

Fries, Sylvia D. "The Ideology of Science during the Nixon Years: 1970-76," *Soc. Stud. Sci.* 14 (1984), 323-341.

Gabbey, Alan. "Huygens et Roberval," in *Huygens et la France*, ed. René Taton, pp. 69-84. Paris: Vrin, 1982.

———. "Philosophia Cartesiana Triumphata: Henry More (1636-1671)," in *Problems of Cartesianism: Studies in the History of Ideas*, ed. Thomas M. Lennon et al., pp. 171-250. Montreal: McGill-Queens University Press, 1982.

Garfinkel, Harold. *Studies in Ethnomethodology.* Englewood Cliffs, N.J.: Prentice-Hall, 1967.

Gargani, Aldo Giorgio. *Hobbes e la scienza.* Turin: Giulio Einaudi, 1971.

Gaukroger, Stephen. *Explanatory Structures: Concepts of Explanation in Early*

Physics and Philosophy. Atlantic Highlands, N.J.: Humanities Press, 1978.

Gee, Henry. "The Derwentdale Plot, 1663," *Trans. Roy. Hist. Soc.*, 3d ser. 11 (1917), 125-142.

Gellner, Ernest. "Concepts and Society," in idem, *Cause and Meaning in the Social Sciences*, ch. 2. London: Routledge and Kegan Paul, 1973.

Geoghegan, D. "Gabriel Plattes' Caveat for Alchymists," *Ambix* 10 (1962), 97-102.

Gillispie, Charles Coulston. *The Edge of Objectivity: An Essay in the History of Scientific Ideas.* Princeton: Princeton University Press, 1960.

———. "The *Encyclopédie* and the Jacobin Philosophy of Science: A Study in Ideas and Consequences," in *Critical Problems in the History of Science*, ed. Marshall Clagett, pp. 255-289. Madison: University of Wisconsin Press, 1959.

Ginzburg, Carlo. *The Night Battles: Witchcraft & Agrarian Cults in the Sixteenth & Seventeenth Centuries*, trans. John and Anne Tedeschi. London: Routledge and Kegan Paul, 1983.

Glover, Willis B. "God and Thomas Hobbes," in *Hobbes Studies*, ed. K. C. Brown, pp. 141-168. Oxford: Basil Blackwell, 1965.

Goldsmith, M. M. *Hobbes's Science of Politics.* New York: Columbia University Press, 1966.

Golinski, Jan V. "Language, Method and Theory in British Chemical Discourse, c. 1660-1770." Ph.D. thesis, University of Leeds, 1984.

Gouk, Penelope. "The Role of Acoustics and Music Theory in the Scientific Work of Robert Hooke," *Ann. Sci.* 37 (1980), 573-605.

Grant, Edward. *Much Ado about Nothing: Theories of Space and Vacuum from the Middle Ages to the Scientific Revolution.* Cambridge: Cambridge University Press, 1981.

Green, I. M. *The Re-establishment of the Church of England, 1660-1663.* Oxford: Oxford University Press, 1978.

Greene, Robert A. "Henry More and Robert Boyle on the Spirit of Nature," *J. Hist. Ideas* 23 (1962), 451-474.

———. "Whichcote, Wilkins, 'Ingenuity,' and the Reasonableness of Christianity," *J. Hist. Ideas* 42 (1981), 227-252.

Grover, R. A. "The Legal Origins of Thomas Hobbes's Doctrine of Contract," *J. Hist. Phil.* 18 (1980), 177-207.

Guenancia, Pierre. *Du vide à Dieu: essai sur la physique de Pascal.* Paris: F. Maspero, 1976.

Guerlac, Henry. *Essays and Papers in the History of Modern Science.* Baltimore: Johns Hopkins University Press, 1977.

———. "Newton's Optical Aether," *Notes Rec. Roy. Soc. Lond.* 22 (1967), 45-57.

Guilloton, Vincent. *Autour de la 'Relation' du voyage de Samuel Sorbière en Angleterre 1663-1664.* Northampton, Mass.: Smith College, 1930.

Guinsburg, Arlene Miller. "Henry More, Thomas Vaughan and the Late Renaissance Magical Tradition," *Ambix* 27 (1980), 36-58.

Gunther, R. T. *Early Science in Oxford*, 15 vols. Oxford: privately printed, 1923-1967.

Hacking, Ian. *The Emergence of Probability: A Philosophical Study of Early Ideas about Probability, Induction and Statistical Inference.* Cambridge: Cambridge University Press, 1975.

———. *Representing and Intervening: Introductory Topics in the Philosophy of Natural Science.* Cambridge: Cambridge University Press, 1983.

Hahn, Roger. *The Anatomy of a Scientific Institution: The Paris Academy of Sciences, 1666-1803.* Berkeley: University of California Press, 1971.

———. "Huygens and France," in *Studies on Christiaan Huygens*, ed. H.J.M. Bos et al., pp. 53-65. Lisse, The Netherlands: Swets & Zeitlinger, 1980.

Hall, A. Rupert. *From Galileo to Newton 1630-1720.* London: Collins, 1963.

———. "Gunnery, Science, and the Royal Society," in *The Uses of Science in the Age of Newton*, ed. John G. Burke, pp. 111-141. Berkeley: University of California Press, 1983.

———. *The Revolution in Science 1500-1750.* London: Longman, 1983.

———. *The Scientific Revolution 1500-1800: The Formation of the Modern Scientific Attitude*, 2d ed. Boston: Beacon Press, 1966.

Hall, A. Rupert, and Hall, Marie Boas. "Philosophy and Natural Philosophy: Boyle and Spinoza," in *Mélanges Alexandre Koyré. II. L'aventure de l'esprit*, ed. René Taton and I. Bernard Cohen, pp. 241-256. Paris: Hermann, 1964.

Hall, Marie Boas. "Boyle, Robert," in *Dictionary of Scientific Biography*, vol. II, pp. 377-382. New York: Charles Scribner's, 1970.

———. "Boyle as a Theoretical Scientist," *Isis* 41 (1950), 261-268.

———. "An Early Version of Boyle's 'Sceptical Chymist'," *Isis* 45 (1954), 156-168.

———. "The Establishment of the Mechanical Philosophy," *Osiris* 10 (1952), 412-541.

———. "Huygens' Scientific Contacts with England," in *Studies on Christiaan Huygens*, ed. H.J.M. Bos et al., pp. 66-81. Lisse, The Netherlands: Swets & Zeitlinger, 1980.

———. *Robert Boyle and Seventeenth-Century Chemistry.* Cambridge: Cambridge University Press, 1958.

———. "Salomon's House Emergent: The Early Royal Society and Cooperative Research," in *The Analytic Spirit: Essays in the History of Science in Honor of Henry Guerlac*, ed. Harry Woolf, pp. 177-194. Ithaca: Cornell University Press, 1981.

———. "Science in the Early Royal Society," in *The Emergence of Science in Western Europe*, ed. Maurice Crosland, pp. 57-77. London: Macmillan, 1975.

Halleux, Robert. "Huygens et les théories de la matière," in *Huygens et la France*, ed. René Taton, pp. 187-195. Paris: Vrin, 1982.

Hannaway, Owen. *The Chemists and the Word: The Didactic Origins of Chemistry*. Baltimore: Johns Hopkins University Press, 1975.

Hanson, Donald W. *From Kingdom to Commonwealth: The Development of Civic Consciousness in English Political Thought*. Cambridge, Mass.: Harvard University Press, 1970.

Harrison, Charles T. "Bacon, Hobbes, Boyle, and the Ancient Atomists," in *Harvard Studies and Notes in Philosophy and Literature*, vol. 15, ed. G. H. Maynadier et al., pp. 191-218. Cambridge, Mass.: Harvard University Press, 1933.

Harrison, John. *The Library of Isaac Newton*. Cambridge: Cambridge University Press, 1978.

Harvey, Bill. "Plausibility and the Evaluation of Knowledge: A Case-Study of Experimental Quantum Mechanics," *Soc. Stud. Sci.* 11 (1981), 95-130.

Havighurst, Alfred F. "The Judiciary and Politics in the Reign of Charles II," *Law Quart. Rev.* 66 (1950), 62-78, 229-252.

Hawes, Joan L. "Newton and the Electrical Attraction Unexcited," *Ann. Sci.* 24 (1968), 121-130.

Heathcote, N. H. deV. "Guericke's Sulphur Globe," *Ann. Sci.* 6 (1950), 293-305.

Heilbron, J. L. *Elements of Early Modern Physics*. Berkeley: University of California Press, 1982.

Heimann, P. M. " 'Nature is a Perpetual Worker': Newton's Aether and Eighteenth-Century Natural Philosophy," *Ambix* 20 (1973), 1-25.

Henry, John. "Atomism and Eschatology: Catholicism and Natural Philosophy in the Interregnum," *Brit. J. Hist. Sci.* 15 (1982), 211-240.

Herschel, J.F.W. *Preliminary Discourse on the Study of Natural Philosophy*, new ed. London: Longman, Rees, 1835.

Hervey, Helen. "Hobbes and Descartes in the Light of Some Unpublished Letters of the Correspondence between Sir Charles Cavendish and Dr. John Pell," *Osiris* 10 (1952), 67-90.

Hesse, Mary B. "Hooke's Development of Bacon's Method," in *Proceedings of the Tenth International Congress of the History of Science, Ithaca, 1962*, 2 vols.; vol. I, pp. 265-268. Paris: Hermann, 1964.

———. "Hooke's Philosophical Algebra," *Isis* 57 (1966), 67-83.

———. "Hooke's Vibration Theory and the Isochrony of Springs," *Isis* 57 (1966), 433-441.

Hexter, J. H. "Thomas Hobbes and the Law," *Cornell Law Rev.* 65 (1980), 471-488.

Heyd, Michael. "The Reaction to Enthusiasm in the Seventeenth Century: Towards an Integrative Approach," *J. Mod. Hist.* 53 (1981), 258-280.

Hill, C. R. "The Iconography of the Laboratory," *Ambix* 22 (1975), 102-110.

Hill, Christopher. *Change and Continuity in Seventeenth-Century England*. Cambridge, Mass.: Harvard University Press, 1975.

———. *The Experience of Defeat: Milton and Some Contemporaries.* London: Faber and Faber, 1984.

———. *God's Englishman: Oliver Cromwell and the English Revolution.* New York: Dial, 1970.

———. *Intellectual Origins of the English Revolution.* Oxford: Oxford University Press, 1965.

———. *Milton and the English Revolution.* Harmondsworth: Penguin, 1979.

———. *Some Intellectual Consequences of the English Revolution.* Madison: University of Wisconsin Press, 1980.

———. "William Harvey and the Idea of Monarchy," in *The Intellectual Revolution of the Seventeenth Century*, ed. Charles Webster, pp. 160-181. London: Routledge and Kegan Paul, 1974.

———. *The World Turned Upside Down: Radical Ideas during the English Revolution.* Harmondsworth: Penguin, 1975.

Hirst, Paul. "Witchcraft Today and Yesterday," *Econ. Soc.* 11 (1982), 428-448.

Hirzel, Rudolf. *Der Dialog: Ein literarhistorischer Versuch*, 2 vols. in 1. Leipzig: S. Hirzel, 1895.

Hodges, Devon Leigh. "Anatomy as Science," *Assays* 1 (1981), 73-89.

Hofmann, Joseph E. *Leibniz in Paris 1672-1676: His Growth to Mathematical Maturity.* Cambridge: Cambridge University Press, 1974.

Home, Roderick W. "Francis Hauksbee's Theory of Electricity," *Arch. Hist. Exact Sci.* 4 (1967), 203-217.

———. "Newton on Electricity and the Aether," in *Contemporary Newtonian Research*, ed. Zev Bechler, pp. 191-213. Dordrecht: Reidel, 1982.

Horne, C. J. "Literature and Science," in *The Pelican Guide to English Literature: 4. From Dryden to Johnson*, ed. Boris Ford, pp. 188-202. Harmondsworth: Penguin, 1972.

Hunter, Michael. "Ancients, Moderns, Philologists, and Scientists," *Ann. Sci.* 39 (1982), 187-192.

———. "The Debate over Science," in *The Restored Monarchy, 1660-1688*, ed. J. R. Jones, pp. 176-195. London: Macmillan, 1979.

———. *John Aubrey and the Realm of Learning.* London: Duckworth, 1975.

———. *The Royal Society and Its Fellows 1660-1700: The Morphology of an Early Scientific Institution.* Chalfont St. Giles: British Society for the History of Science, 1982.

———. *Science and Society in Restoration England.* Cambridge: Cambridge University Press, 1981.

Hutchison, Keith. "Supernaturalism and the Mechanical Philosophy," *Hist. Sci.* 21 (1983), 297-333.

Ivins, William M., Jr. *Prints and Visual Communication.* Cambridge, Mass.: M.I.T. Press, 1969.

Jacob, James R. "Aristotle and the New Philosophy: Stubbe versus the Royal Society," in *Science, Pseudo-Science and Society*, ed. Marsha P.

Hanen et al., pp. 217-236. Waterloo, Ontario: Wilfrid Laurier University Press, for the Calgary Institute for the Humanities, 1980.

———. "Boyle's Atomism and the Restoration Assault on Pagan Naturalism," *Soc. Stud. Sci.* 8 (1978), 211-233.

———. "Boyle's Circle in the Protectorate: Revelation, Politics, and the Millennium," *J. Hist. Ideas* 38 (1977), 131-140.

———. *Henry Stubbe, Radical Protestantism and the Early Enlightenment.* Cambridge: Cambridge University Press, 1983.

———. "Restoration, Reformation and the Origins of the Royal Society," *Hist. Sci.* 13 (1975), 155-176.

———. *Robert Boyle and the English Revolution: A Study in Social and Intellectual Change.* New York: Burt Franklin, 1977.

Jacob, Margaret C. *The Newtonians and the English Revolution, 1689-1720.* Ithaca: Cornell University Press, 1976.

Jacquot, Jean. "Un document inédit: les notes de Charles Cavendish sur la première version du 'De Corpore' de Hobbes," *Thalès* 8 (1952), 33-86.

———. "Notes on an Unpublished Work of Thomas Hobbes," *Notes Rec. Roy. Soc. Lond.* 9 (1952), 188-195.

———. "Sir Charles Cavendish and His Learned Friends," *Ann. Sci.* 8 (1952), 13-27, 175-191.

James, D. G. *The Life of Reason: Hobbes, Locke, Bolingbroke.* London: Longmans, Green, 1949.

Jardine, Lisa. *Francis Bacon: Discovery and the Art of Discourse.* Cambridge: Cambridge University Press, 1974.

Jobe, Thomas Harmon. "The Devil in Restoration Science: The Glanvill-Webster Witchcraft Debate," *Isis* 72 (1981), 343-356.

Jones, Harold Whitmore. "Mid-Seventeenth-Century Science: Some Polemics," *Osiris* 9 (1950), 254-274.

Jones, J. R. *Country and Court: England, 1658-1714.* Cambridge, Mass.: Harvard University Press, 1979.

———. "Political Groups and Tactics in the Convention of 1660," *Hist. J.* 6 (1963), 159-177.

Jones, Richard Foster. *Ancients and Moderns: A Study of the Rise of the Scientific Movement in Seventeenth-Century England*, 2d ed. St. Louis: Washington University Press, 1961.

———. "Science and English Prose Style in the Third Quarter of the Seventeenth Century," *Publ. Mod. Lang. Assoc. America* 45 (1930), 977-1009.

———. "Science and Language in England of the Mid-Seventeenth Century," *J. Eng. Germ. Philol.* 31 (1932), 315-331.

Jones, W. J. *Politics and the Bench: The Judges and the Origins of the English Civil War.* London: Allen and Unwin, 1971.

Judson, Margaret A. *From Tradition to Political Reality: A Study of the Ideas*

Set Forth in Support of the Commonwealth Government in England, 1649-1653. Hamden, Conn.: Archon Books, 1980.

Kaplan, Barbara Beigun. "Greatrakes the Stroker: The Interpretations of His Contemporaries," *Isis* 73 (1982), 178-185.

Kargon, Robert Hugh. *Atomism in England from Hariot to Newton*. Oxford: Clarendon Press, 1966.

———. "Atomism in the Seventeenth Century," in *Dictionary of the History of Ideas*, ed. Philip P. Wiener, vol. 1, pp. 132-141. New York: Scribner's, 1973.

Kauffeldt, Alfons. *Otto von Guericke: Philosophisches über den leeren Raum*. Berlin: Akademie-Verlag, 1968.

Keegan, John. *The Face of Battle: A Study of Agincourt, Waterloo and the Somme*. New York: Viking, 1976.

———. *Six Armies in Normandy: From D-Day to the Liberation of Paris*. New York: Viking, 1982.

Klaaren, Eugene M. *Religious Origins of Modern Science: Belief in Creation in Seventeenth-Century Thought*. Grand Rapids, Mich.: William B. Eerdmans, 1977.

Kocher, Paul H. "Bacon on the Science of Jurisprudence," *J. Hist. Ideas* 18 (1957), 3-26.

Köhler, Max. "Studien zur Naturphilosophie des Th. Hobbes," *Arch. Gesch. Phil.* 9 (1902-1903), 59-96.

Koyré, Alexandre. *Galileo Studies*, trans. John Mepham. Atlantic Highlands, N.J.: Humanities Press, 1978.

Krafft, Fritz. *Otto von Guericke*. Darmstadt: Wissenschaftliche Buchgesellschaft, 1978.

Kuhn, Thomas S. "The Function of Measurement in Modern Physical Science," *Isis* 52 (1961), 161-190.

———. "A Function for Thought Experiments," in idem, *The Essential Tension: Selected Studies in Scientific Tradition and Change*, pp. 240-265. Chicago: University of Chicago Press, 1977.

Kuslan, Louis, and Stone, A. Harris. *Robert Boyle: The Great Experimenter*. Englewood Cliffs, N.J.: Prentice-Hall, 1970.

Labrousse, Elisabeth. "Le démon de Mâçon," in *Scienze, credenze occulte, livelli di cultura*, pp. 249-275. Istituto Nazionale di Studi sul Rinascimento. Florence: Olschki, 1982.

Lacey, Douglas R. *Dissent and Parliamentary Politics in England, 1661-1689*. New Brunswick, N.J.: Rutgers University Press, 1969.

Laird, John. *Hobbes*. London: Ernest Benn, 1934.

Lamont, William M. *Richard Baxter and the Millennium*. London: Croom Helm, 1979.

Latour, Bruno. "Give Me a Laboratory and I Will Raise the World," in *Science Observed: Perspectives on the Social Study of Science*, ed. Karin D. Knorr-Cetina and Michael Mulkay, pp. 141-170. London: Sage, 1983.

Latour, Bruno. *Les microbes: guerre et paix, suivi de irréductions*. Paris: Editions A. M. Métailié, 1984.

Latour, Bruno, and Woolgar, Steve. *Laboratory Life: The Social Construction of Scientific Facts*. Beverly Hills, Calif.: Sage, 1979.

Laudan, Laurens. "The Clock Metaphor and Probabilism: The Impact of Descartes on English Methodological Thought, 1650-65," *Ann. Sci.* 22 (1966), 73-104.

Leach, Edmund. "Melchisedech and the Emperor: Icons of Subversion and Orthodoxy," in *Proceedings of the Royal Anthropological Institute for 1972*, pp. 5-14. London: Royal Anthropological Institute, 1973.

Lennox, James G. "Robert Boyle's Defense of Teleological Inference in Experimental Science," *Isis* 74 (1983), 38-52.

Lenoble, Robert. *Mersenne, ou la naissance du mécanisme*. Paris: Vrin, 1943.

Leopold, J. H. "Christiaan Huygens and His Instrument Makers," in *Studies on Christiaan Huygens*, ed. H.J.M. Bos et al., pp. 221-233. Lisse, The Netherlands: Swets & Zeitlinger, 1980.

Lichtenstein, Aharon. *Henry More: The Rational Theology of a Cambridge Platonist*. Cambridge, Mass.: Harvard University Press, 1962.

Linnell, Charles L. S. "Daniel Scargill: 'A Penitent Hobbist'," *Church Quart. Rev.* 156 (1955), 256-265.

Lupoli, Agostino. "La polemica tra Hobbes e Boyle," *Ann. Fac. Lett. Fil. Univ. Milano* 29 (1976), 309-354.

McAdoo, Henry R. *The Spirit of Anglicanism: A Survey of Anglican Theological Method in the Seventeenth Century*. London: A. & C. Black, 1965.

McClaughlin, Trevor. "Le concept de science chez Jacques Rohault," *Rev. Hist. Sci.* 30 (1977), 225-240.

———. "Sur les rapports entre la Compagnie de Thévenot et l'Académie Royale des Sciences," *Rev. Hist. Sci.* 28 (1975), 235-242.

Macdonald, Michael. "Religion, Social Change and Psychological Healing in England," *Stud. Church Hist.* 19 (1982), 101-126.

Macfarlane, Alan. *Witchcraft in Tudor and Stuart England*. London: Routledge and Kegan Paul, 1970.

MacGillivray, Royce. "Thomas Hobbes's History of the English Civil War: A Study of *Behemoth*," *J. Hist. Ideas* 31 (1970), 179-198.

McGuire, J. E. "Boyle's Conception of Nature," *J. Hist. Ideas* 33 (1972), 523-542.

———. "Force, Active Principles, and Newton's Invisible Realm," *Ambix* 15 (1968), 154-208.

McGuire, J. E., and Tamny, Martin. *Certain Philosophical Questions: Newton's Trinity Notebook*. Cambridge: Cambridge University Press, 1983.

McKeon, Michael. *Politics and Poetry in Restoration England: The Case of Dryden's Annus Mirabilis*. Cambridge, Mass.: Harvard University Press, 1975.

McKeon, Richard. *The Philosophy of Spinoza: The Unity of His Thought*. London: Longmans, Green, 1928.

McKie, Douglas. "Fire and the *Flamma vitalis*: Boyle, Hooke and Mayow," in *Science, Medicine and History*, ed. E. A. Underwood, 2 vols.; vol. 1, pp. 469-488. London: Oxford University Press, 1953.

———. "Introduction" [to facsimile edition of Boyle's *Works*], vol. 1, pp. v*-xx*. Hildesheim: George Olms, 1965.

Mackintosh, Sir James. "Dissertation Second; Exhibiting a General View of the Progress of Ethical Philosophy, Chiefly during the Seventeenth and Eighteenth Centuries," in *Encyclopaedia Britannica*, 7th ed. "Dissertations," vol. 1, pp. 291-429. Edinburgh, 1842.

McLachlan, Herbert. *Socinianism in Seventeenth-Century England.* Oxford: Oxford University Press, 1951.

Macpherson, C. B. "Introduction" [to his edition of *Leviathan*], pp. 9-63. Harmondsworth: Penguin, 1968.

Madden, Edward H. "Thomas Hobbes and the Rationalistic Ideal," in *Theories of Scientific Method: The Renaissance through the Nineteenth Century*, ed. Madden, pp. 104-118. Seattle: University of Washington Press, 1960.

Maddison, R.E.W. *The Life of the Honourable Robert Boyle, F.R.S.* London: Taylor & Francis, 1969.

———. "The Portraiture of the Honourable Robert Boyle, F.R.S.," *Ann. Sci.* 15 (1959), 141-214.

Mandelbaum, Maurice. *Philosophy, Science, and Sense Perception: Historical and Critical Studies.* Baltimore: The Johns Hopkins Press, 1964.

Mandrou, Robert. *Magistrats et sorciers en France au XVIIe siècle.* Paris: Plon, 1968.

Medawar, Peter. "Is the Scientific Paper a Fraud?" in *Experiment: A Series of Scientific Case Histories*, ed. David Edge, pp. 7-12. London: B.B.C., 1964.

Merrill, Elizabeth. *The Dialogue in English Literature.* New York: Henry Holt, 1911.

Merton, Robert K. *The Sociology of Science: Theoretical and Empirical Investigations*, ed. Norman W. Storer. Chicago: University of Chicago Press, 1973.

Mesnard, Jean. "Les premières relations parisiennes de Christiaan Huygens," in *Huygens et la France*, ed. René Taton, pp. 33-40. Paris: Vrin, 1982.

Middleton, W. E. Knowles. *The Experimenters: A Study of the Accademia del Cimento.* Baltimore: The Johns Hopkins Press, 1971.

———. *The History of the Barometer.* Baltimore: The Johns Hopkins Press, 1964.

———. "Science in Rome, 1675-1700, and the Accademia Fisicomatematica of Giovanni Giustino Ciampini," *Brit. J. Hist. Sci.* 8 (1975), 138-154.

———. "What did Charles II Call the Fellows of the Royal Society?" *Notes Rec. Roy. Soc. Lond.* 32 (1977), 13-17.

Miller, David Philip. "Method and the 'Micropolitics' of Science: The Early

Years of the Geological and Astronomical Societies of London," in *The Politics and Rhetoric of Scientific Method: Historical Studies*, ed. John A. Schuster and Richard Yeo, pp. 227-251. Dordrecht: Reidel, 1986.

Millington, E. C. "Studies in Capillarity and Cohesion in the Eighteenth Century," *Ann. Sci.* 5 (1945), 352-369.

———. "Theories of Cohesion in the Seventeenth Century," *Ann. Sci.* 5 (1945), 253-269.

Mintz, Samuel I. "Galileo, Hobbes, and the Circle of Perfection," *Isis* 43 (1952), 98-100.

———. "Hobbes, Thomas," in *Dictionary of Scientific Biography*, vol. VI, pp. 444-451. New York: Charles Scribner's, 1972.

———. *The Hunting of Leviathan: Seventeenth-Century Reactions to the Materialism and Moral Philosophy of Thomas Hobbes.* Cambridge: Cambridge University Press, 1962.

Missner, Marshall. "Skepticism and Hobbes's Political Philosophy," *J. Hist. Ideas* 44 (1983), 407-427.

Mitcham, Carl. "Philosophy and the History of Technology," in *The History and Philosophy of Technology*, ed. G. Bugliarello and D. B. Doner, pp. 163-201. Urbana: University of Illinois Press, 1979.

More, Louis Trenchard. *The Life and Works of the Honourable Robert Boyle.* London: Oxford University Press, 1944.

Muddiman, Joseph G. *The King's Journalist, 1659-1689: Studies in the Reign of Charles II.* London: Bodley Head, 1923.

Multhauf, Robert P. "Some Nonexistent Chemists of the Seventeenth Century: Remarks on the Use of the Dialogue in Scientific Writing," in Allen G. Debus and Multhauf, *Alchemy and Chemistry in the Seventeenth Century*, pp. 31-50. Los Angeles: William Andrews Clark Memorial Library, 1966.

Needham, Joseph. *The Grand Titration: Science and Society in East and West.* London: George Allen & Unwin, 1969.

Nicholas, Donald. *Mr. Secretary Nicholas, 1593-1669: His Life and Letters.* London: Bodley Head, 1955.

Nicolson, Marjorie Hope. "Christ's College and the Latitude-Men," *Mod. Philol.* 27 (1929), 35-53.

———. "Milton and Hobbes," *Stud. Philol.* 23 (1926), 405-453.

———. *Pepys' 'Diary' and the New Science.* Charlottesville: The University Press of Virginia, 1965.

Oakley, Francis. "Jacobean Political Theology: The Absolute and Ordinary Powers of the King," *J. Hist. Ideas* 29 (1968), 323-346.

O'Brien, John J. "Samuel Hartlib's Influence on Robert Boyle's Scientific Development," *Ann. Sci.* 21 (1965), 1-14, 257-276.

Orr, Robert R. *Reason and Authority: The Thought of William Chillingworth.* Oxford: Clarendon Press, 1967.

Pacchi, Arrigo. *Cartesio in Inghilterra: da More a Boyle.* Rome and Bari: Editori Laterza, 1973.

———. *Convenzione e ipotesi nella formazione della filosofia naturale di Thomas Hobbes*. Florence: La Nuova Italia, 1965.

Partridge, Eric. *Origins: A Short Etymological Dictionary of Modern English*. New York: Macmillan, 1958.

Payen, Jacques. "Huygens et Papin: moteur thermique et machine à vapeur," in *Huygens et la France*, ed. René Taton, pp. 197-208. Paris: Vrin, 1982.

Pelseneer, Jean. "Petite contribution à la connaissance de Mariotte," *Isis* 42 (1951), 299-301.

Peters, Richard. *Hobbes*. Harmondsworth: Penguin, 1967.

Pickering, Andrew. "The Hunting of the Quark," *Isis* 72 (1981), 216-236.

Pinch, T. J. "The Sun-Set: The Presentation of Certainty in Scientific Life," *Soc. Stud. Sci.* 11 (1981), 131-158.

———. "Theory Testing in Science—The Case of Solar Neutrinos: Do Crucial Experiments Test Theories or Theorists?" *Phil. Soc. Sci.* 15 (1985), 167-187.

Playfair, John. "Dissertation Third . . . of the Progress of Mathematical and Physical Science," in *Encyclopaedia Britannica*, 7th ed., "Dissertations," vol. I, pp. 431-572. Edinburgh, 1842.

Pocock, J.G.A. *The Ancient Constitution and the Feudal Law: English Historical Thought in the Seventeenth Century*. Cambridge: Cambridge University Press, 1957.

———. "Time, History and Eschatology in the Thought of Thomas Hobbes," in idem, *Politics, Language and Time: Essays on Political Thought and History*, pp. 148-201. London: Methuen, 1972.

Popkin, Richard H. *The History of Scepticism from Erasmus to Spinoza*. Berkeley: University of California Press, 1979.

Powell, Anthony. *John Aubrey and His Friends*. London: Eyre & Spottiswoode, 1948.

Powell, Baden. *History of Natural Philosophy from the Earliest Periods to the Present Time*. London: Longman, 1842.

Price, Derek J. "The Manufacture of Scientific Instruments from *c* 1500 to *c* 1700," in *A History of Technology*, ed. C. Singer et al., vol. III, pp. 620-647. London: Oxford University Press, 1957.

Prior, Moody E. "Joseph Glanvill, Witchcraft and Seventeenth-Century Science," *Mod. Philol.* 30 (1932), 167-193.

Quine, Willard van Orman. *From a Logical Point of View*, 2d ed. Cambridge, Mass.: Harvard University Press, 1964.

Redwood, John. *Reason, Ridicule and Religion: The Age of Enlightenment in England 1660-1750*. Cambridge, Mass.: Harvard University Press, 1976.

Reif, Patricia. "The Textbook Tradition in Natural Philosophy, 1600-1650," *J. Hist. Ideas* 30 (1969), 17-32.

Reik, Miriam M. *The Golden Lands of Thomas Hobbes*. Detroit: Wayne State University Press, 1977.

Reilly, Conor. *Francis Line S.J.: An Exiled English Scientist (1595-1675).* Rome: Institutum Historicum, S.I., 1969.

Reiser, Stanley J. "The Coffee-Houses of Mid-Seventeenth-Century London." Unpubl. typescript, Imperial College, 1966.

Renaldo, John J. "Bacon's Empiricism, Boyle's Science, and the Jesuit Response in Italy," *J. Hist. Ideas* 37 (1976), 689-695.

Robertson, George Croom. *Hobbes.* Edinburgh: William Blackwood, 1886.

———. "Hobbes, Thomas," in *Encyclopaedia Britannica,* 11th ed., vol. XIII, pp. 545-552. Cambridge: Cambridge University Press, 1910.

Robinson, Edward Forbes. *The Early History of Coffee Houses in England. . . .* London: Kegan Paul, Trench, 1893.

Rochot, B. "Comment Gassendi interprétait l'expérience du Puy de Dôme," *Rev. Hist. Sci.* 16 (1963), 53-76.

Roger, Jacques. "La politique intellectuelle de Colbert et l'installation de Christiaan Huygens à Paris," in *Huygens et la France,* ed. René Taton, pp. 41-48. Paris: Vrin, 1982.

Rogers, G.A.J. "Descartes and the Method of English Science," *Ann. Sci.* 29 (1972), 237-255.

Rogers, Philip G. *The Fifth Monarchy Men.* Oxford: Oxford University Press, 1966.

Rorty, Richard. *Philosophy and the Mirror of Nature.* Princeton: Princeton University Press, 1979.

Rosenfeld, Léon. "Newton's Views on Aether and Gravitation," *Arch. Hist. Exact Sci.* 6 (1969), 29-37.

Rosmorduc, Jean. "Le modèle de l'éther lumineux dans le *Traité de la lumière* de Huygens," in *Huygens et la France,* ed. René Taton, pp. 165-176. Paris: Vrin, 1982.

Rowbottom, Margaret E. "The Earliest Published Writing of Robert Boyle," *Ann. Sci.* 6 (1950), 376-389.

Ruestow, Edward G. "Images and Ideas: Leeuwenhoek's Perception of the Spermatozoa," *J. Hist. Biol.* 16 (1983), 185-224.

Sabra, A. I. *Theories of Light from Descartes to Newton,* 2d ed. Cambridge: Cambridge University Press, 1981.

Sacksteder, William. "The Artifice Designing Science in Hobbes." Unpublished typescript.

———. "Hobbes: The Art of the Geometricians," *J. Hist. Phil.* 18 (1980), 131-146.

———. "Hobbes: Geometrical Objects," *Phil. Sci.* 48 (1981), 573-590.

———. "Hobbes: Man the Maker," in *Thomas Hobbes: His View of Man,* ed. J. G. van der Bend, pp. 77-88. Amsterdam: Rodopi, 1982.

———. "Hobbes: Teaching Philosophy to Speak English," *J. Hist. Phil.* 16 (1978), 33-45.

———. "Some Ways of Doing Language Philosophy: Nominalism, Hobbes, and the Linguistic Turn," *Rev. Metaphys.* 34 (1981), 459-485.

———. "Speaking about Mind: *Endeavor* in Hobbes," *Phil. Forum* 11 (1979), 65-79.

Sacret, Joseph H. "The Restoration Government and Municipal Corporations," *Engl. Hist. Rev.* 45 (1930), 232-259.

Sadoun-Goupil, Michelle. "L'oeuvre de Pascal et la physique moderne," in *L'oeuvre scientifique de Pascal*, ed. René Taton, pp. 249-277. Paris: Presses Universitaires de France, 1964.

Salmon, Vivian. "John Wilkins' *Essay* (1668): Critics and Continuators," *Hist. Linguist.* 1 (1974), 147-163.

Schaffer, Simon. "Natural Philosophy," in *The Ferment of Knowledge: Studies in the Historiography of Eighteenth-Century Science*, ed. G. S. Rousseau and Roy Porter, pp. 55-91. Cambridge: Cambridge University Press, 1980.

———. "Natural Philosophy and Public Spectacle in the Eighteenth Century," *Hist. Sci.* 21 (1983), 1-43.

Schmitt, Charles B. "Experience and Experiment: A Comparison of Zabarella's View with Galileo's in *De motu*," *Stud. Renaiss.* 16 (1969), 80-137.

———. "Experimental Evidence for and against a Void: The Sixteenth-Century Arguments," *Isis* 58 (1967), 352-366.

———. "Towards a Reassessment of Renaissance Aristotelianism," *Hist. Sci.* 11 (1973), 159-193.

Schofield, Robert E., ed. *A Scientific Autobiography of Joseph Priestley (1733-1804)*. Cambridge, Mass.: M.I.T. Press, 1966.

Schutz, Alfred. *Collected Papers, Vol. II. Studies in Social Theory*, ed. Arvid Brodersen. The Hague: Nijhoff, 1964.

Scott, J. F. *The Mathematical Work of John Wallis, D.D., F.R.S. (1616-1703)*. London: Taylor & Francis, 1938.

———. "The Reverend John Wallis, F.R.S. (1616-1703)," in *The Royal Society: Its Origins and Founders*, ed. Sir Harold Hartley, pp. 57-67. London: The Royal Society, 1960.

Selden, Raman. "Hobbes and Late Metaphysical Poetry," *J. Hist. Ideas* 35 (1974), 197-210.

Shapin, Steven. "History of Science and Its Sociological Reconstructions," *Hist. Sci.* 20 (1982), 157-211.

———. "Of Gods and Kings: Natural Philosophy and Politics in the Leibniz-Clarke Disputes," *Isis* 72 (1981), 187-215.

Shapin, Steven, and Barnes, Barry. "Head and Hand: Rhetorical Resources in British Pedagogical Writing, 1770-1850," *Oxford Rev. Educ.* 2 (1976), 231-254.

Shapiro, Alan E. "Kinematic Optics: A Study of the Wave Theory of Light in the Seventeenth Century," *Arch. Hist. Exact Sci.* 11 (1973), 134-266.

Shapiro, Barbara J. "Law and Science in Seventeenth-Century England," *Stanford Law Rev.* 21 (1969), 727-766.

———. *Probability and Certainty in Seventeenth-Century England: A Study of the Relationships between Natural Science, Religion, History, Law, and Literature*. Princeton: Princeton University Press, 1983.

Shea, William R. "Descartes and the Rosicrucians," *Ann. Ist. Mus. Stor. Sci. Firenze* 4 (1979), 29-47.

Shepherd, Christine M. "Newtonianism in Scottish Universities in the Seventeenth Century," in *The Origins and Nature of the Scottish Enlightenment*, ed. R. H. Campbell and Andrew S. Skinner, pp. 65-85. Edinburgh: John Donald, 1982.

———. "Philosophy and Science in the Arts Curriculum of the Scottish Universities in the Seventeenth Century." Ph.D. thesis, Edinburgh University, 1975.

Simon, Walter G. "Comprehension in the Age of Charles II," *Church Hist.* 31 (1962), 440-448.

Skinner, Quentin. "Conquest and Consent: Thomas Hobbes and the Engagement Controversy," in *The Interregnum: The Quest for Settlement, 1646-1660*, ed. G. E. Aylmer, pp. 79-98. London: Macmillan, 1972.

———. "History and Ideology in the English Revolution," *Hist. J.* 8 (1965), 158-178.

———. "The Ideological Context of Hobbes's Political Thought," *Hist. J.* 9 (1966), 286-317.

———. "Thomas Hobbes and His Disciples in France and England," *Comp. Stud. Soc. Hist.* 8 (1966), 153-167.

———. "Thomas Hobbes and the Nature of the Early Royal Society," *Hist. J.* 12 (1969), 217-239.

Slaughter, M. M. *Universal Languages and Scientific Taxonomy in the Seventeenth Century*. Cambridge: Cambridge University Press, 1982.

Snelders, H.A.M. "Christiaan Huygens and the Concept of Matter," in *Studies on Christiaan Huygens*, ed. H.J.M. Bos et al., pp. 104-125. Lisse, The Netherlands: Swets & Zeitlinger, 1980.

Spalding, James, and Brown, Maynard. "Reduction of Episcopacy as a Means to Unity in England, 1640-1662," *Church Hist.* 30 (1961), 414-432.

Spiller, Michael R. G. *"Concerning Natural Experimental Philosophie": Meric Casaubon and the Royal Society*. The Hague: Nijhoff, 1980.

Spragens, Thomas A., Jr. *The Politics of Motion: The World of Thomas Hobbes*. Lexington: The University Press of Kentucky, 1973.

Staudenbauer, C. A. "Platonism, Theosophy and Immaterialism: Recent Views of the Cambridge Platonists," *J. Hist. Ideas* 35 (1974), 157-169.

Steneck, Nicholas H. " 'The Ballad of Robert Crosse and Joseph Glanvill' and the Background to *Plus ultra*," *Brit. J. Hist. Sci.* 14 (1981), 59-74.

———. "Greatrakes the Stroker: The Interpretations of Historians," *Isis* 73 (1982), 161-177.

Stephen, Leslie. *Hobbes*. London: Macmillan, 1904.

———. "Hobbes, Thomas," in *Dictionary of National Biography*, vol. IX, pp. 931-939. Cambridge: Cambridge University Press, 1891.

Stewart, M. A. "Introduction," in *Selected Philosophical Papers of Robert Boyle*, ed. Stewart, pp. xvii-xxxi. Manchester: Manchester University Press, 1979.

Stieb, Ernst W. "Robert Boyle's Medicina Hydrostatica and the Detection of Adulteration," in *Proceedings of the Tenth International Congress of the History of Science, Ithaca, 1962*, 2 vols.; vol. II, pp. 841-845. Paris: Hermann, 1964.

Strauss, Leo. *The Political Philosophy of Hobbes: Its Basis and Its Genesis*, trans. Elsa M. Sinclair. Chicago: University of Chicago Press, 1952.

Stroup, Alice. "Christiaan Huygens & the Development of the Air Pump," *Janus* 68 (1981), 129-158.

Syfret, R. H. "Some Early Critics of the Royal Society," *Notes Rec. Roy. Soc. Lond.* 8 (1950), 20-64.

Taylor, A. E. *Thomas Hobbes*. London: Constable, 1908.

Taylor, E.G.R. *The Mathematical Practitioners of Tudor & Stuart England 1485-1714*. Cambridge: Cambridge University Press, 1970.

Thomas, Keith. *Religion and the Decline of Magic*. Harmondsworth: Penguin, 1973.

———. "The Social Origins of Hobbes's Political Thought," in *Hobbes Studies*, ed. K. C. Brown, pp. 185-236. Oxford: Basil Blackwell, 1965.

Thomas, Peter W. *Sir John Berkenhead, 1617-1679: A Royalist Career in Politics and Polemics*. Oxford: Clarendon Press, 1969.

Thorpe, Clarence de W. *The Aesthetic Theory of Thomas Hobbes*. Ann Arbor: University of Michigan Press, 1940.

Tönnies, Ferdinand. *Thomas Hobbes: Leben und Lehre*. Stuttgart: Frommann, 1925.

Tuck, Richard. "*Power* and *Authority* in Seventeenth-Century England," *Hist. J.* 17 (1974), 43-61.

Turberville, A. S. *A History of Welbeck Abbey and Its Owners*, 2 vols. London: Faber and Faber, 1938-1939.

Turnbull, George Henry. *Hartlib, Dury and Comenius: Gleanings from Hartlib's Papers*. Liverpool: Liverpool University Press, 1947.

———. "Peter Stahl, the First Public Teacher of Chemistry at Oxford," *Ann. Sci.* 9 (1953), 265-270.

Turnbull, Herbert Westren, ed. *James Gregory Tercentenary Memorial Volume*. London: G. Bell, for the Royal Society of Edinburgh, 1939.

Turner, H. D. "Robert Hooke and Boyle's Air-Pump," *Nature* 184 (1959), 395-397.

van Helden, Albert. "The Accademia del Cimento and Saturn's Ring," *Physis* 15 (1973), 237-259.

———. " 'Annulo Cingitur': The Solution of the Problem of Saturn," *J. Hist. Astron.* 5 (1974), 155-174.

———. "The Birth of the Modern Scientific Instrument, 1550-1700," in *The Uses of Science in the Age of Newton*, ed. John G. Burke, pp. 49-84. Berkeley: University of California Press, 1983.

van Helden, Albert. "Eustachio Divini versus Christiaan Huygens: A Reappraisal," *Physis* 12 (1970), 36-50.

van Leeuwen, Henry G. *The Problem of Certainty in English Thought 1630-1690.* The Hague: Nijhoff, 1963.

Veall, Donald. *The Popular Movement for Law Reform, 1640-1660.* Oxford: Clarendon Press, 1970.

Verdon, Michel. "On the Laws of Physical and Human Nature: Hobbes' Physical and Social Cosmologies," *J. Hist. Ideas* 43 (1982), 653-663.

von Leyden, W. *Seventeenth-Century Metaphysics: An Examination of Some Main Concepts and Theories.* London: Duckworth, 1968.

Walker, D. P. *The Decline of Hell: Seventeenth-Century Discussions of Eternal Torment.* London: Routledge and Kegan Paul, 1964.

———. *Unclean Spirits: Possession and Exorcism in France and England in the Late Sixteenth and Early Seventeenth Centuries.* Philadelphia: University of Pennsylvania Press, 1982.

Wallace, Karl R. *Francis Bacon on Communication & Rhetoric.* Chapel Hill: University of North Carolina Press, 1943.

Waller, R. W. "Lorenzo Magalotti in England 1668-1669," *Italian Studies* 1 (1937), 49-66.

Wallis, Roy, ed. *On the Margins of Science: The Social Construction of Rejected Knowledge,* Sociological Review Monograph No. 27. Keele: Keele University Press, 1979.

Warner, D.H.J. "Hobbes's Interpretation of the Doctrine of the Trinity," *J. Relig. Hist.* 5 (1969), 299-313.

Warrender, Howard. "Editor's Introduction" [to his edition of Hobbes, *De cive*], pp. 1-67. Oxford: Clarendon Press, 1983.

Watkins, J.W.N. "Confession is Good for Ideas," in *Experiment: A Series of Scientific Case Histories,* ed. David Edge, pp. 64-70. London: B.B.C., 1964.

———. *Hobbes's System of Ideas: A Study in the Political Significance of Philosophical Theories.* London: Hutchinson, 1965.

Webster, Charles. "The Discovery of Boyle's Law, and the Concept of the Elasticity of Air in the Seventeenth Century," *Arch. Hist. Exact Sci.* 2 (1965), 441-502.

———. "English Medical Reformers of the Puritan Revolution: A Background to the 'Society of Chymical Physitians'," *Ambix* 14 (1967), 16-41.

———. *From Paracelsus to Newton: Magic and the Making of Modern Science.* Cambridge: Cambridge University Press, 1982.

———. *The Great Instauration: Science, Medicine and Reform 1626-1660.* London: Duckworth, 1975.

———. "Henry More and Descartes: Some New Sources," *Brit. J. Hist. Sci.* 4 (1969), 359-377.

———. "Henry Power's Experimental Philosophy," *Ambix* 14 (1967), 150-178.

———. "Water as the Ultimate Principle of Nature: The Background to Boyle's 'Sceptical Chymist'," *Ambix* 13 (1965), 96-107.

Weld, C. R. *A History of the Royal Society, with Memoirs of the Presidents*, 2 vols. London: John Parker, 1848.

Western, J. R. *Monarchy and Revolution: The English State in the 1680s.* London: Blandford Press, 1972.

Westfall, Richard S. *The Construction of Modern Science: Mechanisms and Mechanics.* Cambridge: Cambridge University Press, 1977.

———. *Force in Newton's Physics: The Science of Dynamics in the Seventeenth Century.* London: Macdonald, 1971.

———. "Hooke, Robert," in *Dictionary of Scientific Biography*, vol. VI, pp. 481-488. New York: Charles Scribner's, 1972.

———. "Robert Hooke, Mechanical Technology, and Scientific Investigation," in *The Uses of Science in the Age of Newton*, ed. John G. Burke, pp. 85-110. Berkeley: University of California Press, 1983.

———. "Unpublished Boyle Papers Relating to Scientific Method," *Ann. Sci.* 12 (1956), 63-73, 103-117.

Weston, Corinne Comstock, and Greenberg, Janelle Renfrow. *Subjects and Sovereigns: The Grand Controversy over Legal Sovereignty in Stuart England.* Cambridge: Cambridge University Press, 1981.

Westrum, Ron. "Science and Social Intelligence about Anomalies: The Case of Meteorites," *Soc. Stud. Sci.* 8 (1978), 461-493.

Whiteman, Anne O. "The Restoration of the Church of England," in *From Uniformity to Unity, 1662-1962*, ed. Owen Chadwick and Geoffrey F. Nuttall, pp. 21-88. London: S.P.C.K. Books, 1962.

Wiener, Philip Paul. "The Experimental Philosophy of Robert Boyle (1626-91)," *Phil. Rev.* 41 (1932), 594-609.

Wilkinson, Ronald Sterne. "The Hartlib Papers and Seventeenth-Century Chemistry," *Ambix* 15 (1968), 54-69; 17 (1970), 85-110.

Willey, Basil. *The Seventeenth Century Background.* London: Chatto & Windus, 1950.

Willman, Robert. "Hobbes on the Law of Heresy," *J. Hist. Ideas* 31 (1970), 607-613.

Wilson, George. "On the Early History of the Air-Pump in England," *Edinburgh New Philosophical Journal* 46 (1848-1849), 330-354.

———. *Religio chemici.* London: Macmillan, 1862.

Wittgenstein, Ludwig. *On Certainty*, ed. G.E.M. Anscombe and G. H. von Wright, trans. Denis Paul and Anscombe. New York: Harper Torchbooks, 1972.

———. *Philosophical Investigations*, trans. G.E.M. Anscombe. Oxford: Basil Blackwell, 1976.

———. *Preliminary Studies for the "Philosophical Investigations," Generally Known as The Blue and Brown Books*, 2d ed. Oxford: Basil Blackwell, 1972.

———. *Remarks on the Foundations of Mathematics*, ed. G. H. von Wright,

R. Rhees, and G.E.M. Anscombe, trans. Anscombe. Oxford: Basil Blackwell, 1967.

Wood, P. B. "Methodology and Apologetics: Thomas Sprat's *History of the Royal Society*," *Brit. J. Hist. Sci.* 13 (1980), 1-26.

Woolrych, Austin. "Last Quests for a Settlement, 1657-1660," in *The Interregnum: The Quest for Settlement, 1646-1660*, ed. G. E. Aylmer, pp. 183-204. London: Macmillan, 1972.

Worrall, John. "The Pressure of Light: The Strange Case of the Vacillating 'Crucial Experiment'," *Stud. Hist. Phil. Sci.* 13 (1982), 133-171.

Wright-Henderson, Patrick A. *The Life and Times of John Wilkins*. Edinburgh: Blackwood's, 1910.

Yale, D.E.C. "Hobbes and Hale on Law, Legislation and the Sovereign," *Cambridge Law J.* 31 (1972), 121-156.

Yeo, Richard. "Scientific Method and the Image of Science, 1831-1891," in *The Parliament of Science: The British Association for the Advancement of Science 1831-1981*, ed. Roy MacLeod and Peter Collins, pp. 65-88. Northwood, Middx.: Science Reviews Ltd., 1981.

Young, Thomas. *A Course of Lectures on Natural Philosophy and the Mechanical Arts*, ed. Philip Kelland, 2 vols. London: Taylor and Walton, 1845.

Zilsel, Edgar. *Die sozialen Ursprünge der neuzeitlichen Wissenschaft*, ed. and trans. Wolfgang Krohn. Frankfurt am Main: Suhrkamp, 1976.

Zwicker, Steven N. "Language as Disguise: Politics and Poetry in the Later Seventeenth Century," *Annals of Scholarship* 1 (1980), 47-67.

———. *Politics and Language in Dryden's Poetry: The Arts of Disguise*. Princeton: Princeton University Press, 1984.

· INDEX ·

This is a single index of subjects and persons. Emphasis is on subjects, and persons are usually indexed as subheadings rather than as main entries. Where persons' names are found as main entries, these entries are not comprehensive. When used as subheadings, Huygens is always Christiaan Huygens, and More is Henry More. Works by Boyle and Hobbes appear as main entries, and are distinguished by (B) or (H) after their titles.

Conventions used are: f for figure, and n for note. Italicized page numbers denote the main treatment of indexed topic.

LIBRARY OF CONGRESS CATALOGING IN PUBLICATION DATA

Shapin, Steven.

Leviathan and the air-pump.

Bibliography: p. Includes index.
1. Air-pump—History. 2. Physics—Experiments—History. 3. Science—England—History. 4. Hobbes, Thomas, 1588-1679. Leviathan. 5. Boyle, Robert, 1627-1691. I. Schaffer, Simon, 1955- . II. Hobbes, Thomas, 1588-1679. Dialogus physicus de natura aeris. English. 1985. III. Title.

QC166.S47 1985 533′.5 85-42705
ISBN 0-691-08393-2